Internet of Things

Smart Engineering Systems: Design and Applications
Series Editor Suman Lata Tripathi

Internet of Things
Robotic and Drone Technology
Edited by Nitin Goyal, Sharad Sharma, Arun Kumar Rana, Suman Lata Tripathi

For more information about this series, please visit: https://www.routledge.com/

Internet of Things

Robotic and Drone Technology

Edited by

*Nitin Goyal, Sharad Sharma,
Arun Kumar Rana, and Suman Lata Tripathi*

CRC Press is an imprint of the
Taylor & Francis Group, an **informa** business

First edition published 2022
by CRC Press
6000 Broken Sound Parkway NW, Suite 300, Boca Raton, FL 33487-2742

and by CRC Press
4 Park Square, Milton Park, Abingdon, Oxon, OX14 4RN

Library of Congress Cataloging-in-Publication Data

Names: Goyal, Nitin, 1986- editor.
Title: Internet of things : robotic and drone technology / edited by Nitin Goyal, Sharad Sharma, Arun Kumar Rana, Suman Lata Tripathi.
Other titles: Internet of things (CRC Press : 2022)
Description: First edition. | Boca Raton, FL : CRC Press, 2022. |
Series: Smart engineering systems: design and applications | Includes bibliographical references and index.
Identifiers: LCCN 2021016002 (print) | LCCN 2021016003 (ebook) |
ISBN 9780367754532 (hbk) | ISBN 9781032020532 (pbk) |
ISBN 9781003181613 (ebk)
Subjects: LCSH: Autonomous robots. | Micro air vehicles. | Internet of things.
Classification: LCC TJ211.495 .I65 2022 (print) | LCC TJ211.495 (ebook) |
DDC 629.8/924678--dc23
LC record available at https://lccn.loc.gov/2021016002
LC ebook record available at https://lccn.loc.gov/2021016003

ISBN: 978-0-367-75453-2 (hbk)
ISBN: 978-1-032-02053-2 (pbk)
ISBN: 978-1-003-18161-3 (ebk)

DOI: 10.1201/9781003181613

Typeset in Times LT Std
by KnowledgeWorks Global Ltd.

Contents

Preface

This book presents extensive information on the role of IoT and emerging technology in robot and drone networks. The top applications of the Internet of robotics and drones are: infrastructure and building monitoring, fire service systems, insurance investigations, retail fulfillment, agriculture, and forensic evidence collections. Conventional robotic and drone technology is enhanced with the Internet and other emerging technologies such as cloud computing, big data, artificial intelligence, and communication networks, which open up enormous opportunities like on-demand service-oriented and user-friendly IoD applications. It focuses on major research areas of the Internet of robotics-drones and its related applications. It provides a strong knowledge platform toward IoT for graduates, researchers, data scientists, educators, and hobbyists. All chapters are organised and described in a very simple manner to facilitate readability of the chapters.

CHAPTER ORGANIZATION

This book is organized into 18 chapters:

Chapter 1 is a systematic review of smart agriculture using UAV and deep learning.

Chapter 2 is an overview of Internet of Robotic Things (IoRT) and its manifold applications.

Chapter 3 deals with various design issues of smart cities and the use of Internet of Drone Things (IoDT).

Chapter 4 explains the roles of artificial intelligence, machine learning, and IoT on robotic applications.

Chapter 5 explores the IoT-based laser trip wire system for safety and security in industries.

Chapter 6 describes detecting payment fraud using automatic feature engineering with the Harris Grey Wolf deep neural network.

Chapter 7 addresses the real-time agent-based load-balancing algorithm for Internet of Drone (IoD) in cloud computing.

Chapter 8 presents the model of the ultrawide band antenna for wireless communications with applications and challenges.

Chapter 9 attempts to review the state-of-the-art development of Internet of Underwater Things with application and challenges.

Chapter 10 reviews IoT-enabled wireless mobile ad-hoc networks with an introduction, challenges, and applications.

Chapter 11 carries out the IoT hands-on sensing, actuating, and output modules in robotics.

Chapter 12 predicts the fault detection in robotic arms, enabling them to reduce consumption in costly times.

Chapter 13 illustrates how Artificial Intelligence Advanced Technology IOT, 5G robots, and drones helped combat the Covid-19 pandemic.

Chapter 14 explains the architecture and its applications of e-learning with Internet of Things.

Chapter 15 is a review of the concepts, added values, applications, and issues on the Internet of Robotic Things.

Chapter 16 explores a direct relation with GaAs nanostructure-based solar cells with enhanced light harvesting efficiency.

Chapter 17 gives insight into advanced technologies in IoT with development of smart cities.

Chapter 18 explains implementation of soft skills for humanoid robots using artificial intelligence.

Editors

Nitin Goyal is currently working as an associate professor at the Chitkara University Research and Innovation Network (CURIN), Chitkara University, Punjab, India. He specialises in underwater wireless sensor networks (UWSN), having applications in military operations. He has worked on wireless sensor networks (WSN), mobile ad-hoc network (MANET), etc. He has 11 years of teaching and research experience and has published approximately 40 research papers in various international/national journals, books, and conferences, of which five are SCI, four are SCOPUS, and four are SCOPUS-indexed book chapters. He is currently working in the areas of machine learning, IoT, and IoD in communication networks to solve real-world problems.

Sharad Sharma is currently working as head of the Department of Electronics and Communication engineering, Maharishi Markandeshwar (Deemed to be University), India. He has more than 20 years of teaching and research experience. He has conducted workshops on soft computing and its applications in engineering, wireless networks, simulators, etc. He has a keen interest in teaching and implementing the latest techniques related to wireless and mobile communications. His research interests are routing protocol design, performance evaluation, and optimization for wireless mesh networks using nature-inspired computing, Internet of Things, space communication, and more.

Arun Kumar Rana has completed his BTech degree from Kurukshetra University and MTech degree from Maharishi Markandeshwar (Deemed to be University), Mullana, India. Mr. Rana is currently pursuing his PhD from Maharishi Markandeshwar (Deemed to be University), Mullana, India. His areas of interest include image processing, wireless sensor networks, IoT, AI, machine learning and embedded systems. Mr. Rana is currently working as assistant professor at Panipat Institute of Engineering & Technology, Samalkha, and has more than 13 years of experience. He has published approximately 70 SCI, ESCI, and Scopus papers in national and international journals and conferences. He has also published 6 books with national and international publisher's and is a frequent member of Sci Scopus indexed international conferences/symposiums. Also, he has attended 9 workshops and 9 FDPs. He has guided 6 MTech candidates, with 1 candidate in process. He serves as a reviewer for several journals and international conferences. Mr. Rana is also a member of the Asia Society of Research. He also published 6 national and international patents. He's received international awards from various international organization many times, and is listed in the World Scientist Ranking 2021.

Suman Lata Tripathi is presently working as a professor at the School of Electronics and Electrical Engineering, Lovely Professional University, Punjab, India. She has more than seventeen years of teaching and research experience and has published more than 45 research papers in refereed journals and conferences. She was nominated for the "Research Excellence Award" in 2019 at Lovely Professional University, and received the best paper at IEEE ICICS-2018. She has published the edited book, *Recent Advancement in Electronic Device, Circuit and Materials* by Nova Science Publishers. She has also, submitted edited books, *Advanced VLSI Design and Testability Issues* and *Electronic Devices and Circuit Design Challenges for IoT Application* for CRC Press/Taylor & Francis and Apple Academic Press. She is an editor of the book series "Green Energy: Fundamentals, Concepts, and Applications" and "Design and Development of Energy Efficient Systems" to be published by Scrivener Publishing, Wiley (in production). Dr. Tripathi is currently working on two accepted book proposals on biomedical engineering with Elsevier and CRC Press. She is associated with Wiley-IEEE for her multi-authored (ongoing) book in the area of VLSI design with HDLs. Her areas of expertise include micro-electronics device modeling and characterization, low power VLSI circuit design, VLSI design of testing, advance FET design for IoT, embedded system design, and biomedical applications.

List of Contributors

Mohammed Farooq Abdullah
Kumaraguru College of Technology
Coimbatore, India

Mohit Aggarwal
Panipat Institute of Engineering
& Technology
Samalkha Kurukshetra University
Kurukshetra, Haryana, India

Shobhit Aggarwal
Department of Electrical and
Computer Engineering
University of North Carolina at
Charlotte
North Carolina, US

Rakesh Ahuja
Chitkara University Institute of
Engineering and Technology
Chitkara University
Punjab, India

Sapna Arora
Maharishi Markandeshwar
(Deemed to be University)
Mullana, India

Priyesh Arya
Panipat Institute of Engineering
& Technology
Samalkha Kurukshetra University
Kurukshetra, Haryana, India

Saira Banu Atham
Presidency University
Bangalore, Karnataka, India

Sena Kumar Barai
Department of Computer Science
University of Gour Banga
Malda, India

Mohan Kumar Ch
Koneru Lakshmaiah Education
Foundation
Vaddeswaram, AP, India

Gogineni Krishna Chaitanya
Koneru Lakshmaiah Education
Foundation
Vaddeswaram, AP, India

Souvik Das
Indian Institute of Technology
Kharagpur
Kharagpur, West Bengal, India

Radhika G. Deshmukh
Shri Shivaji College of Arts,
Commerce & Science, Akola
India

Rajiv Dey
BML Munjal University
Haryana, India

Sachin Dhawan
Panipat Institute of Engineering and
Technology
Samalkha, Panipat, India

K.R. Don
Department of Oral Pathology
Saveetha Dental College
Saveetha Institute of Medical and
Technical Sciences
Saveetha University
Velappanchavadi, Chennai, Tamil
Nadu, India

Souvik Ganguli
Department of Electrical and
Instrumentation Engineering
Thapar Institute of Engineering and
Technology
Patiala, Punjab, India

Anudeep Goraya
IKG-PTU Jalandhar
Punjab, India

Nitin Goyal
Chitkara University Institute of
  Engineering and Technology,
  Chitkara University
Punjab, India

Sachin Kumar Gupta
School of Electronics &
  Communication Engineering
Shri Mata Vaishno Devi University
Katra, Jammu, Kashmir, India

Shivam Gupta
MRIIRS
Faridabad, Haryana, India

Mubarak Husain
Panipat Institute of Engineering &
  Technology
Samalkha Kurukshetra University
Kurukshetra, Haryana, India

Sanyam Jain
Panipat Institute of Engineering &
  Technology
Samalkha Kurukshetra University
Kurukshetra, Haryana, India

Tarun Jaiswal
Department of Computer Applications
NIT Raipur
Raipur, India

Ajay Jangra
Department of Computer Science &
  Engineering
University Institute of Engineering &
  Technology (U.I.E.T)
Kurukshetra University
Kurukshetra, Haryana, India

Navdeep Kaur Jhajj
Assistant Professor, SBSSTC-Ferozepur
Punjab, India

Sandeep Kajal
North Carolina State University
North Carolina, Raleigh, US

Jappreet Kaur
Department of Computer Science
  and Engineering
Thapar Institute of Engineering
  and Technology
Patiala, Punjab, India

Chandra Sekhar Kolli
Koneru Lakshmaiah Education
  Foundation
Vaddeswaram, AP, India

O. B. Krishna
Indian Institute of Technology
  Kharagpur
Kharagpur, West Bengal, India

Arun Kumar
Panipat Institute of Engineering and
  Technology
Samalkha, Panipat, India

J. Maiti
Indian Institute of Technology
  Kharagpur
Kharagpur, West Bengal, India

K. R. Padmaand
Department of Biotechnology
Sri Padmavati Mahila
VisvaVidyalayam (Women's)
  University
Tirupati, AP, India

Manju Pandey
Department of Computer Applications
NIT Raipur
Raipur, India

Krishna Keshob Paul
Department of Computer Science
University of Gour Banga
Malda, India

D. V. Prashant
Department of Electronics and
    Communication Engineering
PDPM Indian Institute of Information
    Technology Design and
    Manufacturing
Jabalpur, India

Ganeshan Ramasamy
Koneru Lakshmaiah Education
    Foundation
Vaddeswaram, AP, India

Arun Kumar Rana
Panipat Institute of Engineering and
    Technology
Samalkha, Panipat, India

Jishnu Dev Roy
Department of Computer Science
University of Gour Banga
Malda, India

Pankaj Sahu
BML Munjal University
Haryana, India

Savita Saini
Department of Computer Science &
    Engineering
University Institute of Engineering &
    Technology (U.I.E.T)
Kurukshetra University
Kurukshetra, Haryana, India

Dip Prakash Samajdar
Department of Electronics and
    Communication Engineering
PDPM Indian Institute of Information
    Technology Design and
    Manufacturing
Jabalpur, India

Agniv Sarkar
Indian Institute of Technology
    Kharagpur
Kharagpur, West Bengal, India

Sourav Sarkar
Department of Computer Science
University of Gour Banga
Malda, India

Sharad Sharma
Maharishi Markandeshwar
    (Deemed to be University)
Mullana, India

Gurpreet Singh
Department of Computer Science
    & Engineering
University Institute of Engineering
    & Technology (U.I.E.T)
Kurukshetra University
Kurukshetra, Haryana, India

Manwinder Singh
School of Electronics and Electrical
    Engineering
Lovely Professional University
Phagwara, Jalandhar, Punjab,
    India

Ashutosh Srivastava
JNPS OxyJoy Pvt. Ltd. (OXY)
BHU, Varanasi, India

Yashonidhi Srivastava
Department of Electrical and
    Instrumentation Engineering
Thapar Institute of Engineering and
    Technology
Patiala, Punjab, India

Abu Sufian
Department of Computer
    Science
University of Gour Banga
Malda, India

Priyanka Tripathi
Department of Computer
    Applications
NIT Raipur
Raipur, India

Suman Lata Tripathi
School of Electronics and Electrical
    Engineering
Lovely Professional University
Phagwara, Punjab, India

Nitin Yadav
Panipat Institute of Engineering &
    Technology
Samalkha Kurukshetra University
Kurukshetra, Haryana, India

Sahil Virk
Department of Electrical and
    Instrumentation Engineering
Thapar Institute of Engineering and
    Technology
Patiala, Punjab, India

1 Smart Agriculture Using UAV and Deep Learning

A Systematic Review

Krishna Keshob Paul*, Jishnu Dev Roy*, Sourav Sarkar*, Sena Kumar Barai*, Abu Sufian*, Sachin Kumar Gupta#, and Ashutosh Srivastava†
*Department of Computer Science,
University of Gour Banga, Malda, India
#School of Electronics & Communication
Engineering, Shri Mata Vaishno Devi University,
Katra, Jammu, Kashmir, India
†JNPS OxyJoy Pvt. Ltd. (OXY), BHU, Varanasi, India

CONTENTS

1.1 INTRODUCTION

Technical advances in agriculture increase the supply of needs for an increasing population. Over the last few years, the volume of investments is increasing in agriculture to meet cultivation growth of at least 70% by 2050 [1–3] because cultivation areas will be decreased. Information and communication technology (ICT) like Internet of Things (IoT) devices, modern data analysis, unmanned aerial vehicles (UAVs), etc. can improve precision agriculture and smart farming in a long-term manner. Today, smart agriculture is a buzzword integrated with the latest ICT to grow better food with purity. ICT can provide automation in farming. IoT devices can monitor fields, crops, and the environment; UAVs can help irrigation processes to map large fields and monitor crops in large fields. Smart agriculture not only helps

DOI: 10.1201/9781003181613-1

1

supply food for increasing population but also promotes recent trends like organic farming, etc. [4, 5].

UAVs or drones are disposable or recoverable flying objects that can fly on their own without a pilot, either by remote control on the ground or by pre-arrival programs. UAVs are equipped with ICT tools like embedded systems and different sensors like red-green-blue (RGB) cameras, LIDARs, RADARs, infrared cameras, global positioning system (GPSs), which can perform different tasks and make UAVs smart [6]. UAVs can be used for hazardous and risky missions for humans. Initially invented for military applications, today's UAVs are successfully used in several civil applications, such as agriculture, policing, surveillance, recreational purposes, and so on [7]. Drones are used for aerial photography, monitoring civilian movements in social gatherings, monitoring gatherings, and making announcements about lockdown rules in COVID-affected areas to resist the COVID-19 pandemic [8, 9].

With the advancement of UAV technologies, UAVs can now do things that previously required helicopters. The advantages of UVAs in smart agriculture mainly provide mobility in farming and can work under variable weather conditions. Using these UAVs to scan the agricultural field from the above offers many possibilities. However, examining these photographs by humans is cumbersome. Deep learning comes into play in these areas. Convolutional neural networks (CNNs) are probably the most widely used deep learning architecture [10]. The recent upsurge in deep learning is due to the huge popularity and efficiency of ConvNets. Recent interest in CNN began with AlexNet in 2012, and it has grown exponentially since then [11]. Through ConvNets object detection [12] and image segmentation [13] can easily be done. Deep learning is not implemented much in the agricultural field as of now, but things are changing rapidly [14]. This chapter examines the literature for plant detection and classification, plant health assessment, smart pest and herb control, and field analysis and yield estimation. We systematically reviewed different approaches using deep learning on UAV-captured spatial data to solve these challenges. Here, we also discuss technical backgrounds and future scopes.

1.2   BACKGROUND DETAILS

Smart Agriculture: Smart agriculture is a concept combining ICT with agriculture. This means managing farms using modern ICT to achieve higher productivity outcomes while optimizing the human workforce required [15, 16].

Recent ICTs used in modern agricultural environments include:

- Robotics: UAV, UGV, automated manufacturing machines, etc.
- Sensors: For sensing water, temperature, soil, humidity, light, etc.
- Data analytics: Modern data analysis, image-processing algorithms, etc.
- Software: Dedicated software solutions that target specific farm types or use platforms that can manage various agricultural processes and systems without knowing the details of the system.
- Connectivity: Wi-Fi, 5G, LTE, cellular networks, satellites, GPS, etc.

IoT: The IoT set of computing devices are connected through the Internet, and all digital and mechanical components are provided unique identifiers (UIDs) and it can transfer data through a network without human-to-computer or human-to-human interaction [17]. Sensors, connectivity, data processing, and user-interface are the pillars of IoT. IoT devices sense the environment through sensors (e.g., camera sensors), send data directly to the controller or the cloud, and receive responses from cloud servers' controller and do the work accordingly.

UAVs are also IoT devices, as they are equipped with cameras, GPS, etc., and controlled by the user or remote server using an Internet connection.

UAVs: UAVs or drones are disposable or recoverable flying objects that can fly on their own without a pilot, either by remote control on the ground or by pre-arrival programs. UAVs are equipped with different sensors that could perform different tasks that make UAVs smart like RGB cameras, LIDARs, RADARs, infrared cameras, etc. [6].

UAVs are categorized according to their shape, weight, range, velocity, and other factors. Here, UAVs are categorized and summarized by the shape of their wings.

- Fixed-wing UAVs are suitable for large fields; they are less affected by wind, long flights, and high altitude; they require vertical fire.
- Rotary-wing UAVs are suitable for small surfaces for take-off and landing, requiring steady fire and fire at various angles.

Deep Learning: This is a subset of machine learning algorithms, such as learning models. These models can learn how to perform pattern analysis or classifications directly from pictures, texts, sounds, or videos. These models sometimes achieve human-level accuracy. A large set of labeled data is used to train these models; also neural network architectures may have many layers on them [18]. Deep learning is the main technology that is behind the success of driverless cars as UAVs. It is used in speech control devices like hand-free speakers (Alexa, Google Home, etc.), tablets, phones, TVs, turbots, etc. Deep learning achieves results that were not possible previously.

There are many possibilities with deep learning. Some examples are virtual assistants, translators, vision for driverless delivery trucks, drones, autonomous cars, facial recognition, etc. [19, 20]. Therefore, in the same way, it is very effective in agriculture domains [14].

1.3 STATE-OF-THE-ART LITERATURE STUDY

In these subsections, we are going to discuss various reported works in this domain. Strictly, we have pointed out four domains, plant detection and classification, plant health assessment, smart pest and herb control, and field analysis and yield estimation.

1.3.1 PLANT DETECTION AND CLASSIFICATION

Detecting plants like grass and weeds is very difficult for humans in large agricultural fields. If actual crops and weeds are not detected and classified, then it can affect profit, as well as the environment due to excessive use of fertilizers, pesticides, and

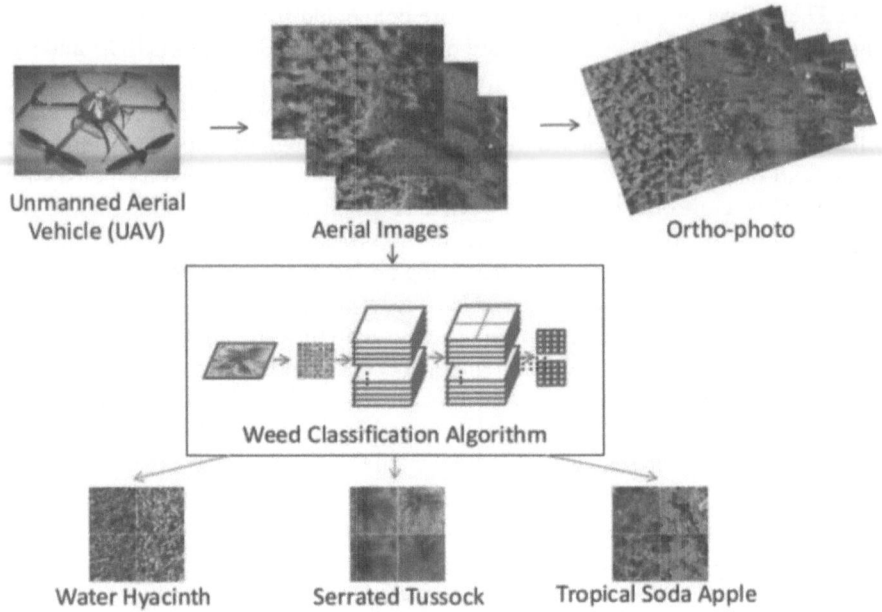

FIGURE 1.1 Detection of plants using UAV images.

insecticides. Many plants are the raw material of medical industries, and identifying these plants is a tedious job for humans. Many rare plants need to be preserved and identified. We can use UAVs for gathering information, and using machine learning should be very easy, economical, and less time-consuming.

Figure 1.1 is a perfect explanation of plant classification by UAVs. UAVs capture aerial images, then these images go through various plant classification algorithms, then plants are classified accordingly [21].

In this area of plant classification, many significant works have already explored this, Bing Lu et al. [22] studied how UAV images are used for vegetation studies for identifying grass in grassland. Wan et al. developed a UAV-based dual-camera platform [23] and, using k-means classification, estimated flower numbers of oilseed rape. Hung et al. developed an algorithm based on learning methods [21] for identifying weeds in fields using UAV imagery. They claimed that they achieved 90% accuracy. Sandino et al. proposed an integrated pipeline methodology [24] using gradient boosted decision trees to map grasses and vegetation in arid lands achieving around 96% accuracy. Pena et al. achieved an improvement for classifying summer crops using machine learning methods [25]. They proposed that hierarchical models of machine learning methods give more performance. Hula Yilin et al. proposed a CNN architecture [26] for classifying plants from image sequences. They achieve 97.47% accuracy for classifying 16 types of plants. C. Arun Priya et al. proposed an algorithm [27] for plant leaf reorganization using a support vector machine. They claimed that their proposed method could classify 32 types of plants. Dyrmann et al. developed a CNN from scratch [28]. They tested on 22 plant species, including weed and crops, and achieved 86.2% average accuracy. Kaya et al. experimented [29] with four different deep learning models with four open datasets: Flavia, Swedish Leaf, UCI Leaf, and PlantVillage. Grinblat et al. demonstrated an example of a deep learning application using CNN [30] and identified plants through their leaf vein morphological patterns. Zhiyu Liu et al. [31] demonstrated that hybrid deep learning models can give better performance than pure CNN, autoencoder (AE), or support vectors machines (SVMs) for leaf classification. Andres Miliot et al. implemented a system [32] that can discriminate between weeds and sugar beets using CNN. Yu Li et al. implemented a system to recognize rice plants using a capsule network (cabinet). Lottes et al. introduced a vision-based plant classification system [33] using UAV images. Their system can discriminate between sugar beets and weeds, and they train their model using the Phantom dataset. Ghazi et al. proposed a system for an open-set plant identification problem, the Plant CLEF 2006 campaign [34]; they claimed that their model achieved an average precision of 0.738 with the Plant CLEF 2016 test dataset. Yu Sun et al. proposed the BJFU100 dataset and deep learning model ResNet26 [35] and achieved 91.78% accuracy and published their dataset for academic use. Natesan et al. achieved 80% accuracy [36] of classifications of plants using ResNet. Anderson et al. experimented with R-CNN, YOLOv3 [37], and Retina Net and showed that Retina Net achieved the most accuracy with 92.64% average precision. Zortea et al. proposed a CNN based methodology [38] to detect citrus trees among high-density orchards. Csillik et al. demonstrated a system using simple CNN and simple linear iterative clustering [39] to identify citrus trees with 96.24% overall accuracy

In this section, we study some significant works. Many other works are there to study in areas of interest based on deep learning with UAV-captured images. There are many advanced works yet to be achieved, including developing more accurate and less resource-consuming algorithms, as well as UAVs that detect plants using their own hardware, etc.

1.3.2 PLANT HEALTH ASSESSMENT

The continuously increasing population demands increases in food production. This has inspired the development of the agriculture system over past decades. As technology continues to improve, it also needs to implement automation systems in the agricultural fields. To develop high-tech agricultural systems, we also have to focus on plant health. As our primary objective is to develop an automated farming system that can be easily and cost-effectively used by farmers, also increases the production and the profitability of farming by the use of commonly available resources. In this section, we discuss some significant works done in this area of plant health assessment. This section is divided into three subcategories:

Crop health monitoring
Crop disease and damage detection
Fruit grading and shape analysis of agricultural products

Crop Health Monitoring: In 2019, Dat Do et al. [40] develop a machine learning technique that uses UAV-based digital images to assess citrus plant health assessment. The authors' work presents the use of machine learning to develop a methodology that analyzes digital images of the citrus plants collected from the UAVs. Ground-based sensors including a water potential meter, chlorophyll meter, and spectroradiometer are used to monitor the plant conditions. Here, they used precision agriculture farming technique via this method. The collected images and ground truth data are then used as training data for the machine learning models. For this proposed model, they also evaluate other machine learning techniques from simple linear regression to CNN. In 2015 Eermas et al. [41] implemented an automation solution for the evolution of plant health in cornfields. They employ small UAVs and computational vision algorithms that work with information in the RGB visual spectrum. This provides numerous results on the use of commercial RGB sensors to provide farmers with vital information on the condition of their fields. The main motive of their research is to reduce fertilizers and improve crop yields. They achieved performance as high as 82%.

Crop Disease and Damage Detection: Huang et al. [42] planted wheat at Xinxiang, Henan province, China, and selected two fields for the experiment. One was infected with Helminthosporium leaf blotch (HLB) disease, and the other was normal. HLB infection turned the leaves yellow at the initial stage. Phantom 4 (a UAV) with an onboard RGB camera was used to capture 4000- ×3000-pixel images from a height of 80 m. The CNN method achieved an accuracy of 91.43%, and error was 0.83%, outperforming others in terms of accuracy and stability. Su et al. [43] experimented on the Caoxinzhuang experimental field of Northwest A and F University, Yangling, Shanxi Province, China with a DJI S1000 Octocopter and a five-band multispectral camera. The system was tested with winter wheat infected with yellow rust inoculum. A UAV captured aerial multispectral images from 16 to 24 m with a 1- to 1.5-cm/pixel ground resolution. The developed system achieved average precision of 89.2%, recall of 89.4%, and accuracy of 89.3%. Abdulridha et al. [44] collected hyperspectral data using a DJI Matrice 600Pro Hexacopter and a hyperspectral camera. They used ARI and TCARI 1, which detected plants with canker infection accurately. Dang et al. [45] experimented in different areas in Korea with two UAVs equipped with RGB sensors (12MP). They captured 40 images of dimension 4000 × 3000 pixels. Using GoogLeNet, they detected Fusarium wilt with over 90% accuracy. In 2012 Bashir and Sharma [46] implemented automated plant disease detection using image processing for remote areas. They implemented an effective image segmentation algorithm for color and texture analysis. In 2015 Khirade and Patil implemented plant disease detection using an image processing system [47]; they used the RELIEF-F algorithm for different disease detection. They discuss various techniques for segmenting the infected part of the plant with ANN classification techniques. The flowchart in Figure 1.2 shows the steps for plant disease detection and disease classification [47].

Ferentinos [48] developed a CNN model for detecting and diagnosing plant diseases using leaf images of healthy and diseased plants using deep learning methodologies. He trained this model with an open database which contains 87,848 images of 25 different plants, sets of 58 distinct classes of (plant, disease) combinations, and achieved a 99.53% success rate. In 2015 Rastogi [49] et al. developed a system to detect different leaf diseases from plant leaves using computer vision

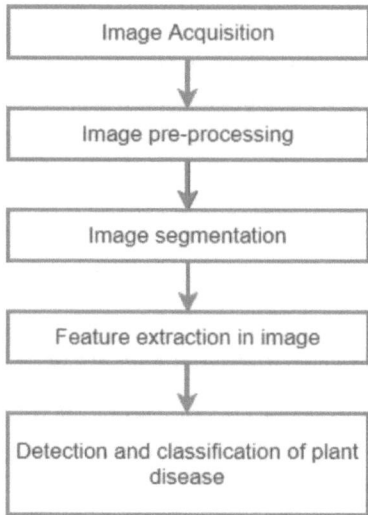

FIGURE 1.2 Basic steps for plant disease classification and detection.

technology and fuzzy logic. Bhange and Hingoliwala researched Talya, a bacterial blight disease affecting fruit [50]. They developed a web-based image processing system that quickly resizes the image. Then features are extracted on the basis of color, morphology, and CCV clustering algorithms with k-means. SVM is used to detect whether fruit is infected or non-infected. Overall a system accuracy of 82% is achieved. Islam et al. developed an image processing and machine learning system [51] to diagnose diseases of potato plants. They developed an easily accessible and automated system using image segmentation with multiclass SVM. They use a dataset called PlantVillage and achieve an overall accuracy of 95%. Pujaria et al. implemented a technique [52] to identify and classify fungal diseases. They studied four different agriculture/horticulture crops: vegetable crops, fruit crops, cereal crops, and commercial crops. The proposed work aims to develop a methodology for classification and identification of fungal disease symptoms affected by horticulture and agriculture crops.

Fruit Grading and Shape Analysis of Agricultural Products: Seema et al. [53] reviewed different classification techniques based on image processing for classifying and sorting fruit. Corrado Costa et al. [54] reviewed some papers on shape analysis of agricultural products. S. Cuberocite et al. [55] developed post-harvest vegetable and fruit grading systems for packing houses that can grade fruits and vegetables automatically using computer vision technology. Al-Marakeby et al. [56] developed a vision-based model to sort food products. To train this model they use a dataset of 1000 images; this model achieved accuracy of 97%.

In this section, we examined significant works of plant health assessment. Many other researches also include its study, an area of interest in works based on deep learning with UAV-captured images. In the future more accurate and less resource-consuming algorithms will likely be developed and UAVs should detect disease and record damages to crops accurately.

1.3.3   SMART PEST AND HERB CONTROL

Pests are the main roadblock in farming. Efficient use of pesticides is necessary for the environment as well as crops. Excessive use of pesticides is also not economical. Spraying pesticides using a UAV has the advantage of speeding up the process and preventing the pesticides from being excessively absorbed into the soil. However, a high amount of pesticides may drift to surrounding fields due to wind speed and direction. With increased precision, it is possible to limit the loss of pesticides while spraying from above.

Faial et al. [57] proposed an architecture that employs a UAV with a system of coupled spray that can communicate with a wireless sensor network. This system aims to send feedback on weather conditions and how spraying is falling in the target crop field. Based on the information, the UAV appropriately applies an optimized route. Figure 1.3 shows a UAV with a sprayer [58].

Wang et al. [59] experimented with water-sensitive paper and rhodamine B tracer to compare the droplet parameters' deposition of electric air pressure (EAP) sprayers and UAVs under different spray volumes. It was discovered that an electric air-pressure knapsack sprayer leads to run-off and lower deposition. UAV sprayers perform better

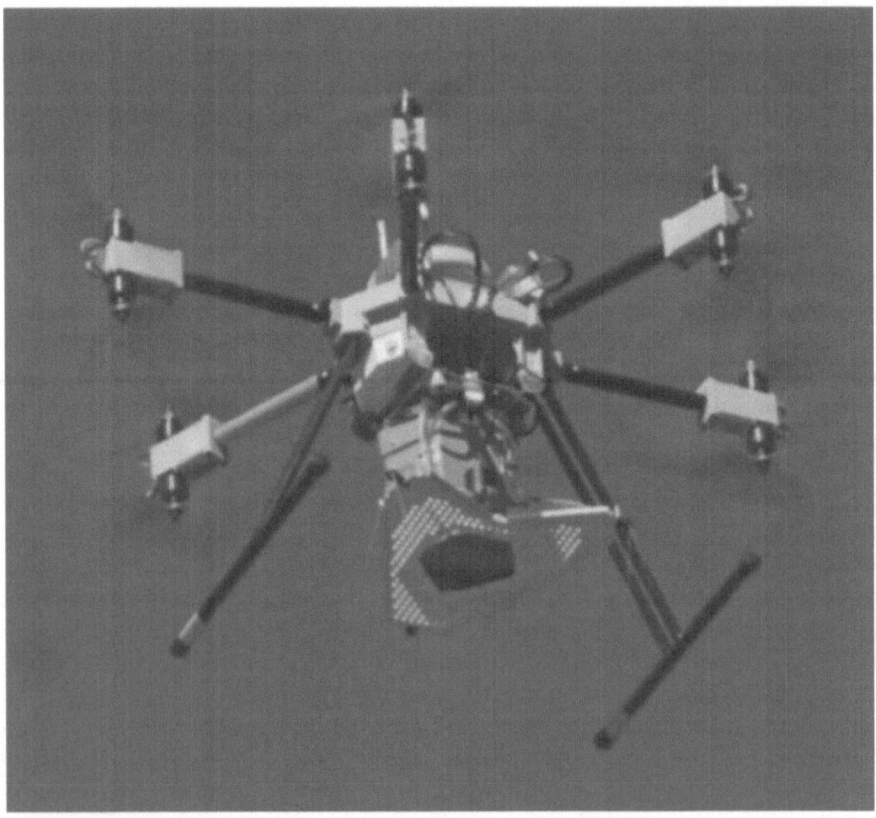

FIGURE 1.3   UAV by CNR for treatment of processionary moth.

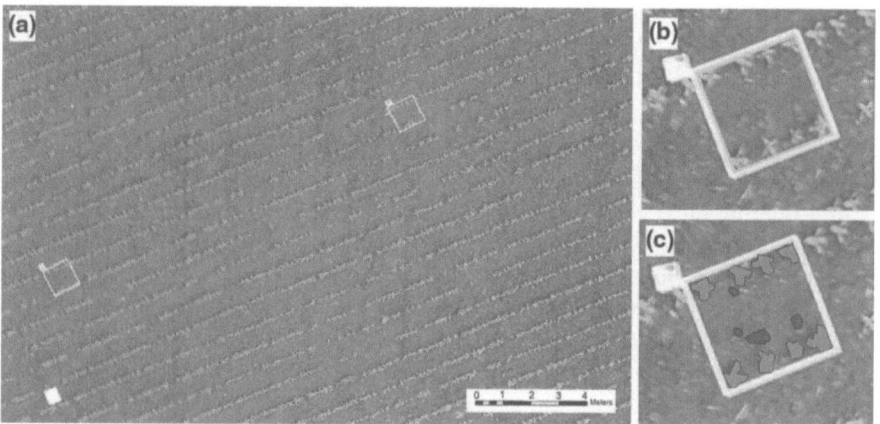

FIGURE 1.4 (a) Image taken from UAV-mounted camera; (b) detail vector created; (c) detail vector filled with green sunflower crop and violet depicting weeds.

in comparison; they're more cost-effective and wasteless. They achieve an optimal control efficacy using the UAV at 16.8 L ha^{-1} (with LU120-02, -03 nozzles) with a systemic insecticide and at 28.1 L ha^{-1} with contact insecticide and fungicide. Tetila et al. [60] did a comparative study among VGG-19, Resnet-50, Inception-v3, VGG-16, and Xception deep learning architectures with 5000 soybean pest images and found that Resnet-50 architecture reached an accuracy of up to 93.82%. Yang et al. [61] proposed a system for recognizing pests based on salience analysis of the image and a deep learning model to classify insect species in Chinese tea fields. Machado et al. [62] invented a mobile application named BioLeaf that can measure the damage of soybean leaves, and the application is supported by two techniques, Otsu segmentation and Bezier curves. The application has been tested on real-world images gathered from soybean fields, and the tool can be used for different width and narrow leaf cultures.

Lopez et al. [63] performed a study on early monitoring of grass weeds in maize fields using UAV images and OBIA algorithms. They have done another study [64] on sunflower fields early-season weed mapping using the same algorithm. Figure 1.4 shows an example of a UAV captured image of a sunflower field with image analysis, it separates weeds from sunflowers [64]. Castro et al. performed a study on weed detection on cereals (wheat) and legumes (broad bean and pea) fields and found that multi-spectral aerial images can be used to map site-specific treatment for weeds and they work better for crops at green stages.

Automated pesticide spraying UAVs are very useful for farming as well as for the environment. There are more works yet to be discussed, and future UAVs should be more accurate and spray pesticides only on affected areas.

1.3.4 FIELD ANALYSIS AND YIELD ESTIMATION

Estimating crop yield is an important task in managing and marketing commodities. Accurate yield forecasting assists farmers in improving the quality of their crops. It also helps reduce operating costs by making better decisions about the intensity of harvest and labor required. In general, crop yield estimates are done using previous data, with workers manually counting fruit at selected locations in the field. There are two types of computer-based methods for estimating performance: regional methods and counting methods.

Wang et al. [65] developed an automatic system to estimate apple crop yield, in which images of apples orchards are captured at night for reducing unpredictable natural lighting during the day. Linker et al. [66] used color images for their model to count the number of apples in orchards under natural illumination, the model can detect apples with more than 85% accuracy. Yan Gong et al. [67] developed a technique to estimate rapeseed yield with remotely sensed data by spectral mixture analysis. Figures 1.5 and 1.6 show the results.

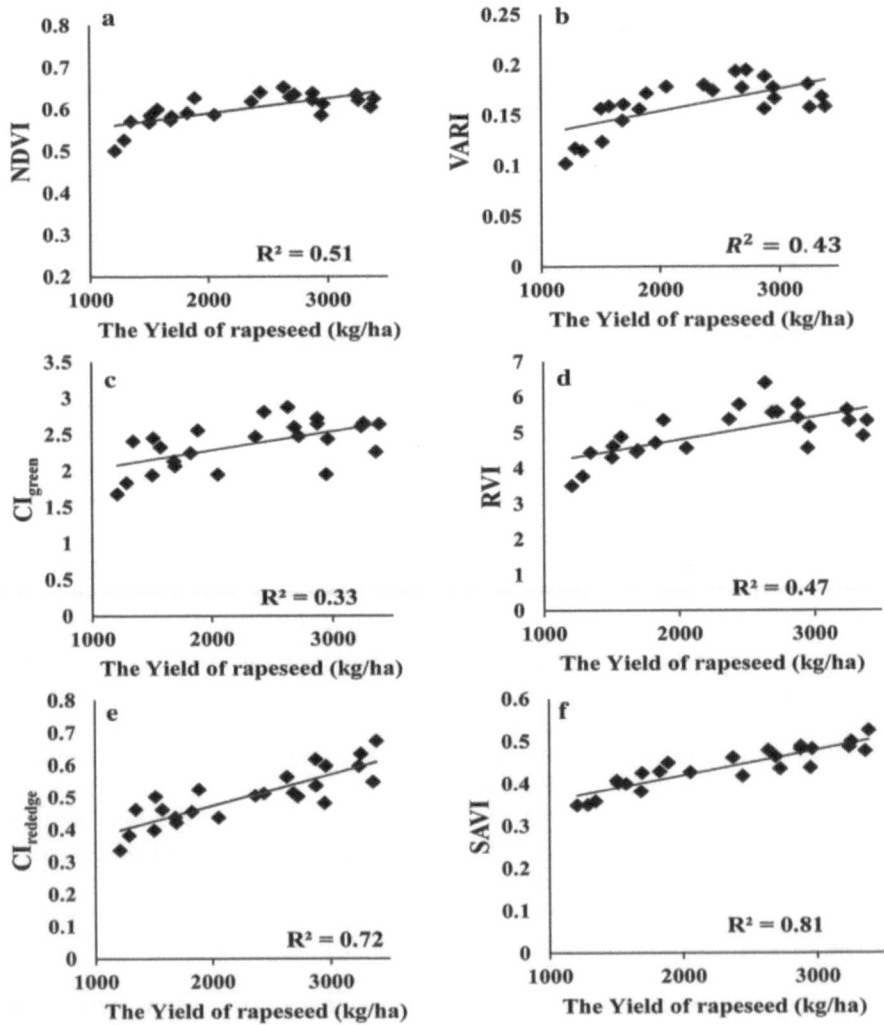

FIGURE 1.5   The yield relationships.

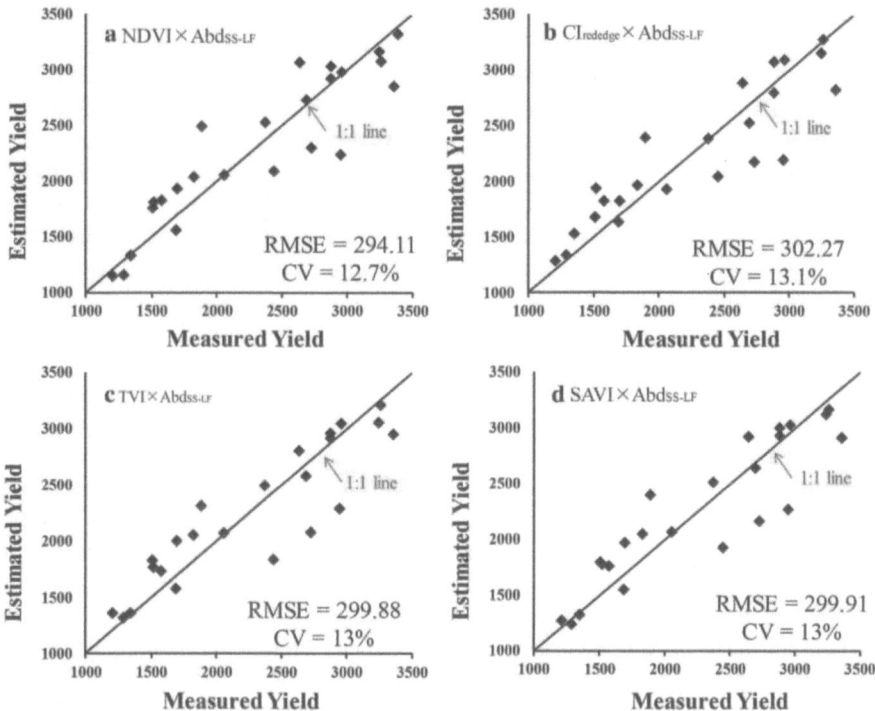

FIGURE 1.6 Validation of algorithms for estimation of rapeseed yield.

These results showed that the product of short-stalk-leaf abundance and a normalized difference vegetation index was most accurate for estimating rapeseed yield under different nitrogen treatments with estimation errors lower than 13% [67]. Kuwata et al. [68] discovered the convolutional architecture for fast feature embedding (CAFFE) model of deep learning developed by Berkeley Vision and Learning Center (BVLC) to solve crop yield estimation-related regression problems.

Yield estimation is an important task in agriculture; more accurate estimation makes more profit. In the future UAVs should estimate resources and crops automatically and more accurately in a way that is less time-consuming and more profitable.

1.4 DISCUSSION AND FUTURE SCOPE

As mentioned, food production must increase with the increase in world populations, so technological advances are required for fulfilling such requirements. In this chapter, we present studies on existing deep learning and UAV-based technologies for smart agriculture. From the surveys in this chapter, we conclude that vast research has been going on in this domain, but most of the research techniques are either cloud-based or fully autonomous equipped systems. In the first case, latency and data security are a big problem, whereas in the latter case, it is not feasible to provide a fully autonomous system for every farmer.

Therefore, edge computing [69], where computation is performed locally, in association with a cloud server, could be an optimized strategy, where data not required to be sent to a remote cloud server for processing; instead computing shall be done locally. As a result, data security and latency shall be mitigated. For lightweight devices such as UAVs where computing resources are lesser, transfer learning [70] or few-shot learning [71] could be adopted. Therefore, a UAV may perform edge computing easily. In addition, transfer learning or few-shot learning also mitigates the requirement of a large-scale ground labeled dataset for training the used algorithm. Through this type of system setup, some pilot studies should be carried out at suitable agricultural grounds.

1.5 CONCLUSION

As technologies become more advanced, and it impacts our lifestyles and work, conventional methods are always being changed. For example, the use of UAVs in agricultural fields has evolved. UAVs with machine learning can do works automatically like spray pesticide, fruit counting, weed detection, etc. In this literature review, we tried to present a study on smart agriculture using UAVs and deep learning methods. We analyzed different areas of agricultural farming, how UAVs could be used, and what works were done in that area previously. There are many works on effective and smart agricultural farming, but the use of UAVs is still has a huge scope for smart farming. Machine learning or deep learning techniques could be the key enablers for smart farming and give us a better understanding and management of agricultural firms.

REFERENCES

1. T. Reardon, R. Echeverria, J. Berdegué, B. Minten, S. Liverpool-Tasie, D. Tschirley, D. Zilberman, Rapid transformation of food systems in developing regions: highlighting the role of agricultural research & innovations, Agricultural Systems 172 (2019), 47–59.
2. Food and Agricultural Organization of the UN. FAO declaration of the world summit on food security, FAO: Rome, Italy, 2009.
3. D.C. Tsouros, S. Bibi, P.G. Sarigiannidis, A review on UAV-based applications for precision agriculture, Information 10 (11) (2019), 349.
4. A. Khatri-Chhetri, P.K. Aggarwal, P.K. Joshi, S. Vyas, Farmers' prioritization of climate-smart agriculture (CSA) technologies, Agricultural Systems 151 (2017), 184–191.
5. P.K.R. Maddikunta, S. Hakak, M. Alazab, S. Bhattacharya, T.R. Gadekallu, W.Z. Khan, Q.-V. Pham, Unmanned aerial vehicles in smart agriculture: applications, requirements and challenges, arXiv preprint arXiv:2007.12874.
6. S. Lee, Y. Choi, Reviews of unmanned aerial vehicle (drone) technology trends and its applications in the mining industry, Geosystem Engineering 19 (4) (2016), 197–204.
7. M. Mozaffari, W. Saad, M. Bennis, Y.-H. Nam, M. Debbah, A tutorial on UAVs for wireless networks: Applications, challenges, and open problems, IEEE Communications Surveys & Tutorials 21 (3) (2019), 2334–2360.
8. V. Chamola, V. Hassija, V. Gupta, M. Guizani, A comprehensive review of the covid-19 pandemics and the role of IoT, drones, AI, blockchain, and 5G in managing its impact, IEEE Access 8 (2020), 90225–90265.
9. A. Sufian, D.S. Jat, A. Banerjee, Insights of artificial intelligence to stop spread of covid-19, in: Big Data Analytics and Artificial Intelligence Against COVID-19: Innovation Vision and Approach, Springer, 2020, pp. 177–190.

10. A. Ghosh, A. Sufian, F. Sultana, A. Chakrabarti, D. De, Fundamental concepts of convolutional neural network, in: Recent Trends and Advances in Artificial Intelligence and Internet of Things, Springer, 2020, pp. 519–567.

11. F. Sultana, A. Sufian, P. Dutta, Advancements in image classification using convolutional neural network, in: 2018 Fourth International Conference on Research in Computational Intelligence and Communication Networks (ICRCICN), IEEE, 2018, pp. 122–129.

12. F. Sultana, A. Sufian, P. Dutta, A review of object detection models based on convolutional neural network, in: Intelligent Computing: Image Processing Based Applications, Springer, 2020, pp. 1–16.

13. F. Sultana, A. Sufian, P. Dutta, Evolution of image segmentation using deep convolutional neural network: a survey, Knowledge-Based Systems (2020), 106062.

14. A. Kamilaris, F.X. Prenafeta-Boldu´, Deep learning in agriculture: A survey, Computers and Electronics in Agriculture 147 (2018), 70–90.

15. N. Gondchawar, R. Kawitkar, et al., IoT-based smart agriculture, International Journal of Advanced Research in Computer and Communication Engineering 5 (6) (2016), 838–842.

16. M. Ayaz, M. Ammad-Uddin, Z. Sharif, A. Mansour, E.-H.M. Aggoune, Internet-of-things (IoT)-based smart agriculture: toward making the fields talk, IEEE Access 7 (2019), 129551–129583.

17. S. Niˇzeti´c, P. Sˇoli´c, D.L.-D.-I. Gonzalez-de, L. Patrono et al., Internet of things (IoT): Opportunities, issues and challenges towards a smart and sustainable future, Journal of Cleaner Production 274 (2020), 122877.

18. I. Goodfellow, Y. Bengio, A. Courville, Deep Learning, MIT Press, 2016, http://www.deeplearningbook.org.

19. Y. LeCun, Y. Bengio, G. Hinton, Deep learning, Nature 521 (7553) (2015), 436–444.

20. J. Sultana, M.U. Rani, M. Farquad, An extensive survey on some deep-learning applications, in: Emerging Research in Data Engineering Systems and Computer Communications, Springer, 2020, pp. 511–519.

21. C. Hung, Z. Xu, S. Sukkarieh, Feature learning based approach for weed classification using high resolution aerial images from a digital camera mounted on a UAV, Remote Sensing 6 (12) (2014), 12037–12054.

22. B. Lu, Y. He, Species classification using unmanned aerial vehicle (UAV)-acquired high spatial resolution imagery in a heterogeneous grassland, ISPRS Journal of Photogrammetry and Remote Sensing 128 (2017), 73–85.

23. L. Wan, Y. Li, H. Cen, J. Zhu, W. Yin, W. Wu, H. Zhu, D. Sun, W. Zhou, Y. He, Combining UAV-based vegetation indices and image classification to estimate flower number in oilseed rape, Remote Sensing 10 (9) (2018), 1484.

24. J. Sandino, F. Gonzalez, K. Mengersen, K.J. Gaston, UAVs and machine learning revolutionising invasive grass and vegetation surveys in remote arid lands, Sensors 18 (2) (2018), 605.

25. J.M. Penˆa, P.A. Guti´errez, C. Herv´as-Mart´ınez, J. Six, R.E. Plant, F. Lopez Granados, Object-based image classification of summer crops with machine learning methods, Remote Sensing 6 (6) (2014), 5019–5041.

26. H. Yalcin, S. Razavi, Plant classification using convolutional neural networks, in: 2016 Fifth International Conference on Agro-Geoinformatics (Agro-Geoinformatics), IEEE, 2016, pp. 1–5.

27. C. A. Priya, T. Balasaravanan, A. S. Thanamani, An efficient leaf recognition algorithm for plant classification using support vector machine, in: International Conference on Pattern Recognition, Informatics and Medical Engineering (PRIME-2012), 2012, pp. 428–432, doi: 10.1109/ICPRIME.2012.6208384

28. M. Dyrmann, H. Karstoft, H.S. Midtiby, Plant species classification using deep convolutional neural network, *Biosystems Engineering* 151 (2016), 72–80.

29. A. Kaya, A.S. Keceli, C. Catal, H.Y. Yalic, H. Temucin, B. Tekinerdogan, Analysis of transfer learning for deep neural network based plant classification models, Computers and Electronics in Agriculture 158 (2019), 20–29.

30. G.L. Grinblat, L.C. Uzal, M.G. Larese, P.M. Granitto, Deep learning for plant identification using vein morphological patterns, Computers and Electronics in Agriculture 127 (2016), 418–424.

31. Z. Liu, L. Zhu, X.-P. Zhang, X. Zhou, L. Shang, Z.-K. Huang, Y. Gan, Hybrid deep learning for plant leaves classification, in: International Conference on Intelligent Computing, Springer, 2015, pp. 115–123.

32. A. Milioto, P. Lottes, C. Stachniss, Real-time blob-wise sugar beets vs weeds classification for monitoring fields using convolutional neural networks, ISPRS Annals of the Photogrammetry, Remote Sensing and Spatial Information Sciences 4 (2017), 41.

33. P. Lottes, R. Khanna, J. Pfeifer, R. Siegwart, C. Stachniss, UAV-based crop and weed classification for smart farming, in: 2017 IEEE International Conference on Robotics and Automation (ICRA), IEEE, 2017, pp. 3024–3031.

34. M. Mehdipour Ghazi, B. Yanıkoğlu, E. Aptoula, Open-Set Plant Identification Using an Ensemble of Deep Convolutional Neural Networks, CLEF, 2016.

35. Yu Sun, Yuan Liu, Guan Wang, Haiyan Zhang, Deep Learning for plant identification in natural environment, Computational Intelligence and Neuroscience, 2017, Article ID 7361042, 6 pages (2017). https://doi.org/10.1155/2017/7361042

36. S. Natesan, C. Armenakis, U. Vepakomma, Resnet-based tree species classification using UAV images, International Archives of the Photogrammetry, Remote Sensing & Spatial Information Sciences, 2019.

37. A.A.D. Santos, J. Marcato Jr., et al., Assessment of CNN-based methods for individual tree detection on images captured by RGB cameras attached to UAVs, Sensors 19 (16) (2019), 3595.

38. D. Koc-San, S. Selim, N. Aslan, B.T. San, Automatic citrus tree extraction from UAV images and digital surface models using circular through transform, Computers and Electronics in Agriculture 150 (2018), 289–301.

39. O. Csillik, J. Cherbini, R. Johnson, A. Lyons, M. Kelly, Identification of citrus trees from unmanned aerial vehicle imagery using convolutional neural networks, Drones 2 (4) (2018), 39.

40. D. Do, F. Pham, A. Raheja, S. Bhandari, Machine learning techniques for the assessment of citrus plant health using UAV-based digital images, in: Autonomous Air and Ground Sensing Systems for Agricultural Optimization and Phenotyping III, Vol. 10664, International Society for Optics and Photonics, 2018, p. 1066400.

41. D. Zermas, D. Teng, P. Stanitsas, M. Bazakos, D. Kaiser, V. Morellas, D. Mulla, N. Papanikolopoulos, Automation solutions for the evaluation of plant health in corn fields, in: 2015 IEEE/RSJ International Conference on Intelligent Robots and Systems (IROS), IEEE, 2015, pp. 6521–6527.

42. H. Huang, J. Deng, Y. Lan, A. Yang, L. Zhang, S. Wen, H. Zhang, Y. Zhang, Y. Deng, Detection of helminthosporium leaf blotch disease based on UAV imagery, Applied Sciences 9 (3) (2019), 558.

43. J. Su, C. Liu, M. Coombes, X. Hu, C. Wang, X. Xu, Q. Li, L. Guo, W.-H. Chen, Wheat yellow rust monitoring by learning from multispectral UAV aerial imagery, Computers and Electronics in Agriculture 155 (2018), 157–166.

44. J. Abdulridha, O. Batuman, Y. Ampatzidis, UAV-based remote sensing technique to detect citrus canker disease utilizing hyperspectral imaging and machine learning, Remote Sensing 11 (11) (2019), 1373.

45. L. M. Dang, S. I. Hassan, I. Suhyeon, A. Kumar Sangaiah, I. Mehmood, S. Rho, S. Seo, H. Moon, UAV based wilt detection system via convolutional neural networks, Sustainable Computing: informatics and Systems, 28, 2020, 100250.

46. S. Bashir, N. Sharma, Remote area plant disease detection using image processing, IOSR Journal of Electronics and Communication Engineering 2 (6) (2012), 31–34.

47. S.D. Khirade, A. Patil, Plant disease detection using image processing, in: 2015 International Conference on Computing Communication Control and Automation, IEEE, 2015, pp. 768–771.

48. K.P. Ferentinos, Deep learning models for plant disease detection and diagnosis, Computers and Electronics in Agriculture 145 (2018), 311–318.

49. A. Rastogi, R. Arora, S. Sharma, Leaf disease detection and grading using computer vision technology & fuzzy logic, in: 2015 2nd International Conference on Signal Processing and Integrated Networks (SPIN), IEEE, 2015, pp. 500–505.

50. M. Bhange, H. Hingoliwala, Smart farming: pomegranate disease detection using image processing, Procedia Computer Science 58 (2015), 280–288.

51. M. Islam, A. Dinh, K. Wahid, P. Bhowmik, Detection of potato diseases using image segmentation and multiclass support vector machine, in: 2017 IEEE 30th Canadian Conference on Electrical and Computer Engineering (CCECE), IEEE, 2017, pp. 1–4.

52. J.D. Pujari, R. Yakkundimath, A.S. Byadgi, Image processing based detection of fungal diseases in plants, Procedia Computer Science 46 (2015), 1802–1808.

53. A. Kumar, G. Gill, et al., Automatic fruit grading and classification system using computer vision: a review, in: 2015 Second International Conference on Advances in Computing and Communication Engineering, IEEE, 2015, pp. 598–603.

54. C. Costa, F. Antonucci, F. Pallottino, J. Aguzzi, D.-W. Sun, P. Menesatti, Shape analysis of agricultural products: a review of recent research advances and potential application to computer vision, Food and Bioprocess Technology 4 (5) (2011), 673–692.

55. S. Cubero, N. Aleixos, F. Albert, A. Torregrosa, C. Ortiz, O. García-Navarrete, J. Blasco, Optimised computer vision system for automatic pre-grading of citrus fruit in the field using a mobile platform, Precision Agriculture 15 (1) (2014), 80–94.

56. A. Al-Marakeby, A. A. Aly, F. A. Salem, Fast quality inspection of food products using computer vision, International Journal of Advanced Research in Computer and Communication Engineering 1 (2013), 2.

57. B.S. Faical, G. Pessin, P. Geraldo Filho, A.C. Carvalho, G. Furquim, J. Ueyama, Fine-tuning of UAV control rules for spraying pesticides on crop fields, in: 2014 IEEE 26th International Conference on Tools with Artificial Intelligence, IEEE, 2014, pp. 527–533.

58. M. Bacco, A. Berton, et al., Smart farming: opportunities, challenges and technology enablers, in: 2018 IoT Vertical and Topical Summit on Agriculture Tuscany (IOT Tuscany), IEEE, 2018, pp. 1–6.

59. G. Wang, Y. Lan, H. Qi, P. Chen, A. Hewitt, Y. Han, Field evaluation of an unmanned aerial vehicle (uav) sprayer: effect of spray volume on deposition and the control of pests and disease in wheat, Pest Management Science 75 (6) (2019), 1546–1555.

60. E.C. Tetila, B.B. Machado, G. Astolfi, N.A. de Souza Belete, W.P. Amorim, A.R. Roel, H. Pistori, Detection and classification of soybean pests using deep learning with UAV images, Computers and Electronics in Agriculture 179 (2020), 105836.

61. G. Yang, Y. Bao, Z. Liu, Localization and recognition of pests in tea plantation based on image saliency analysis and convolutional neural network, Transactions of the Chinese Society of Agricultural Engineering 33 (6) (2017), 156–162.

62. B.B. Machado, J.P. Orue, et al., Bioleaf: a professional mobile application to measure foliar damage caused by insect herbivory, Computers and Electronics in Agriculture 129 (2016), 44–55.

63. F. Lopez-Granados, J. Torres-Sánchez, A.-I. De Castro, A. Serrano-Pérez, F.-J. Mesas-Carrascosa, J.-M. Peña, Object-based early monitoring of a grass weed in a grass crop using high resolution UAV imagery, Agronomy for Sustainable Development 36 (4) (2016), 67.

64. F. Lopez-Granados, J. Torres-Sanchez, A. Serrano-P´erez, A.I. de Castro, F.-J. MesasCarrascosa, J.-M. Pena, Early season weed mapping in sunflower using UAV technology: variability of herbicide treatment maps against weed thresholds, Precision Agriculture 17 (2) (2016), 183–199.
65. Q. Wang, S. Nuske, M. Bergerman, S. Singh, Automated crop yield estimation for apple orchards, in: Experimental Robotics, Springer, 2013, pp. 745–758.
66. R. Linker, A procedure for estimating the number of green mature apples in nighttime orchard images using light distribution and its application to yield estimation, Precision Agriculture 18 (1) (2017), 59–75.
67. Y. Gong, B. Duan, S. Fang, R. Zhu, X. Wu, Y. Ma, Y. Peng, Remote estimation of rapeseed yield with unmanned aerial vehicle (uav) imaging and spectral mixture analysis, Plant Methods 14 (1) (2018), 70.
68. K. Kuwata, R. Shibasaki, Estimating crop yields with deep learning and remotely sensed data, in: 2015 IEEE International Geoscience and Remote Sensing Symposium (IGARSS), IEEE, 2015, pp. 858–861.
69. W. Shi, J. Cao, Q. Zhang, Y. Li, L. Xu, Edge computing: vision and challenges, IEEE Internet of Things Journal 3 (5) (2016), 637–646.
70. F. Zhuang, Z. Qi, K. Duan, D. Xi, Y. Zhu, H. Zhu, H. Xiong, Q. He, A comprehensive survey on transfer learning, Proceedings of the IEEE, 109 (1), 43–76.
71. Y. Wang, Q. Yao, J.T. Kwok, L.M. Ni, Generalizing from a few examples: a survey on few-shot learning, ACM Computing Surveys (CSUR) 53 (3) (2020), 1–34.

2 Internet of Robotic Things (IoRT) and Its Manifold Applications

An Overview

Yashonidhi Srivastava*, Sahil Virk*, Souvik Ganguli*, and Suman Lata Tripathi‡
*Department of Electrical and Instrumentation Engineering, Thapar Institute of Engineering and Technology, Patiala, Punjab, India
‡School of Electronics and Electrical Engineering, Lovely Professional University, Phagwara, Punjab, India

CONTENTS

2.1 INTRODUCTION

With the use of the Internet of Things (IoT), a large amount of data can be transferred over a network without using human-to-human interaction or human-to-computer interaction. On the other hand, robots have been employed in many industries to perform specialized, tedious, critical, and sometimes dangerous tasks. Internet of Robotic Things (IoRT) is an embryonic model that brings together the platform of IoT with the evolving robotic technology. Researchers and scholars are tirelessly working on integrating the developments in the robotic systems field with the network technologies to extend the functions and values of these robots [1].

DOI: 10.1201/9781003181613-2

In IoT, the robot can be connected as a thing. With the help of IoT, we will now have "from anytime, anyplace connectivity for anyone," to connectivity for anything. IoT applications are widely spread in the business, society, and environment domains, and if IoT can be clubbed together with the robotic technology in these particular domains, it can result in the burgeoning of the novel technologies in the contemporary world. Programming these robots and instructing them to follow someone or to patrol a particular area and collecting the data while patrolling can ease the task of the police force of a particular region to maintain law and order. But this technology may raise some privacy concerns of the citizens of that state [1].

IoRT systems can also be employed in agriculture fields to improve the quality of soil, hence increasing the productivity of the fields and gaining more profit. Task-specific robots can be assigned different jobs in various domains. The main advantage of the IoRT systems would be that the robots can operate for full 24/7. With the use of these systems, human labor will reduce drastically, which will thus reduce the costs and, in turn, increase the quality of the products and therefore the profits. IoRT systems can be engaged in the harvesting season and the selective harvesting of the mushrooms, tomatoes, sweet peppers, strawberries, and raspberries. They can also be utilized in the gentle handling and storage of the crops [2].

IoRT systems are a boon to the health sector. They provide real-time health information and also help to quickly diagnose patients' conditions. Such systems also help track ambulances, staff members, and doctors. In addition, they can automatically gather data regarding different drugs in reducing the ailments. These systems can treat individuals suffering from stroke and mental disabilities. This technology finds its place in streaming large amounts of data using drones or unmanned air vehicles (UAVs). It is also used in military applications [3].

The remaining part of the chapter is organized with the help of the following subsections. In Section 2.2, IoRT use is explored in intelligent production plants. Robot technologies in the area of healthcare are also explored with the IoT in Section 2.3. In Section 2.4, robot arm movement applications and its control are taken into account in the IoRT area. In Section 2.5, the semantic and web applications of IoRT are taken into account. There is also ample weightage of the advancement of sensing technology and IoRT applications discussed in Section 2.6. The emergence of the IoRT based on the artificial neural network (ANN) is discussed in Section 2.7. It should also be noted that IoRT's cloud applications play an important role deliberated in Section 2.8. The use of IoRT is also found in intelligent space applications, which are discussed in Section 2.9. In addition, a variety of different IoRT applications are addressed in Section 2.10, such as in education and smart cities. These are outlined in the conclusion, Section 2.11, with some guidelines for further implementation of IoRT.

2.2 IoRT IN FACTORY AUTOMATION AND SMART MANUFACTURING APPLICATIONS

The utilization of IoRT in manufacturing industries, its evolution, and the scopes of further innovation are discussed in the following section. Fields associated with the monitoring and safety of workers, as well as warehouse management has been elaborated upon in addition to the introduction of novel frameworks based on IoRT.

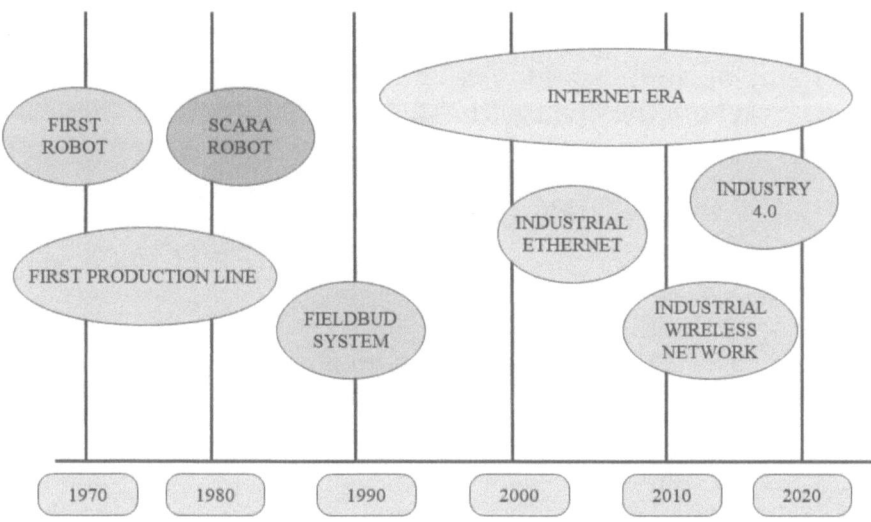

FIGURE 2.1 The evolution of communication technologies, robotics, and essential landmarks.

The interdependence of IoT and IoRT, as well as the systems developed utilizing the two is presented [4–6].

A historical outlook on industrial robotics and its evolution is discussed in this chapter. Industrial communication systems are studied and their roles are discussed in the process of enhancing and developing industrial robots in the future for smart factories. The evolutions of robots along with the upgrades in the communication infrastructure are produced with an emphasis on the discussion of hurdles in the upcoming future for robotics. An intricate link is found between the success of industrial robots and the evolution of communication systems [4]. Figure 2.1 highlights how either of the fields has grown in the last few decades, as well as their interdependence.

A system for determining the health and well-being of indoor plants has been developed with the inclusion of social media messaging services like WhatsApp. The target of this chapter is to combine IoT and IoRT to entirely automate the environment. The machine learning models inherit the readings from the robotic arm to ensure the health of the plants. The data hence received gets stored in the cloud environment of Ericsson's Apple Internet of Things (APPIoT) platform. The notifications sent to the professionals through messaging platforms require immediate attention. Post several trials, the system turned out to be affordable and trusty [7]. The working has also been visually presented in Figure 2.2 below that interlinks data with mobile robot and social media platforms.

Transport and storage services had a huge contribution in the logistics domain, especially in warehouses that depended on labor and had not experienced complete automation. To modify the preceding methods facilitated by the adoption of novel techniques, a warehouse management system was presented. Technologies of Industry 4.0 encompassing autonomous robots, Industrial Internet of Things (IIoT),

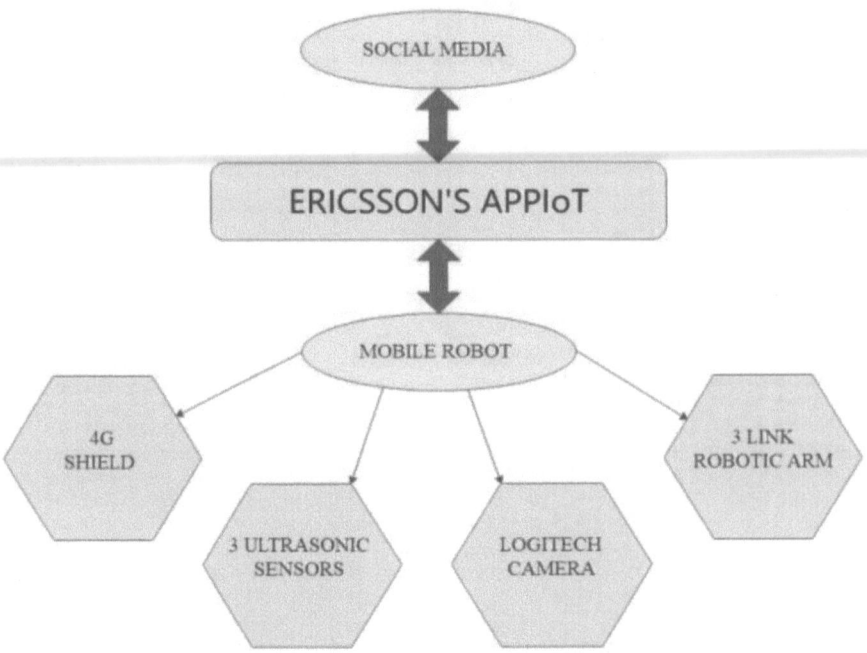

FIGURE 2.2   Structure of the system to ensure the well-being of plants.

and cloud computing were adopted to come up with smart robotic warehouse management systems. The aspects of the robots to segregate between goods to be delivered and the ones to be put away have been presented. Efficient usage of the floor area and labor to enhance logistics has been targeted [5]. A diagram has been presented in the form of Figure 2.3, which highlights the slotting and re-slotting procedures followed by the mobile robots to segregate goods that are to be transported based on the time of their delivery.

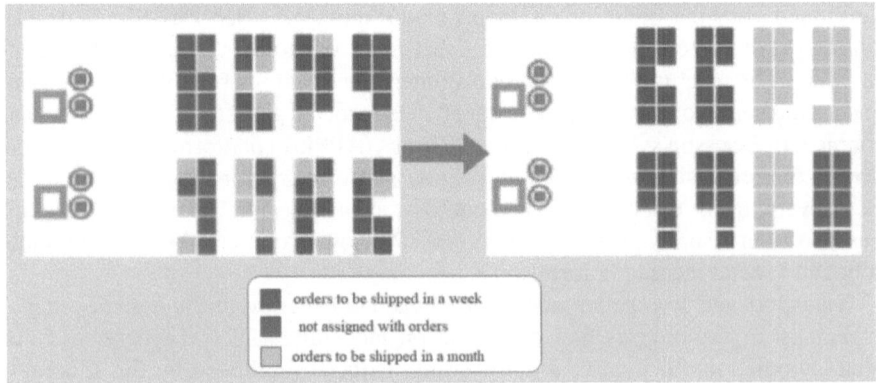

FIGURE 2.3   A representation of segregation and re-slotting of goods in the warehouse.

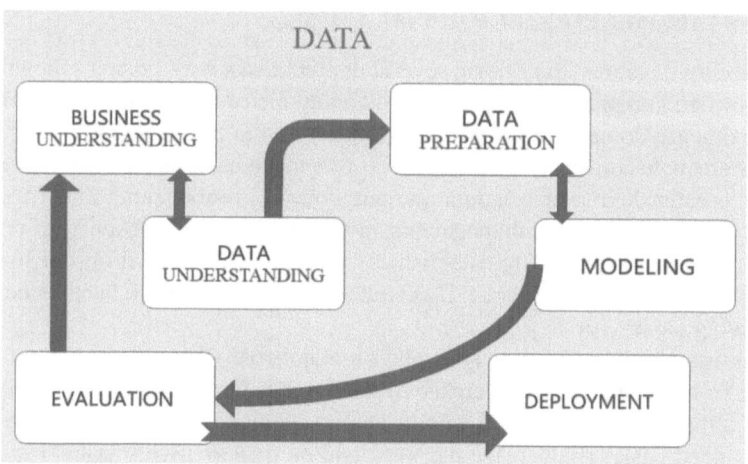

FIGURE 2.4 Process of data mining.

The evolution of technology in the fields of application technology, storage technology, and network technology had resulted in the accumulation of huge amounts of data. Data mining hence became a necessity to fish crucial information and discover critical knowledge. The difference between the big data received from the industries in contrast to the traditional ones was discussed based on the internet environment. Analytical methods, therefore, become inapplicable to such vast quantities of data. A novel algorithm for data mining, based on IoRT was introduced. The process followed included basic steps like business understanding, data preparation, deployment, evaluation, and modeling. The proposed algorithm presented efficient and trusty results [8]. The process of data mining is showcased in Figure 2.4.

Safety of workers at construction sites has been a factor of concern for many years. Recent advancements in technology have significantly improved the systems used and have made sites safer. This was facilitated by initializing communication between databases, sensors, and other smart devices with the help of IoT. To prevent accidents from taking place at multiple construction sites, smart technology was proposed for the system that stores the data accumulated at construction sites. The data received further gets transported to the cloud. It further gets utilized to warn the workers working in the proximity of dangerous sites or locations [6].

The connection of people, devices, and objects is imperative and has been finding more utilization than ever due to IoT. Positive enhancements have been experienced in various fields, such as healthcare, agriculture, and robotics. Primarily in the field of robotics, IoT had developed and has had positive impacts in the realm if IoRT as well. Assessing the usage of IoRT in fields of machine learning, cloud computing, and artificial intelligence, an architectural framework for IoRT has been introduced. The need for digitalization and the promotion of IoRT has been discussed along with its implementation. The shortcomings associated with the framework have also been listed in addition to the scope and areas of further research [9].

2.3   IoRT IN HEALTHCARE APPLICATIONS

The capability of robots to perform several distinct tasks with greater efficiency and precision when compared to humans played a key factor in the inclusion of robotics in the healthcare domain. With recent advancements in the healthcare sector, there has been an inclusion of a variety of sensors and devices. Hence, factors of possessing a greater degree of freedom and autonomy in robots came under the radar of consideration. Post the enhancements in the field of IoT, robots were provided with the position of "things" to establish connections with other things utilizing the Internet. The challenges in this field, as well as its prospects, have been discussed in the following paper [10].

The assistance of the nursing staff plays a major role when taking care of people suffering from dementia. The increase in the number of patients had made it tricky for the healthcare workers to deal with as the requirements of every patient varied on a need-to-need basis. To decrease the workload on the caregivers at nursing homes, the inclusion of Internet of Robotics was discussed. The concepts of robotic intervention strategy discoursed along with its interdependence on healthcare workers. The reliability as well as the alleged dependence of the healthcare workers on the robots was examined and presented [11].

For the application in the domain of exoskeletons and IoT, the construction of electromyography (EMG) sensors was introduced. The proposed sensor was compared with the existing ones being utilized commercially, based on signal-to-noise ratio (SNR). The relevant comparisons and subsequent validations were included in the following paper. It was concluded that the proposed sensors were feasible in the field of IoT. However, it was not found to be viable in terms of dealing with the diagnosis or the monitoring of musculoskeletal subjects [12].

Several novel robotic innovations were introduced to enhance the process of patient care. The applicability and the facilitation of robotic technology in terms of services in ICUs, surgeries, and general care were discussed. Its facilitation to reduce the risk for patients, as well as doctors, were also included. The application for collecting as well as storing medical samples has also been discussed, which played a key role in its preservation as well. The approach of the Internet of Medical Robotics Things (IoMRT) and the conception of including robots as "things" was presented to enhance its connection with other things over the Internet. An overview of robots during the conduction of surgeries and other services was included along with a description of the challenges being faced during the proceedings as well as the relevant solutions [13].

The possibility and the horizons of robots being termed as "things" were presented with the help of IoT and its recent advancements. The network infrastructure required for such communication was described, which became essential to connect the IT systems with the physical world. The increased accuracy and efficiency of the system was also presented post the proposed integration [14].

Robots vary from regular machines in terms of what is expected from them by their users. The users require robots to behave and interact with them like they do with other humans. Therefore, it became essential for robots to adapt and learn more human characteristics to be more acceptable and welcome in several different places depending on the applications. A model that had acquired information from several resources to create a better interactive experience for its users was presented. The features of assessing the user's expressions and the factors that make up their speech, including its tone and rate, were discussed to create the best possible interactive experience. The possible inclusion of parameter learning strategy to further improve the model was also included [15].

2.4   IoRT IN ROBOTIC MOVEMENT AND CONTROL

The interconnection between the IoT and robotics was discussed to extend its applications in various domains. The challenges, corresponding issues, and the targeted applications of the robotic domain post its amalgamation with IoT was presented. The facilitation and the functioning of a robotic arm were majorly discussed along with its mechanism. Rotational motion as well as linear displacement was a key factor of discussion for the wrist and the elbow movements. A microcontroller was brought under utilization as it could be connected with a Wi-Fi module, which in turn would expand its controllability to mobile, as well as PC users [16]. The general idea behind the project has been represented with the help of Figure 2.5 provide below.

To make robots more familiar to humans and consequent interactions, the paper attempted to validate the concept of the Internet of Things–based Robotic Car Control. To understand the integration of robotic things with IoT, many hardware, communication protocols, actuators, and sensors were brought together. The time taken by the IoRCar was measured and presented for better understanding [17].

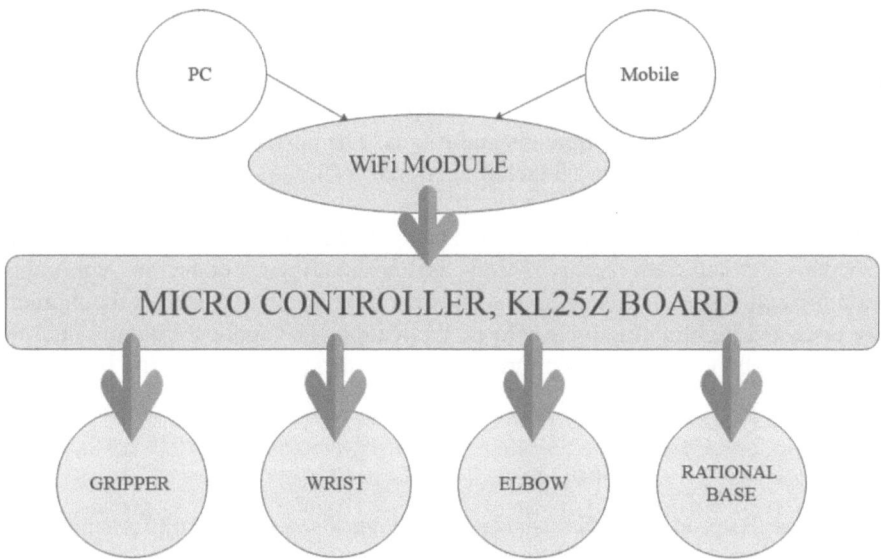

FIGURE 2.5   The flow of the project represented through block diagrams.

The procedure of building a robotic arm, utilizing IoT was presented. The controlling and monitoring of the arm were proposed to be done by the utilization of Internet facilities. Multiple servo motors were included with pulse width modulation (PWM) outputs. The utilization of node-red as a server made it possible for it to be controlled using smartphones. The results and its verification were also included in the form of practical experiments [18].

The novel conception of the IoRT and its similarity to mechanical smart intelligence was presented focusing on the development and the applications of a robotic arm. The required programming to get it as close to a human arm as possible was discussed, which was represented by five degrees of freedom (5 DOF). The development of a robotic arm as an IoRT device was introduced along with its ability to be controlled by several other devices [19].

2.5   IoRT IN SEMANTIC AND WEB APPLICATIONS

The manipulation of entities, engagement in conversation, and the ability to displace them was discussed in regard to assistive and companion robots. The robots and their assistance of a cloud backend were also presented along with its need for analyzing data, determination of execution and the provision of necessary support. A case study based on personal interaction systems was introduced along with the architectural design of the proposed IoRT system [20].

A semantic framework for IoRT-based systems to enhance the development of applications was proposed. The applications targeted to successfully monitor and manage IoRT systems. For contextual modeling, a framework called Smart Rules was presented, which worked essentially on reactive reasoning. To encompass the effectiveness of the proposed system, an elderly person was monitored during their daily activities to assess the working in ambient assisted living (AAL) [21].

A declarative ontological model and a middleware platform for building services were introduced for social robots functioning in smart IoT environments. A data-driven and modular workflow were introduced to determine the time and content for the interactions between humans and robots. The performance of the interfaces was explained in three scenarios including a smart office, a smart home, and a smart medical facility [22].

Several real-world examples were used to study the applications of IoT in the field of robotics attained from research. Many systems and their architecture, communication methodologies, hardware and software systems being utilized at the moment were analyzed highlighting their area of application. The applicability and facilitation brought about by Web of Things (WoT) were discussed concerning the field of robotics [23].

2.6   IoRT WITH SENSING TECHNOLOGIES

To optimize the process of data sharing and diagnosis, a framework for smart sensor interface was presented. To highlight the applications of the proposed design, various implementation schemes and verification experiments were discussed. The proposed system provided valuable results when detecting deviations from normal

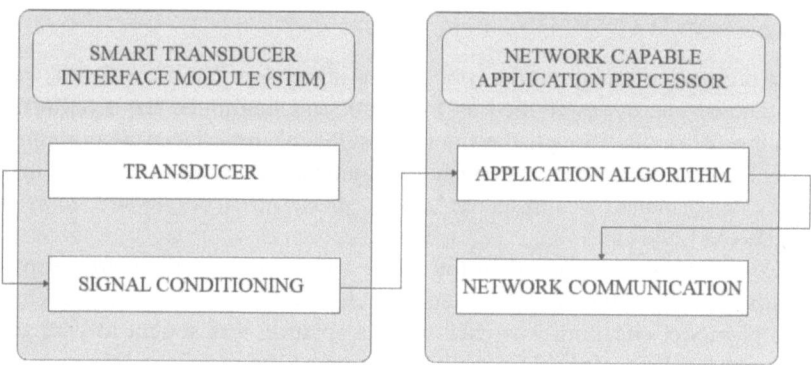

FIGURE 2.6 The partitioning of the smart sensors.

behaviors [24]. The module presented could be utilized for network communications as represented in Figure 2.6.

A system to fuse sensor data, monitor events and act to control several mobile robots was proposed based on the recent advancements in the field of IoRT. The division and a description of the host controller, as well as the multiple robot layers, were included. The proposed system targeted on the fulfilment of the following tasks—avoiding obstacles, creation of a map, moving towards waypoint stability, planning of paths, and localizing robots. Simulation results were also included to highlight the effectiveness of the proposed algorithm. The results included the localization as well as other successes of two robots under consideration [25].

A multi-terrain navigational system was introduced to monitor the environment. The implementation design along with wireless sensor network (WSN) and Internet related technology (IRT) was discussed. The robot measured and further analyzed four physical as well as four chemical parameters to test the air quality. An artificial visual device was also included in the proposed system [26].

2.7 ANN-BASED IoRT

The coverage as well as the interconnectivity among multiple IoRT robots was discussed. To maintain the connectivity among the robots, two methodologies were proposed: IoRT-based and a neural network control scheme. The proposed methods attempted to find a middle ground between collective coverage and the quality of communication. In terms of consumption, energy, and convergence, the system provided effective results [27].

An IoRT-based control system was proposed to maintain the connectivity on a global level when taking into consideration several mobile robots to reach a certain level of quality of service (QoS). The connectivity was proposed to be achieved by utilizing the virtual force algorithm. The inclusion of a neural network-based controller enhanced the system greatly and happened to be very inclusive of IoT-based approaches [28].

2.8 IoRT AND CLOUD APPLICATIONS

A novel protocol to integrate the robot operating system (ROS) with IoT was proposed. The protocol was termed as ROSLink and facilitated the monitoring and controlling of robots through the Internet. ROSLink introduced a communication protocol through the cloud to establish communication between the robots and their users. Its performance was tested on a cloud platform. Accurate and trusty results were received [29].

A framework was proposed in the paper which utilized the platform of cloud computing to rectify the challenges encountered while using model checking. The method of model checking was used when a solution was sought to post an issue being faced by the system. The implementation of the proposed framework along with the validation has been presented with the help of several experiments [30].

2.9 IoRT IN SMART SPACE APPLICATIONS

The proposals in regard to smart spaces and related awareness were brought about in [31]. The awareness in the proposed field could be increased by providing the robots with more refined data. The availability of more data also facilitated the robot with better decision-making abilities. The integration of robots with smart spaces by the utilization of a network communication protocol was discussed. The interaction of the robots with smart space has been highlighted with the help of Figure 2.7.

The robot operating system (ROS) was utilized as a framework to involve smart spaces and robots to expand the horizon of sensory information. The involvement of sensory information becomes crucial as it plays a significant role in avoiding collisions. The IoRT architecture was also discussed in terms of the association of robots with sensors and its interdependence with the network layer as well as

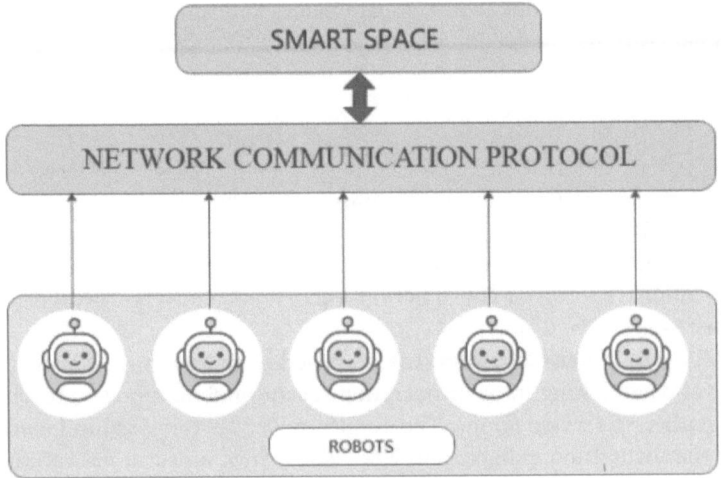

FIGURE 2.7   The interaction between smart space and robots.

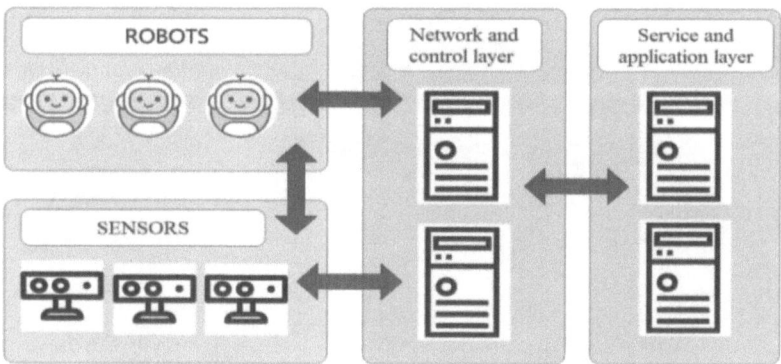

FIGURE 2.8 The architecture of IoRT.

the service and application layers [31]. The architecture of IoRT is explained in Figure 2.8.

The advancements of IoRT and its impact in the field of research were discussed in this paper. The challenges of the integration of robots and corresponding technologies into smart spaces were addressed. The applicability of IoRT technologies in real life was highlighted along with the need for more research in the field. The development of cyber-physical system (CPS) was discussed encompassing applications ranging from power grids, transportation to health care services [3]. The applications have been explored in Figure 2.9.

2.10 IoRT IN OTHER MISCELLANEOUS APPLICATIONS

The utilization of educational robots as smart mobile components was discussed. They found their need in the educational field as they could communicate, facilitate learning, and had computational capabilities. The visualization of knowledge as well as student engagement was discussed with an emphasis on instant feedback. The

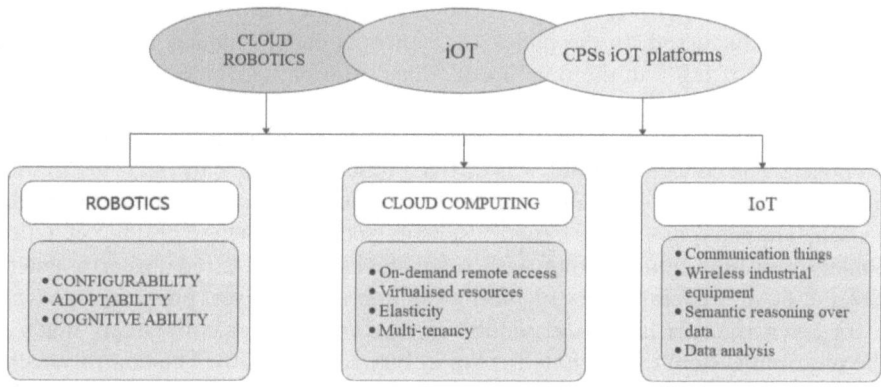

FIGURE 2.9 The applications of cloud, IoT, and CPS in IoRT.

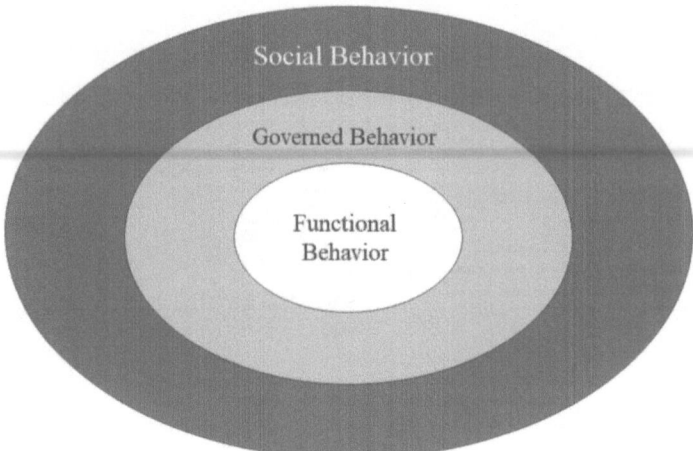

FIGURE 2.10   Robotic things and their layers of behavior.

proposed system was assessed and evaluated using a four-phase model of interest development [32].

Several issues and challenges being brought up by robotic technologies were discussed and the need for fixed regulations was brought about. The importance and requirement of cyber institutions for robotic things have discoursed [33]. The concept of layers of behavior in robotic things was introduced and has been represented with the help of Figure 2.10.

A method for information security was proposed in the sharing domain of the Web of Robotics Things (WoRT), which essentially focused on the preservation of privacy in terms of visual data. Segmentations based on neural networks get utilized to protect the content in the database at the level that is open to access by the users. The data stored is kept anonymous to add a further layer of protection. Results showed that the proposed system is capable of keeping the data of users secure [34].

The application of IoT-based technology in the field of smart homes and spaces was discussed. In the paper, the concepts of IoT majorly highlighted the applications of IoRT in the domain of transportation, agriculture, robotics. and logistics [35].

The development of robotic behavior control to facilitate a smart city was proposed. The concept of IoRT was utilized based on deep learning. The implications of the control methods for the model to react with its environment were discussed. The interconnection between sensors and controls in regard to behavioral control was highlighted along with other necessities required to carry out the tasks at hand. Several layers including the application layer, physical layer, network layer, internet layer, and robotic infrastructure layer collectively made up the required system to facilitate the smart city services [36]. The control mechanisms associated with the robotic system in smart city services are provided in Figure 2.11.

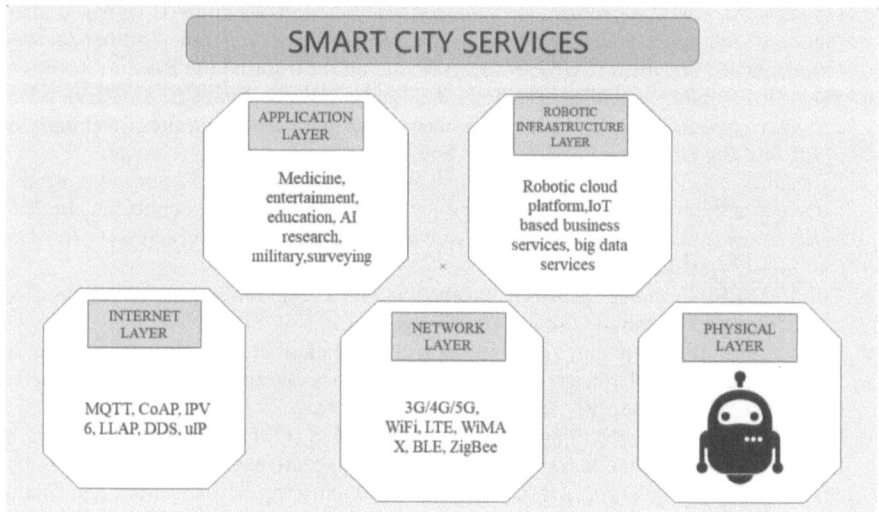

FIGURE 2.11 The behavior of the robotic control system.

2.11 CONCLUSIONS AND PROSPECTS

This chapter provided an overview of the Internet and its applications in various areas of robotics. IoRT has been used in intelligent production facilities. The robotic applications in the health sector are often debated with the Internet (IoT). Robot arm movement applications and its control were taken into account in the IoRT area. In this chapter too, the semantic and web applications of IoRT were taken into account. There is also enough weightage of the advancement of sensing technology and IoRT applications. Also considered was the development of the IoRT based on ANN. It should also be noted that IoRT's cloud applications play an important role. The use of IoRT is also found in intelligent space applications. In addition, a variety of different IoRT applications are addressed, such as in education and intelligent cities. This chapter gives researchers a clear basis for understanding the IoRT principle to think out-of-the-box in real-world applications and to incorporate IoT-based robotic systems.

REFERENCES

1. Schubotz, R., Vogelgesang, C., Antakli, A., Rubinstein, D., & Spieldenner, T. (2017). Requirements and specifications for robots, linked data and all the REST. In *SEMANTICS Workshops*.
2. Beck, M. A., Liu, C. Y., Bidinosti, C. P., Henry, C. J., Godee, C. M., & Ajmani, M. (2020). An embedded system for the automated generation of labeled plant images to enable machine learning applications in agriculture. *arXiv preprint arXiv:2006.01228*.
3. Romeo, L., Petitti, A., Marani, R., & Milella, A. (2020). Internet of Robotic Things in smart domains: Applications and challenges. *Sensors, 20*(12), 3355.
4. Grau, A., Indri, M., Bello, L. L., & Sauter, T. (2017, October). Industrial robotics in factory automation: From the early stage to the Internet of Things. In *IECON 2017-43rd Annual Conference of the IEEE Industrial Electronics Society* (pp. 6159–6164). IEEE.

5. Lee, C. K. M. (2018). Development of an Industrial Internet of Things (IIoT) based smart robotic warehouse management system. In *International Conference on Information Resources Management (CONF-IRM)*. Association for Information Systems, China.

6. Singh, R., Gehlot, A., Gupta, D., Rana, G., Sharma, R., & Agarwal, S. (2019). XBee and Internet of Robotic Things based worker safety in construction sites. In *Handbook of IoT and Big Data* (pp. 81–107). CRC Press.

7. Saravanan, M., Perepu, S. K., & Sharma, A. (2018, November). Exploring collective behavior of Internet of Robotic Things for indoor plant health monitoring. In *2018 IEEE International Conference on Internet of Things and Intelligence System (IOTAIS)* (pp. 148–154). IEEE.

8. Cui, L. (2019). Complex industrial automation data stream mining algorithm based on random Internet of robotic things. *Automatika*, 60(5), 570–579.

9. Masuda, Y., Zimmermann, A., Shirasaka, S., & Nakamura, O. Internet of Robotic Things with digital platforms: Digitization of robotics enterprise. In *Human Centered Intelligent Systems* (pp. 381–391). Springer, Singapore.

10. Patel, A. R., Patel, R. S., Singh, N. M., & Kazi, F. S. (2017). Vitality of robotics in healthcare industry: An Internet of Things (IoT) perspective. In *Internet of Things and Big Data Technologies for Next Generation Healthcare* (pp. 91–109). Springer, Cham.

11. Ongenae, F., De Backere, F., Mahieu, C., De Pestel, S., Nelis, J., Simoens, P., & De Turck, F. (2017). Personalized robotic intervention strategy by using semantics for people with dementia in nursing homes. In *ESWC2017, the Extended Semantic Web Conference; Lecture Notes in Computer Science* (pp. 21–30) Springer, Germany.

12. González-Mendoza, A., Pérez-SanPablo, A. I., López-Gutiérrez, R., & Quiñones-Urióstegui, I. (2018, September). Validation of an EMG sensor for Internet of Things and robotics. In *2018 15th International Conference on Electrical Engineering, Computing Science and Automatic Control (CCE)* (pp. 1–5). IEEE.

13. Guntur, S. R., Gorrepati, R. R., & Dirisala, V. R. (2019). Robotics in healthcare: An Internet of Medical Robotic Things (IoMRT) perspective. In *Machine Learning in Bio-Signal Analysis and Diagnostic Imaging* (pp. 293–318). Academic Press.

14. Katona, J., Ujbanyi, T., Sziladi, G., & Kovari, A. (2019). Electroencephalogram-based brain-computer interface for internet of robotic things. In *Cognitive Infocommunications, Theory and Applications* (pp. 253–275). Springer, Cham.

15. Rossi, S. (2019). Introduction to Internet of intelligent robotic things for healthy living and active ageing.

16. Gawli, K., Karande, P., Belose, P., Bhadirke, T., & Bhargava, A. (2017). Internet of things (IoT) based robotic arm. *International Research Journal of Engineering and Technology*, 4(03), 757–759.

17. Ray, P. P., Chettri, L., & Thapa, N. (2018, March). IoRCar: IoT supported autonomic robotic movement and control. In *2018 International Conference on Computation of Power, Energy, Information and Communication (ICCPEIC)* (pp. 077–083). IEEE.

18. Ishak, M. K., Roslan, M. I., & Ishak, K. A. (2018). Design of robotic arm controller based on Internet of Things (IoT). *Journal of Telecommunication, Electronic and Computer Engineering*, 10(2–3), 5–8.

19. Arefin, S. E., Heya, T. A., & Uddin, J. (2018, June). Real-life implementation of Internet of Robotic Things using 5 DoF heterogeneous robotic arm. In *2018 Joint 7th International Conference on Informatics, Electronics & Vision (ICIEV) and 2018 2nd International Conference on Imaging, Vision & Pattern Recognition (icIVPR)* (pp. 486–491). IEEE.

20. Simoens, P., Mahieu, C., Ongenae, F., De Backere, F., De Pestel, S., Nelis, J., ... & Jacobs, A. (2016, October). Internet of robotic things: Context-aware and personalized interventions of assistive social robots (short paper). In *2016 5th IEEE International Conference on Cloud Networking (Cloudnet)* (pp. 204–207). IEEE.

21. Sabri, L., Bouznad, S., Rama Fiorini, S., Chibani, A., Prestes, E., & Amirat, Y. (2018). An integrated semantic framework for designing context-aware Internet of Robotic Things systems. *Integrated Computer-Aided Engineering, 25*(2), 137–156.
22. Mahieu, C., Ongenae, F., De Backere, F., Bonte, P., De Turck, F., & Simoens, P. (2019). Semantics-based platform for context-aware and personalized robot interaction in the internet of robotic things. *Journal of Systems and Software, 149*, 138–157.
23. Kamilaris, A., & Botteghi, N. (2020). The penetration of Internet of Things in Robotics: Towards a Web of Robotic Things. *arXiv preprint arXiv:2001.05514.*
24. Mahakalkar, N., Pethe, R., Kimmatkar, K. M., & Tiwari, H. D. (2019, May). Smart interface development for sensor data analytics in Internet of Robotic things. In *2019 International Conference on Intelligent Computing and Control Systems (ICCS)* (pp. 957–962). IEEE.
25. Dai, Y., & Lee, S. G. (2020). Multiple Internet of Robotic Things robots based on LiDAR and camera sensors. *International Journal of Advanced Robotic Systems, 17*(2), https://doi.org/10.1177/1729881420913769
26. Singh, R., Gehlot, A., Samkaria, R., & Choudhury, S. (2020). An intelligent and multi-terrain navigational environment monitoring robotic platform with wireless sensor network and internet of robotic things. *International Journal of Mechatronics and Automation, 7*(1), 32–42.
27. Razafimandimby, C., Loscri, V., & Vegni, A. M. (2016, April). A neural network and IoT based scheme for performance assessment in internet of robotic things. In *2016 IEEE first international conference on internet-of-things design and implementation (IoTDI)* (pp. 241–246). IEEE.
28. Akshay, P., Tabassum, N., Fathima, S., & Ahmed, I. (2018). Artificial neural network and IoT based scheme in internet of robotic things. *Perspectives in Communication, Embedded-systems and Signal-processing-PiCES, 2*(6), 126–130.
29. Koubaa, A., Alajlan, M., & Qureshi, B. (2017). Roslink: Bridging ROS with the internet-of-things for cloud robotics. In *Robot Operating System (ROS)* (pp. 265–283). Springer, Cham.
30. Gomez, M. A., Chibani, A., Amirat, Y., & Matson, E. T. (2018). IoRT cloud survivability framework for robotic AALs using HARMS. *Robotics and Autonomous Systems, 106*, 192–206.
31. Uchechukwu, D., Siddique, A., Maksatbek, A., & Afanasyev, I. (2019, November). ROS-based integration of smart space and a mobile robot as the Internet of Robotic Things. In *2019 25th Conference of Open Innovations Association (FRUCT)* (pp. 339–345). IEEE.
32. Plauska, I., & Damaševičius, R. (2014, October). Educational robots for Internet-of-Things supported collaborative learning. In *International Conference on Information and Software Technologies* (pp. 346–358). Springer, Cham.
33. Loke, S. W. (2018, October). Are we ready for the Internet of Robotic Things in public spaces? In *Proceedings of the 2018 ACM International Joint Conference and 2018 International Symposium on Pervasive and Ubiquitous Computing and Wearable Computers* (pp. 891–900), New York.
34. Abbasi, M. H., Majidi, B., Eshghi, M., & Abbasi, E. H. (2019). Deep visual privacy preserving for internet of robotic things. In *2019 5th Conference on Knowledge Based Engineering and Innovation (KBEI)* (pp. 292–296). IEEE.
35. Dmitriev, A. S., Ryzhov, A. I., & Popov, M. G. (2019). Direct chaotic communications and active RFID tags for Internet of Things and Internet of Robotic Things.
36. Liu, Y., Zhang, W., Pan, S., Li, Y., & Chen, Y. (2020). Analyzing the robotic behavior in a smart city with deep enforcement and imitation learning using IoRT. *Computer Communications, 150*, 346–356.

3 Smart Cities and the Use of Internet of Drone Things (IoDT)

Yashonidhi Srivastava*, Sahil Virk*, Souvik Ganguli*, and Suman Lata Tripathi#
*Department of Electrical and Instrumentation Engineering, Thapar Institute of Engineering and Technology, Patiala, Punjab, India
#School of Electronics and Electrical Engineering, Lovely Professional University, Phagwara, Punjab, India

CONTENTS

3.1 INTRODUCTION

Smart cities are the future of the upcoming generation. A smart city uses smart technology to improve infrastructure, culture, lifestyle, capital, and behaviors. More than 50% of the world population lives in cities, and this number is expected to increase to 70% by 2050 [1]. Smart cities integrated with the Internet of Drone Things (IoDT) could be extensively employed to overcome the obstacles related to the growing population and fast urbanization; this will ultimately be going to benefit both the government and the masses. This combined technology would aim to enhance the life quality and the socioeconomic development of cities, and hence will make them more sustainable and efficient [2].

Smart cities using unmanned aerial vehicles (UAVs), that is drones, can extensively boost agricultural production and will take primary sector activities to a new level. Because of the aforementioned technology, farmers can optimally control the growth of crops, which will help to improve crop quality. As extreme climate changes can be predicted, there will be more control over weeds and pests' diseases [3].

DOI: 10.1201/9781003181613-3

33

Moreover, IoDT-based farm technologies and solutions will be affordable and low cost to end users. The technology will aid in irrigation scheduling to monitor soil moisture and weather conditions. IoDTs could be employed over fields to collect the information of soil nutrients for predicting crop health. This move could lead to the rise in food production to feed the expanding global population [4].

Smart cities can employ UAVs for disaster management and rescue applications. Drones are proposed for localization and tracking technology for firefighters helping them in a more precise search operation. For the early detection of accidents, sensors can be employed in IoDT technology in cities. The main advantage of the drone is that these UAVs can enter the high temperature and thick smoke zone where the firefighters could not enter to gather the internal information of the affected area. The best example of this mechanization is the employment of a mobile rescue robot in the nuclear accident of Fukushima power plant for exploration and monitoring operations [5].

In the health sector, drones are commonly used in medical emergencies. Drone-based COVID-19 medical service (DBCMS) is operated to consider the safety issues associated with the medical employees who are more likely to be susceptible to the COVID-19 disease. Furthermore, this technology can also aid in the treatment of COVID-19 patients. The proposed mechanism has categorized the spread of the virus into the three stages: the imported stage, local transmission stage, and the community transmission stage. In the first stage, the travel history of COVID-19-positive patients can be detected. In the next stage, if this infected person stays in close contact with other persons or transfers the virus, then the travel history of that infected person can be obtained. In the last stage, it is difficult to detect the source of transmission because of the community transfer of the virus and is called the epidemic stage. But using this approach, actions can be taken before stage 3 arrives [6].

The remainder of the chapter is constructed in the following manner. The applications of the Internet in drone technology for smart cities is explored in Section 3.2. The use of IoDT in unmanned vehicles is discussed in Section 3.3. Section 3.4 addresses the security and communication issues pertaining to IoDT. The usage of IoT drone technology in agriculture is discussed in Section 3.5. Other applications of IoDT in different disasters have been discussed in Section 3.6, while Section 3.7 discusses its other diverse applications. Section 3.8 outlines the conclusions and points out some future developments in this field.

3.2   IoDT IN SMART CITIES AND SMART GRID APPLICATIONS

A taxonomy to present an overview of IoT and its applications was proposed in the field of integrated ICT, smart cities, and network types. The attempts being made thus far by organizations were also included to cover a broader perspective on the subject. To enhance smart cities and improve the applications of IoT, current efforts were also included, along with a few case studies to highlight the challenges and possible solutions to the same. Traffic management turned out to be a key factor in terms of managing smart cities. A general layout of the modeling has been presented through a diagram in Figure 3.1. Initiatives adopted on a global scale were discussed along with issues that need to be resolved to improve and redirect upcoming research work [7].

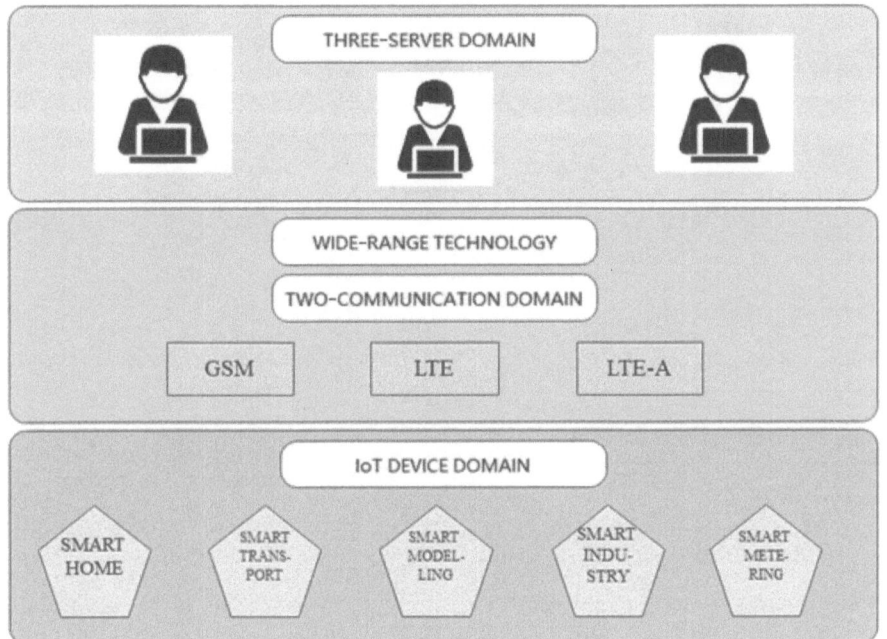

FIGURE 3.1   System of traffic modeling for smart cities.

A model in regards to UAV coalition was presented along with an algorithm attained from the basics of geometry. The objective of the algorithm to find a path from the Dubins curve was discussed. The presented algorithm was further validated with the incorporation of some coalition formation algorithms. The model performance, as well as the results offered, was tested on several parameters. The number and types of resources were varied in accordance with the sensor ranges to test the accuracy and practicality of the model hence presented [8].

With the recent advancements in the field of IoT, several techniques adopted to enhance smart cities were discussed. A survey on the methodologies being adopted as well as the application of collaborative drones was presented. The survey presented had a wider scope and was significantly inclusive. The factors determining the smartness of smart cities were highlighted. The key factors included disaster management, energy consumption, the safety of the public, quality of life being lived, and collection of data and security [9], which are represented with the help of Figure 3.2.

The networking of the alliance of drones with the Internet of Public Safety Things (IoPST) was discussed. The importance of drones in terms of deliveries and wireless communication was presented. The aspects of exploration of areas that are hard to access, the safety of the public, as well as the issue of connectivity were covered. To successfully meet the safety requirements of smart cities, the need for the collaboration of drones and IoPST for real-time monitoring, real-time analytics, and a strong decision-making ability were presented. The objective of the paper to improve safety regulations in terms of traffic control, crime control, and several other areas were discussed [10]. The collaboration and affected fields are presented in Figure 3.3.

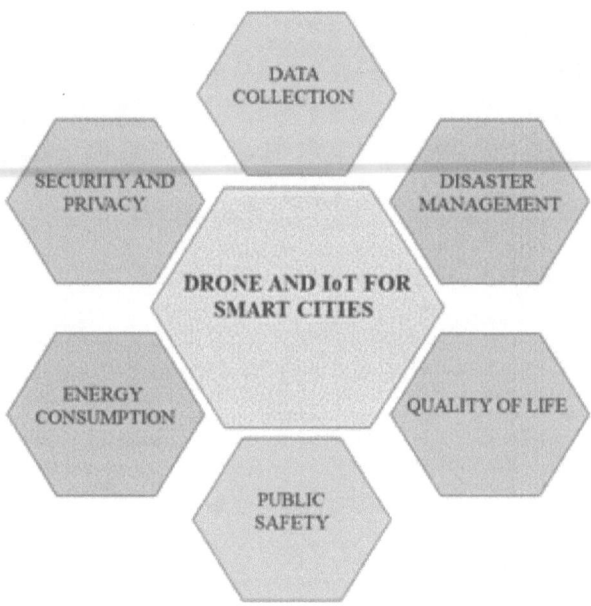

FIGURE 3.2 IoDT and its different aspects.

An analysis regarding the needs of visual IoT was presented in terms of smart cities. A novel IoT architecture named A-VIoT was proposed to improve end-to-end performance. The system possessed the intelligence to decipher complex environments, the ability to analyze videos, crowd coordination, the most optimum utilization of resources, and the availability of controls that promote adaptation according

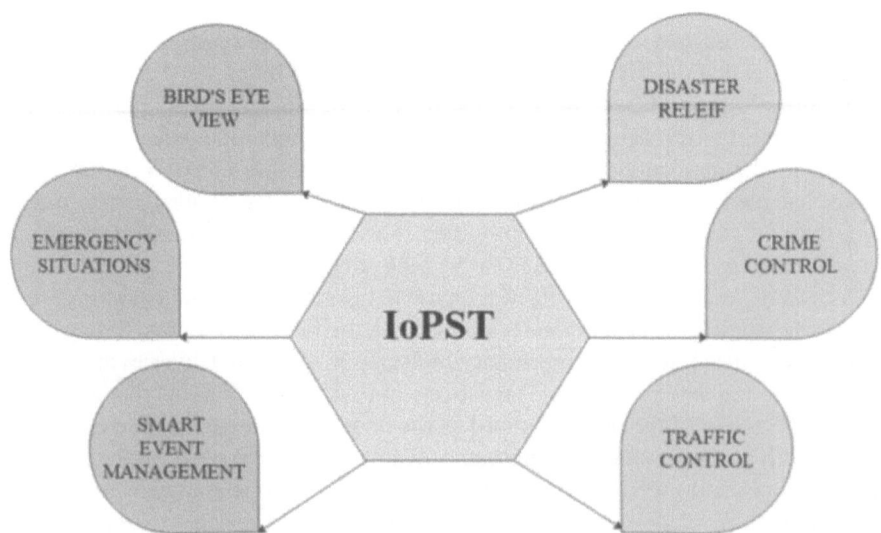

FIGURE 3.3 IoPST and its related components.

to the requirements. Several challenges being faced in the field were highlighted, and the scope for further research was also discussed [11].

Two major hurdles in the field of smart grids, namely the management and communication, were presented along with appropriate solutions. To enhance good communication in a smart grid network, a delay-tolerant system along with the IoDT technology was presented. The system was designed to enable the sharing of information and offloading in the IoT cloud base. A load forecasting method was presented, which turned out to be an amalgamation between boosting strategies and the concept of deep learning. The architecture was developed for improved predictions despite the presence of noise and faulty transmissions in terms of the physical layer [12].

3.3   IoDT IN UNMANNED VEHICLE APPLICATIONS

The recent development in the domain of UAVs was discussed addressing its scope in the market. Its deployment on a global scale was also presented. Crediting the quality of high mobility, its applications in terms of delivery, rescue operations, farming, and the mitigation of pollution were addressed along with a possible role in the field of IoT vision. The issues in terms of the deployment of UAVs were also put forward, which included cases of physical collisions, privacy violations, and instances of smuggling. A survey on the issues was presented in addition to the architecture designed for delivery purposes [13].

A drone management system based on global standards was presented. It was inspired by the interworking proxy entity (IPE), based primarily on drones, and its working. The system hence proposed was demonstrated, and the implementation process was also included. The architectural model for the system was also presented, which included several common services entities. Domains of location, device management, subscriptions, notifications, security, registration, and discovery were encompassed by the CSE structure [14]. The reference model for the layered architecture has been showcased in Figure 3.4 for inquisitive readers.

A methodology for expanding the network connectivity was proposed to achieve the required standard of services arisen from the integration of IoT and UAVs. Predictions in terms of the strength of the signal, as well as fading conditions, were discussed along with adaptive transmissions. The transmissions and its benefits were put forward in regards to the end-users. The ability of ANN to accurately predict the strength of the signal was addressed along with the results that covered the distortion of signals, as well as techniques to resolve them [15].

3.4   IoDT: SECURITY AND COMMUNICATION ISSUES

Concerns regarding drone security and the areas of vulnerability of WiFi-enabled drones were addressed. Multiple varieties of attacks were identified and implemented on commercially available drones. Each attack that was implemented had its method stated along with the process of its execution that was highlighted during the process of implementation. The current status of drone security along with its management was discussed. Its control, resilience, and other concerns were also highlighted to make the working even more secure [16].

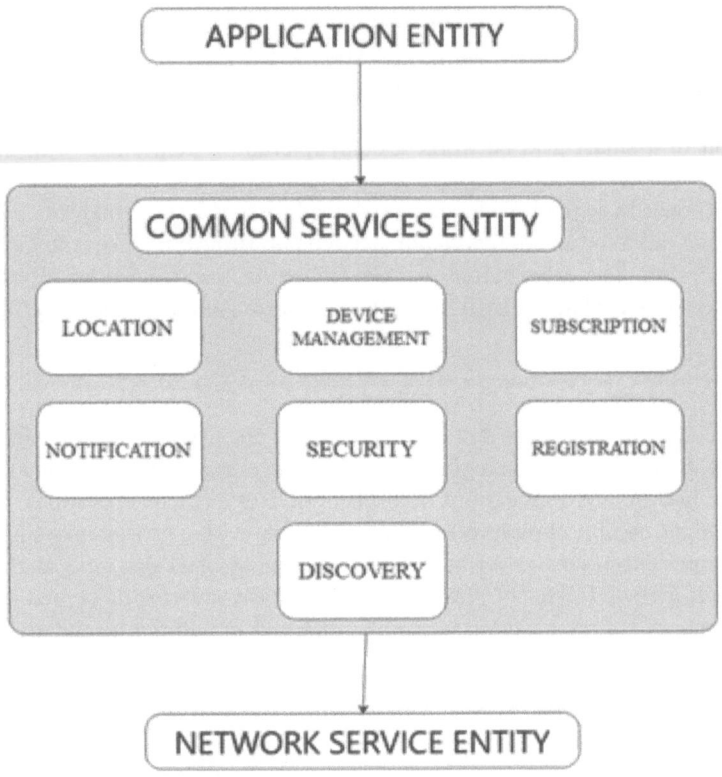

FIGURE 3.4   A reference model of the layered architecture.

Multiple real-time security vulnerabilities and latency issues and their conse-quences were discussed. To overcome the mentioned challenges, an authentication model was introduced based on blockchain technology. The objective of the pro-posed architecture was to make the system more secure and minimize latency. A comparative study with other existing models has also been presented along with experimental results. It was concluded that the architecture presented had high throughput and low delay [17].

A multi-hop airborne method was presented utilizing the latest Bluetooth tech-nology. The objective was to connect several drones for real-time transfer of data. To successfully execute communication among drones and between drones and the ground, an onboard Bluetooth transceiver (OBT) was presented. To assess real-time flight trajectories and data delivery, a graphical user interface (GUI) was introduced. An overview of the system is represented in Figure 3.5. Experimental results were also discussed to test flight control systems [18]. A replica of the suggested network is provided in Figure 3.5.

An information-searching algorithm was presented to extract and segregate infor-mation from the ratings received in regard to services rendered in the field of IoDT. The presented algorithm was based on mechanisms used in drone networks to exploit information. Experimental results were discussed, which therefore compelled the conclusion of the algorithm being efficient, accurate, and fast [19].

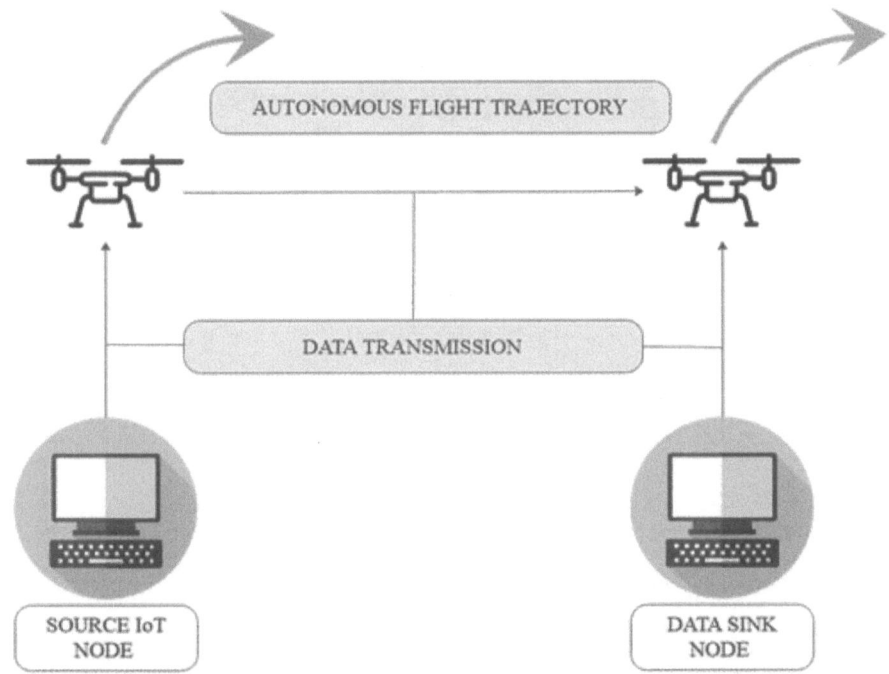

FIGURE 3.5　An overview of an IoDT system.

3.5 APPLICATIONS OF IoDT IN SMART AGRICULTURE

Multiple applications of IoT in the field of farming were discussed along with the challenges associated with them. The requirement of wireless communication and other needs for the devices to be able to function effectively were also analyzed and addressed. IoT systems enabled with sensors to facilitate smart agriculture were studied. Several IoT-based systems were then reviewed, and the flaws were pointed out. Areas of improvement were highlighted along with the provision of a road map for further work [4].

The demand and need for UAVs in agriculture were highlighted. The increase in efficiency post the usage of drones with sensors, cameras, and integrating modules was discussed. It was concluded that a possible integration with ML and IoT-based technologies could be significantly transformative. The proposed solution included the usage of Raspberry Pi 3B module in collaboration with the GPS module. With the help of the red-green-blue (RGB) and the gas sensors, the model provided efficient results [20]. The proposed model is shown in Figure 3.6.

A do-it-yourself (DIY) technique was discussed to develop an agricultural drone. The applications of computer-guided tools, their implementation, and the design requirements were addressed. The technology pertaining to IoT and cloud computing was utilized to successfully achieve the task at hand. Post the assessment of the results received, it was concluded that a drone could be built for the required agricultural assistance. Future aspects of the drone and possible amendments in its design were also included [21].

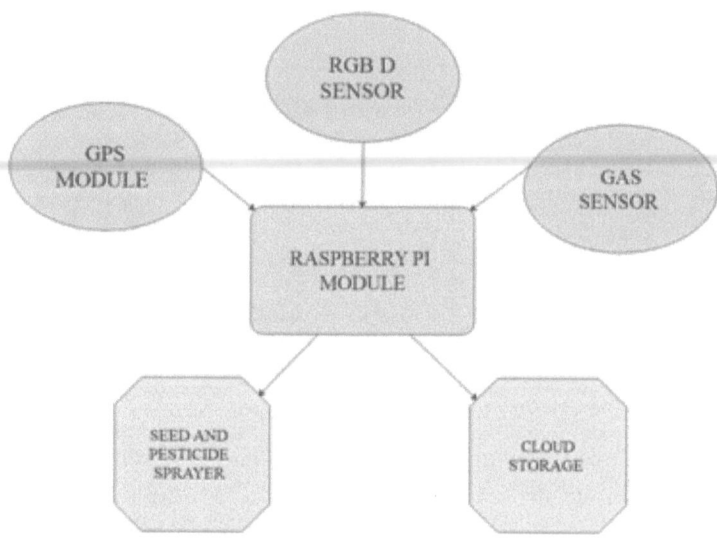

FIGURE 3.6   A diagrammatic representation of the proposed model.

Several challenges that have arisen in the domain of agriculture due to rising population, extreme weather conditions, and climatic transformations were addressed. The concept of smart farming and the utilization of IoT-based techniques were presented. The impacts of the usage of mentioned technologies to promote growth, reduce wastage, and improvements in the quality of plantations were discussed. The hardware as well as the software requirements for smart farming were highlighted along with a description of the results hence achieved [22].

The research presented provided on overview of agricultural solutions based on the technology required for the provision of crucial insights. The health of the crop as well as the requirements to improve the same were discussed. The same was achieved by extraction of complementary features from datasets as well as the minimization of crop ground survey. The mentioned factors played a major role when the land under consideration was a large area [23].

A method to predict rainfall based on the genetic algorithm was proposed. It was a decision-making algorithm to decide whether or not a manual supply of water was needed. The rectification of the proposed method was done by sensing the moisture content in the soil of that particular locality. A system based on the concept of terrace gardening was also included, which utilized a pump to provide water [24]. The representation of the drones during the process of deployment is illustrated in Figure 3.7.

3.6   IoDT IN DISASTER MANAGEMENT AND COVID-19 APPLICATIONS

The creation of an accident monitoring system based on IoDT-based technology was discussed. The presented design consisted of a subsystem based on IoT, three drone-based subsystems, and a single wired subsystem. Several sensors and systems were

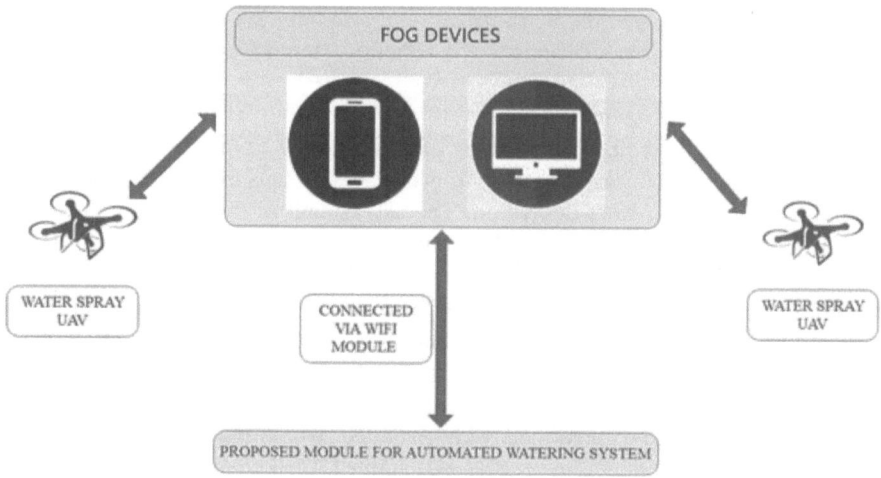

FIGURE 3.7   The representation of drones during deployment.

taken under consideration during the development of block diagrams. Post the development of block diagrams, reliability models of the system were developed. The possibilities of a system free of failures were analyzed and addressed [25].

The utilization of UAVs for the prevention of accidents was discussed. IoT-based technology was utilized to successfully launch from fire control units. The procedure of the system including the working of the sensors and alert calls being sent to the control unit was described in detail along with the possibilities of actions to be taken at the time of a distress call. The ability of the machine to land and reload to further rectify the situation was also explained. The proposed method and the expectations to improve the quality of service was also described [26].

A mechanism was proposed catering to the healthcare workers, who are more susceptible to COVID-19 infections. The system was termed DBCMS. The proposed method was expected to significantly improve the treatment procedure of those suffering from the infection. The utilization of drone-based services to minimize contact and to hence keep medical workers safe was described. The importance of social distancing was elaborated upon to prevent the need for being admitted in a hospital [6].

3.7 IoDT IN OTHER MISCELLANEOUS APPLICATIONS

The presented model in [27] evaluated the behavior of drones and also assessed the potential of loopholes and vulnerabilities. The verifications being carried out for the model were described along with the presentation of simulations. The presented model was significantly successful in the procedure of drone selection. The probability of the selection of the appropriate drone increased critically along with a decline in the number of accidents [28].

The concepts and forms of AVUS were clarified in [29]. In addition to that, the possible methods and devices needed to originate solutions were discussed. The possible challenges that could be faced while the facilitation of the system was also highlighted. The feasible solutions to the concerns stated were also included in the article [29].

A drone positioning measuring system in the field of Internet-aided drone topology was introduced. The abilities of every drone to be connected, features that enabled the sharing of location, as well as the measurement of the distance between other drones, were described. The assessment of the position of obstacles was also discussed. Simulation results were included to emphasize the fact that proposed methodology could significantly improve the process of the determination of position-utilizing technology based on IoDT [27].

A drone-enabled relay system was presented based on IoT technology. The objective was to provide environmental monitoring as well as fast data collection. Long-range (LoRa) technologies were adopted for better utilization of high-powered modules. Results from experiments were included, which made it easy to conclude the fact that the data collection process was stable [30]. The architecture of the proposed system has been represented in the form of a diagram, given in Figure 3.8.

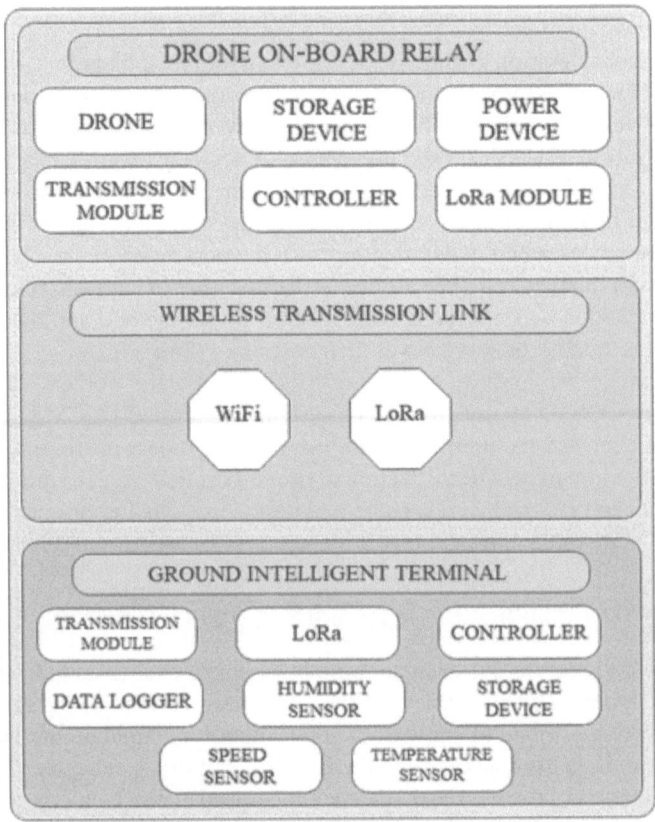

FIGURE 3.8 The architecture of the Internet-supported drone technology used for data collection.

3.8 CONCLUSIONS

This chapter dealt with a broad overview of the Internet and advances of drone technology in various areas of our everyday lives, with a specific emphasis on smart cities. More precisely, the chapter discussed drone technology with Internet applications in the smart city scenario. The use of IoDT in unmanned vehicles was also discussed. The chapter also dealt with security and communication problems related to IoDT. In various agricultural uses, additional IoDT implementations were presented. The use of IoT-based drone technology to monitor flood and fire outbreaks has also been pointed out. Thus these drone applications can benefit mankind in manifold ways and may lead to the development of smart cities. Some specific applications like traffic management, controlling unmanned vehicles, aiding in the emergency corridor, smart manufacturing, as well as agricultural developments, disaster management, and last but not the least managing COVID outbreaks are the areas that can be improved by drone technology supported by IoT framework.

REFERENCES

1. Alshekhly, I. F. (2012). Smart cities: Survey. *Journal of Advanced Computer Science and Technology Research Vol*, *2*(2), 79–90.
2. Joshi, S., Saxena, S., & Godbole, T. (2016). Developing smart cities: An integrated framework. *Procedia Computer Science*, *93*, 902–909.
3. Al-Turjman, F., & Altiparmak, H. (2020). Smart agriculture framework using UAVs in the Internet of Things era. In Fadi Al-Turjman (Ed.), *Drones in Smart-Cities* (pp. 107–122). Elsevier, USA.
4. Ray, P. P. (2017). Internet of things for smart agriculture: Technologies, practices and future direction. *Journal of Ambient Intelligence and Smart Environments*, *9*(4), 395–420.
5. Kim, Y. D., Son, G. J., Kim, H., Song, C., & Lee, J. H. (2018). Smart disaster response in vehicular tunnels: Technologies for search and rescue applications. *Sustainability*, *10*(7), 2509.
6. Angurala, M., Bala, M., Bamber, S. S., Kaur, R., & Singh, P. (2020). An internet of things assisted drone based approach to reduce rapid spread of COVID-19. *Journal of Safety Science and Resilience*, *1*(1), 31–35.
7. Mehmood, Y., Ahmad, F., Yaqoob, I., Adnane, A., Imran, M., & Guizani, S. (2017). Internet-of-things-based smart cities: Recent advances and challenges. *IEEE Communications Magazine*, *55*(9), 16–24.
8. Ismail, A., Bagula, B. A., & Tuyishimire, E. (2018). Internet-of-things in motion: A uav coalition model for remote sensing in smart cities. *Sensors*, *18*(7), 2184.
9. Alsamhi, S. H., Ma, O., Ansari, M. S., & Almalki, F. A. (2019). Survey on collaborative smart drones and internet of things for improving smartness of smart cities. *IEEE Access*, *7*, 128125–128152.
10. Alsamhi, S. H., Ma, O., Ansari, M. S., & Gupta, S. K. (2019). Collaboration of drone and internet of public safety things in smart cities: An overview of qos and network performance optimization. *Drones*, *3*(1), 13.
11. Ji, W., Xu, J., Qiao, H., Zhou, M., & Liang, B. (2019). Visual IoT: Enabling internet of things visualization in smart cities. *IEEE Network*, *33*(2), 102–110.
12. Mukherjee, A., Mukherjee, P., De, D., & Dey, N. (2020). iGridEdgeDrone: Hybrid mobility aware intelligent load forecasting by edge enabled internet of drone things for smart grid networks. *International Journal of Parallel Programming*, *49*(3), 285–325.

13. Motlagh, N. H., Taleb, T., & Arouk, O. (2016). Low-altitude unmanned aerial vehicles-based internet of things services: Comprehensive survey and future perspectives. *IEEE Internet of Things Journal, 3*(6), 899–922.

14. Choi, S. C., Sung, N. M., Park, J. H., Ahn, I. Y., & Kim, J. (2017, July). Enabling drone as a service: OneM2M-based UAV/drone management system. In *2017 Ninth International Conference on Ubiquitous and Future Networks (ICUFN)* (pp. 18–20). IEEE.

15. Alsamhi, S. H., Ma, O., & Ansari, M. S. (2018). Predictive estimation of the optimal signal strength from unmanned aerial vehicle over internet of things using ANN. *arXiv preprint arXiv:1805.07614.*

16. Westerlund, O., & Asif, R. (2019, February). Drone hacking with raspberry-pi 3 and wifi pineapple: Security and privacy threats for the internet-of-things. In *2019 1st International Conference on Unmanned Vehicle Systems-Oman (UVS)* (pp. 1–10). IEEE.

17. Yazdinejad, A., Parizi, R. M., Dehghantanha, A., Karimipour, H., Srivastava, G., & Aledhari, M. (2020). Enabling drones in the internet of things with decentralized blockchain-based security. *IEEE Internet of Things Journal, 8*(8), 6406–6415.

18. Li, K., Lu, N., Zhang, P., Ni, W., & Tovar, E. (2020, April). Multi-drone assisted internet of things testbed based on bluetooth 5 communications. In *2020 19th ACM/IEEE International Conference on Information Processing in Sensor Networks (IPSN)* (pp. 345–346). IEEE.

19. Rehman, A., Paul, A., Ahmad, A., & Jeon, G. (2020). A novel class based searching algorithm in small world internet of drone network. *Computer Communications, 157,* 329–335.

20. Saha, A. K., Saha, J., Ray, R., Sircar, S., Dutta, S., Chattopadhyay, S. P., & Saha, H. N. (2018, January). IOT-based drone for improvement of crop quality in agricultural field. In *2018 IEEE 8th Annual Computing and Communication Workshop and Conference (CCWC)* (pp. 612–615). IEEE.

21. Suhas, M. V., Tejas, S., Yaji, S., & Salvi, S. (2018, October). AgrOne: An agricultural drone using internet of things, data analytics and cloud computing features. In *2018 4th International Conference for Convergence in Technology (I2CT)* (pp. 1–5). IEEE.

22. Mat, I., Kassim, M. R. M., Harun, A. N., & Yusoff, I. M. (2018, November). Smart agriculture using Internet of Things. In *2018 IEEE Conference on Open Systems (ICOS)* (pp. 54–59). IEEE.

23. Shafi, U., Mumtaz, R., Iqbal, N., Zaidi, S. M. H., Zaidi, S. A. R., Hussain, I., & Mahmood, Z. (2020). A multi-modal approach for crop health mapping using low altitude remote sensing, Internet of Things (IoT) and machine learning. *IEEE Access, 8,* 112708–112724.

24. Roy, S. K., & De, D. (2020). Genetic algorithm based internet of precision agricultural things (IopaT) for agriculture 4.0. *Internet of Things*, 100201, DOI:10.1016/j. iot.2020.100201.

25. Fesenko, H., Kharchenko, V., Sachenko, A., Hiromoto, R., Kochan, V., Kor, A., & Rucinski, A. (2018). An Internet of Drone-based multi-version post-severe accident monitoring system: Structures and reliability. In Vyacheslav Kharchenko, Ah Lian Kor, & Andrzej Rucinski (Eds.), Dependable IoT for Human and Industry Modeling, Architecting, Implementation (pp. 197–217). River Publishers, Denmark, The Netherlands.

26. Jayapandian, N. (2019, November). Cloud enabled smart firefighting drone using internet of things. In *2019 International Conference on Smart Systems and Inventive Technology (ICSSIT)* (pp. 1079–1083). IEEE.

27. Lee, C. Y. (2020, January). Cooperative drone positioning measuring in internet-of-drones. In *2020 IEEE 17th Annual Consumer Communications & Networking Conference (CCNC)* (pp. 1–3). IEEE.

28. Sharma, V., Choudhary, G., Ko, Y., & You, I. (2018). Behavior and vulnerability assessment of drones-enabled industrial internet of things (IIoT). *IEEE Access, 6*, 43368–43383.
29. Ahmadhon, K., Al-Absi, M. A., Lee, H. J., & Park, S. (2019, February). Smart flying umbrella drone on internet of things: AVUS. In *2019 21st International Conference on Advanced Communication Technology (ICACT)* (pp. 191–195). IEEE.
30. Zhang, M., & Li, X. (2020). Drone-enabled internet of things relay for environmental monitoring in remote areas without public networks. *IEEE Internet of Things Journal, 7*(8), 7648–7662.

4 The Roles of Artificial Intelligence, Machine Learning, and IoT in Robotic Applications

Jappreet Kaur, Souvik Ganguli#,
and Suman Lata Tripathi†*
*Department of Computer Science and
Engineering, Thapar Institute of Engineering
and Technology, Patiala, Punjab, India
#Department of Electrical and Instrumentation
Engineering, Thapar Institute of Engineering
and Technology, Patiala, Punjab, India
†School of Electronics and Electrical Engineering, Lovely
Professional University, Phagwara, Punjab, India

CONTENTS

DOI: 10.1201/9781003181613-4

4.1 INTRODUCTION

Artificial intelligence (AI) and machine learning (ML) are opening new opportunities in virtually all industries, thus making commonly used equipment more proficient. No wonder that AI and machine learning are often applied to robots to improve them. AI and robotics are a leading arrangement for automating chores. Lately, the existence of AI has progressively developed and is now very common in robotics, bringing together flexibility and learning competencies in formerly inflexible applications [1, 2].

Currently, AI is used in online advertising, driving, aeronautics, medication, and image recognition. The modern success of AI has caught the attention of both the scientific community and the public. An example of this is vehicles equipped with an automated steering system, also referred to as autonomous cars. Each vehicle is equipped with a series of LIDAR sensors and cameras, which allow recognition of its three-dimensional environment and provides the ability to make intelligent decisions on real traffic road situations. One more example is the Alpha-Go, established by Google DeepMind, to play the board game Go. Sedol became the very first appliance to beat a specialized competitor, and it recently triumphed against the current world champion, Ke Jie, who is from China. Also the actual count of potential games in Go is projected to be 10^{761} and assuming the thrilling difficulty of the game, the best AI researchers suppose that it will take years for this to happen. This, in turn, has led to both the eagerness and terror in many that AI will beat humans in all the turfs it walks into [1, 2].

ML is a branch of learning in AI that provides computers with the capability to study from statistics. To do this, ML determines the extension of prototypes that are capable of learning and predicting from accessible data. These prototypes work with the help of algorithms that can estimate based on the statistics instead of ambiguous programs. Consequently, ML is frequently used in a variety of difficulties where designing precise systems is not convenient. Thus, machine learning can substitute for human proficiency in data management. Also, ML delivers the algorithmic gears for allocating with data and delivering estimates [3, 4].

ML technology is drastically changing how robots work and dramatically increasing the range of their abilities. ML is just one facet of upgraded robotics. Robotics has challenging statistical necessities, and those are being aided by developments in multicore processing power. Furthermore, variations in instrument expertise and even a motor-powered mechanism have influenced development. Robots are now being equipped with RADARs to deliver them with more awareness of their surroundings [4].

Robotics is undoubtedly an exciting application of ML skills. It is categorized as a direct interconnection with the actual physical world, physical response, and acomplicated controlling scheme. Recently, numerous methods have been tried to relate ML to precise robotics chores. Yet, to date, we are quite far away from a completely automated robot control system with the capability of learning [5, 6].

In recent years, the robotics community and the IoT community have together come up with the Internet of Robotic Things (IoRT). The idea of IoRT is one in which intellectual appliances are capable of monitoring the proceedings taking place in the surroundings, and then their sensor data is fused. The appliances then use native and dispersed intellect to choose the sequence of actions that are to take place and then act to employ those ideas in the actual world [7].

Until now, the IoRT community has been focused on varied but at the same time extremely linked goals. IoT aimed at subsidiary facilities for persistent detecting, observing, and tracing, whereas the robotic community aimed at manufacturing accomplishment, communication, and independent behavior. These two communities combined have added immense value to many fields [8]. Further in this chapter, we are going to discuss the inventions and wonders bought to the world by combining AI, ML, and IoT with robotics.

The remaining chapter is structured in the following fashion. Section 4.2 highlights some contributions of AI to the field of robotics. The role of ML and its applications in robotics are deliberated in Section 4.3. Section 4.4 points out the association of IoT with robots. In Section 4.5, the future trends of robotics are discussed. Section 4.6 concludes the chapter, giving an outline for future trends.

4.2 ARTIFICIAL INTELLIGENCE AND ROBOTICS

4.2.1 CHOOSING CORRECT PATHS

AI plays an extremely valuable role in the application of robotic assembly. When AI collaborates with advanced vision systems, the systems benefit with the correction of real-time courses, which is most suitable for intricate industrialized sectors, such as aerospace. AI can also aid in helping robots learn on their own and decide the best paths for the courses that are to be performed [5].

Robotic process automation (RPA) is the expertise that permits its user to design software for the computer or in this case a robot, to implement any business idea by assimilating and competing with the activities of a human interrelation along with digital electronics. Similar to humans, RPA robots also use the interface of the operator to deploy the applications and note data. These RPA robots construe, generate replies, and interact with the rest of the systems to complete massive ranges of monotonous activities [2, 3]. RPA is explained with the help of Figure 4.1.

The RPA-based robot implements its share of the processor, and at the same time another robot does its minute task of aligning the assembly workers, and thus the process is completed. A normal robot just does whatever it is told; it never makes any improvisations nor does it learn anything. On the other hand, when we talk about an AI robot, it is altogether a different thing. This particular robot after being trained once becomes better in the given task with time [3, 4].

FIGURE 4.1   The RPA.

4.2.2   ARTIFICIAL INTELLIGENCE AND ITS USAGE IN PACKAGING INDUSTRIES

AI has a very strong capability to bring forward transformational proficiencies into packaging operations. AI is also capable of assisting with keener examinations of products, to ensure that the deliveries are precise and that each customer gets what they ordered. Many AI-based tools, such as the AI power vision system, are used to execute quality examination, sorting, and instantaneous tracing making sure that there are minimal human errors.

One of the most important aspects of the packaging industry is the quality examination and the machine vision systems that are used to sense and recognize irregularities. These machine vision systems cannot learn on their own, however, the systems are reprogrammed depending on the object that has to be examined. The supply chain thus can be enhanced with the capability to give an instantaneous tracing status of the object. This ability later leads to better efficiency in the business [4].

The packaging industry uses diverse systems of AI along with robotics for faster, comparatively cheaper, and much more precise packaging. AI and robotics aid in saving certain steps and also help in the regular refinement of the process of packaging, which in return makes the process much easier and efficient [4, 5].

4.2.3 APPLICATION OF ARTIFICIAL INTELLIGENCE IN THE SORTING OF RECYCLED GOODS

Sorting recycled goods is an application that barely comes into our mind whenever we talk about the packaging industry and the application of AI within it. But this is actually a perfect application for the industry. Considering the large quantity of waste which is generated and thrown into the seas or landfills every single day, we need to work on and upgrade the reprocessing and recycling procedures. To do this, we need to make sure that the products that are being recycled and also their by-products are discarded properly. If all this is executed with a perfect plan with proper use of AI and robotics, then this tiring, arduous, and monotonous job will become easier and much less time-consuming.

4.2.4 APPLICATIONS OF ARTIFICIAL INTELLIGENCE IN THE FOOD PACKAGING SECTOR

The Carlton Council of North America along with AMP Robotics bought a revolutionary inventive in the Food Packaging industry with the incorporation of AI and Robotics into it. They came up with a robot by the name of CLARKE which was capable of sorting different sorts of products like boxes, plastic containers, and bottles.

After collecting all the recyclable waste from North America, it is then taken to the materials recovery facility (MRF). This is where all the collectables are sorted and then distributed into batches, and then later these batches are distributed among other facilities where they are then recycled. After the distribution process is done by MRF, then it's time for CLARKE to take charge. CLARKE is a robot that scans all the products through its camera. With its ability to scan a huge amount of products, it helps the AI technology to learn and decide which products are to be recycled and which are to be thrown out.

According to the latest data collected, CLARKE is capable of identifying 150 different types of cartons, and it is also capable of picking up those cartons using its spider-like grippers. Also, CLARKE picks up 60 times in one minute, whereas a typical human worker can only pick around 40 cartons per minute.

4.2.5 ARTIFICIAL INTELLIGENCE-POWERED VISION SYSTEMS FOR INSPECTION

AI is capable of assisting with keener scrutiny. AI can assist with smarter inspection of goods, to ensure that shipments are accurate with respect to what is being delivered, to whom, and in what quantity. The use of AI-powered tools like AI-powered vision systems can perform superior scrutiny, sorting, and live tracing, making sure that there is no human error [1].

One of the most important components of the packaging industry is scrutinizing the quality of products, and the system that is used for identifying the irregularities is a machine vision system. This system is not capable of learning on its own, therefore it has to be reprogrammed every time depending on the product that has to be scrutinized. However, because of the capability of giving live tracking of the records, the chain of supply can be enhanced. Also, the availability of live tracking helps in the all-round development of the industry [3].

One of the finest case studies showing how the packaging industry has been affected by AI is that of the Packaging Distributors of America (PDA). This particular company was running a lighting system business, and it was going through crisis. They went through a remarkable customer complaint rate of around 75%, which kept on rising. Customers used to complain about receiving packages with missing products. To resolve this problem, PDA came up with a solution and installed AI–powered cameras in the packaging machines. ML technology was used by the cameras to keep a check on what goes into which package. This solved their problem. There were no more customers complaining, and the complaint rate reduced from 75% to zero [9, 10].

4.2.6 ARTIFICIAL INTELLIGENCE FOR WAREHOUSE AUTOMATION

Another game-changer for several industries these days is the AI–powered warehouse automation system. There is near incorporation among deep technology, as it has successfully brought about remarkably effective enhancements and at the same time also maintained a check on business expenses. Warehouse management is an intricate procedure, and it has its own distinctive set of difficulties. There exists a close link among automation and warehouse systems; better results are observed if it is more simplified and associated with other machines and procedures. IoT-powered warehouses are known as smart warehouses. They employ AI and several other technologies like ML and RFID, which assimilate data from different bases all over the warehouse to progress and enhance the procedures on their own without the requirement of human interference or any type of programming [6].

Packaging is the key player for current AI and smart warehousing organizations. AI technologies are capable of gathering data from all the available data points and then using it for suggesting which packaging is to be used. One of the examples for such a system is given by the Tetra Pack Index Survey which took place in 2018. According to this survey, 74% of consumers from the United States and 80% from South Korea decided to avoid using products with multiple layers of packaging as a step toward protecting the environment. This step acted as a call for revisiting the regulations and customs of packaging and bring up alternatives that could meet the demands of the consumers [6, 7].

4.2.7 ARTIFICIAL INTELLIGENCE IN CUSTOMER SERVICE

Today, customer service capacity in large-scale shops, malls, and the hotel industry are using robots all over the world. The majority of these robots being used are powered by AI natural language dispensing capabilities, so that they can communicate with visitors similarly as humans would so. Also, there is a special ability of these AI-powered robots, which is that the more they interact, the more data they collect, the more they learn [3, 4]. The process of customer service powered by AI is provided in Figure 4.2.

FIGURE 4.2   AI-powered customer service process.

4.2.8   SELF-DRIVING CARS

Self-driving cars are also known to many as autonomous vehicles (AVs), connected and autonomous vehicles (CAVs), driverless cars, and robocars. These robocars can sense their surroundings and move around safely with little, or in most cases no, human interference [7, 8].

Robocars utilize several different types of sensors, such as SONAR, GPS, odometer, and RADAR to sense their environment. To interpret the information gathered by the installed sensors, categorize the suitable navigation directions, and identify the hindrances, advanced control systems are used.

These connected and autonomous vehicles have some firm features just like digital electronics, which extricate their technology from other types of cars. Due to these technologies, these vehicles are capable of becoming more responsive and transformative towards changes. Some of these technologies are decoupling, modularity, cross-map reading, and vehicle communication systems.

To make a vehicle 95% as secure as one with a driver, around 275 million autonomous miles are needed failure-free, while numerous billions of miles are required to make them 10% or 20% safer than humans.

Apple is one of the major companies testing these robocars. They started with three cars in April 2017 then increased to 27 cars in January 2018 and then further increased to 45 in March 2018. According to the latest records, they are currently having 69 driverless robocars on-road [8, 11].

4.2.8.1   Challenges Being Faced by Robocars

The possible assistance from improved automation may be restricted by predictable difficulties like arguments over accountability, the time required to revenue the prevailing stock of cars from non-autonomous to autonomous, and therefore an extended era of humans and self-directed cars sharing the streets, conflict by people regarding surrendering control of their vehicles, fears about security, and the application of a legal framework and steady universal government guidelines for self-directed vehicles.

Some other difficulties are low level of driving skill for dealing with possibly hazardous circumstances and irregularities. Ethical difficulties arise where the software of the autonomous car is enforced during an unescapable crash to pick amid many injurious sequences of action. Other problems would be that this will leave a mass population of drivers unemployed. Further, there is a possibility for added invasive mass scrutiny of situation- and travel-utilizing forces and intelligence agency. Thus, the AI will be able to have access to a larger dataset, which is generated by the sensors. Also, it will be very difficult for the police to recognize the sounds of the non-verbal clues in case there is an accident [9, 12].

Another problem is that the sensors are not able to detect some of the animals, such as kangaroos, which act like hindrances. Also, these automated self-driving vehicles require extremely high-graded maps, which are specialized so that there are no anomalies. These maps would need to be updated again and again whenever any route is out of date or a new route is built. Another big challenge is that the road infrastructure existing in the world today may need to be changed to make it compatible with the robocars [12, 13].

4.3   MACHINE LEARNING APPLICATIONS IN ROBOTICS

4.3.1   DATE LABELING AND MACHINE LEARNING

ML has been a game-changer for business industries. It has been able to provide precise data, scrutiny reports, and perceptions on the processes taking place. The packaging industry is also one of the industries that have highly profited by implementing ML technologies into their industry.

ML has helped this industry by saving them from customer dissatisfaction by the implementation of this technology in the labeling process by helping in better standardization. If the products are not labeled correctly it can lead to failures in scrutiny and dissatisfaction among consumers. Today, the implementation of ML has become a necessity for the labeling process to ensure that there are no errors and the process is efficient.

4.3.2   ASSISTIVE AND MEDICAL TECHNOLOGIES

Jaffe [14] states that a robot is a device that has capabilities similar to a human brain; it can process information and perform actions, which can help human beings with different sorts of disabilities. There are movement robots, which are capable of giving therapeutic benefits to human beings. These technologies are frequently reserved for medical labs, as they are very costly for most of the hospitals all over the world.

MICO is a robotic arm, which is one of the latest inventions among the ML-powered robotic technology, and it was developed by collaborating autonomy with assistive machines at Northwestern University.

The difficulties faced while developing these robots is much more complex than one can conceive, although smart robots are capable of making the required adjustments as per the users' requirements. In comparison to some other industries, the healthcare industry is the one that is making the best use of ML technologies and their application in robotics. An association between the researchers from several universities along with a group of physicians have led to the invention of a smart tissue–automated robot through the Center for Automation and Learning for medical robotics [9, 12].

4.3.2.1 The MICO Robot

The MICO Robotic Arm is segmental and effective with a load-to-weight proportion that is perfect for daily actions, impeccable for progressive management research. A standard 24-volt battery is used to power the arm, making it convenient for users. The MICO robot can carefully manage the daily regimes of the person by using the two under-actuated fingers. It has 6-degree freedom and permits supple, harmless, and active contact in nearly any situation [13, 10, 15].

4.3.3 IMITATION LEARNING

Imitation learning systems attempt to imitate human behavior in any given chore. The concept of training by imitation has been part of the teaching concept for a long time, though this field of imitation learning is achieving consideration lately because of the developments in sensing and computing and also due to growing demands for intellectual claims. The model of gaining knowledge by imitating is achieving great admiration because it simplifies instructing complicated chores with negligible professional knowledge. Generally, imitation-learning techniques can decrease the difficulty of training a chore to that of giving instructions, without the requirement for unambiguous programming or scheming return purposes precise for the chore. The recent sensors available are capable of collecting and transmitting a large volume of data speedily, and computers with extraordinary computational supremacy permit reckless processing that plots the data collected by sensors to movements in an appropriate way. This brings up the opportunity for countless possible AI applications that need live-time insight and response such as human-like robots, self-directed cars, the interaction of humans with computers, and video games. Nevertheless, dedicated procedures are required to efficiently and vigorously study models as learning by the process of imitation has its own set of difficulties [15, 16, 17].

Imitation learning is not different from learning by observing, i.e., observational learning. This type of learning comes under the reinforcement learning category. Today, imitation learning has become an important part of the robotic industry. It is used in almost every industry be it agricultural industry, military, health sector, and many others yet to name. So, for these situations, the manual programming of robots

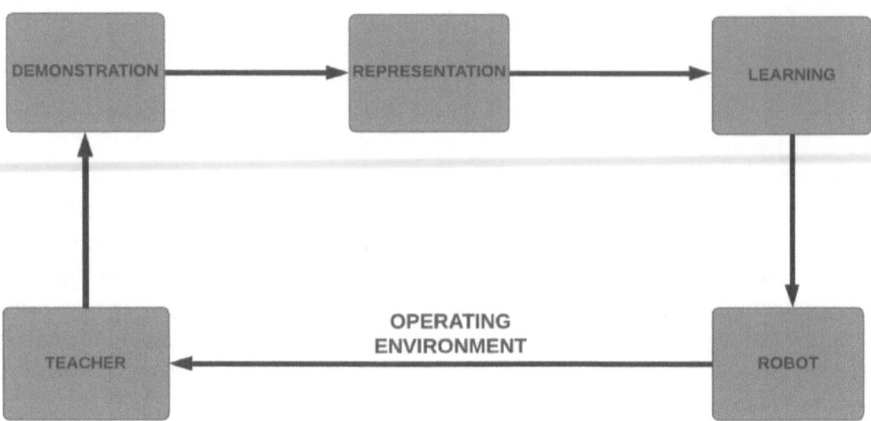

FIGURE 4.3   The imitation learning process by a robot.

is tedious and very challenging. As an alternative, collective approaches for instance programming by the means of demonstrating are used in combination with ML to program in the field [9, 12]. The imitation learning process is demonstrated with the help of Figure 4.3.

4.3.4   Pepper—Social Humanoid Robot

Pepper, the world's premier social humanoid robot, is capable of recognizing human faces and distinguishing some elementary human reactions. Pepper was enhanced and developed for being able to interact with humans and involve itself with people through the means of communication and touchscreen [7].

Pepper is accessible at the moment for industries and universities. More than 2,000 corporations all over the globe have accepted Pepper as an associate to welcome, update, and escort guests in an inventive manner [9, 12].

4.3.4.1   Commercial Applications of Pepper Robots

Several offices in the UK are using Pepper as their receptionist, and it is capable of identifying visitors using its facial recognition technology. It also sends alert messages to the organizers in case there is any meeting and to the hospitality department for arranging drinks for the visitors. Pepper is capable of chatting with the potential clients autonomously. The very first operative Pepper receptionist was installed in London at Brainlabs in the UK, and it was delivered by the SoftBank distributor [9, 18].

Pepper is also being used in many North American airports one of them being the Pierre Elliott Trudeau International Airport in Montreal, Canada. Pepper is used to welcome travelers, and present them with menus and references [7, 8].

4.3.4.2   Sports-Based Applications

On July 9, 2020, a team made of Pepper robots performed at the baseball match between the Rakuten Eagles and the Fukuoka Soft Bank Hawks as cheerleaders [11].

4.3.4.3   Academic Applications

Pepper is accessible in many schools and universities for research purposes, or for teaching programming languages and researching the interaction of humans with a robot. One international team in the year 2017 started research in using Pepper as an adaptable robot for helping in taking care of elderly people in care houses and shelters [11].

CARESSES is a project that intended at the evolving world's premier socially-proficient robot; this project received a subsidy of around 2 million euros and one of their donors was the government of Japan. Many renowned universities were involved in this project. In October 16, 2018, one of the Pepper robots cited the CARESSES project while confirming the Education Committee of the House of Commons of the Parliament of UK [11, 19].

4.4   IoT-AIDED ROBOTICS APPLICATIONS

4.4.1   IoT Allows Robots to Learn

In addition to working with extraordinary efficiency, robots also have the capability of learning and evolving as better robots with time using the data that they keep on gathering. This capability endorses an atmosphere of recurrent enhancement, where the schemes involved are not going with the planned strategy. ML technologies are used to power all these approaches because it is believed to be a more reasonable methodology to process the data modeled [13].

4.4.2   IoT Powers Co-Bots

It is only because of IoT that more collaborative robots (Co-Bots) have been able to take up responsibilities together with their human complements, which has allowed additional convenient scenarios where humans are seen working along with robots rather than completely replacing them. Co-Bots have not taken over laborers' work completely; instead they have allowed humans to work at places that are more important and leave the recurring work for the robots to do [10, 20].

This has also endorsed a harmless atmosphere by substitution. Now laborers no longer have to carry such heavy loads on their backs and walk for long distances; instead the robots are doing that [16].

4.5   WHAT THE FUTURE HOLDS

AI has always created a massive impact on almost all industries. And it will continue to do so even in the field of robotics. AI is tremendously impacting the robotics industry, and it doesn't seem like that impact is ending anytime soon [21].

It is said that soon robots' AI powers will be capable of achieving complicated and dangerous goals. Though this statement may appear very fictional, AI is going to take over many human activities. These AI-powered robots in the future will be capable of taking over dangerous works, for instance managing radioactive materials or deactivating bombs. Moreover, AI robots can endure working in extreme conditions like enormously loud circumstances, roasting temperature, and poisonous atmospheres. Therefore, AI powered-robots will be able to save many lives [22].

Gradually, robots will become cleverer and more proficient; consequently it will become safer to work with them in the same place. AI will play a critical role in the advancements of the robotics industry, and this revolution will enable robots to perform extraordinary tasks. AI and ML along with IoT are ready to completely transform the robotics industry. It is very evident from the applications of these technologies that this is just the beginning of a huge transformation that is yet to come [23, 24].

4.6 CONCLUSIONS

The chapter provides a complete overview of the application of AI, ML, and IoT in the robotics industry. It addresses in detail the applications of each of these technologies into robotics and how these are helping the robotics industry to grow and revolutionize. It tells us how applying these technologies is helping the robotics industry come up with robots that can replace and reduce human efforts. Robotics has begun a revolutionary journey; right now it is in the infancy stage, but if these applications are incorporated into the robotics industry completely, then we are not far from developing robots that can handle hazardous substances.

REFERENCES

1. Perez, J. A., Deligianni, F., Ravi, D., & Yang, G. Z. (2018). Artificial intelligence and robotics. *arXiv preprint arXiv:1803.10813*.
2. Mosavi, A., & Varkonyi, A. (2017). Learning in robotics. *International Journal of Computer Applications, 157*(1), 8–11.
3. Bredenfeld, A., Hofmann, A., & Steinbauer, G. (2010). Robotics in education initiatives in Europe-status, shortcomings and open questions. In *Proceedings of International Conference on Simulation, Modeling and Programming for Autonomous Robots (SIMPAR 2010) Workshops* (pp. 568–574).
4. Kreuziger, J. (1992). Application of machine learning to robotics—An analysis. In *Proceedings of the Second International Conference on Automation, Robotics, and Computer Vision (ICARCV'92)* (pp. 1–7).
5. Fan, Y. J., Yin, Y. H., Da Xu, L., Zeng, Y., & Wu, F. (2014). IoT-based smart rehabilitation system. *IEEE Transactions on Industrial Informatics, 10*(2), 1568–1577.
6. Cho, J. E., & Nezhat, F. R. (2009). Robotics and gynecologic oncology: Review of the literature. *Journal of Minimally Invasive Gynecology, 16*(6), 669–681.
7. Biswas, S., Kinbara, K., Niwa, T., Taguchi, H., Ishii, N., Watanabe, S., ... & Aida, T. (2013). Biomolecular robotics for chemomechanically driven guest delivery fuelled by intracellular ATP. *Nature Chemistry, 5*(7), 613–620.
8. Huang, M. H., & Rust, R. T. (2020). Engaged to a robot? The role of AI in service. Journal of Service Research, 24(1), 30–41, 1094670520902266.
9. McCabe, J., Monkiewicz, M., Holcomb, J., Pundik, S., & Daly, J. J. (2015). Comparison of robotics, functional electrical stimulation, and motor learning methods for treatment of persistent upper extremity dysfunction after stroke: A randomized controlled trial. *Archives of Physical Medicine and Rehabilitation, 96*(6), 981–990.
10. Fan, Y. J., Yin, Y. H., Da Xu, L., Zeng, Y., & Wu, F. (2014). IoT-based smart rehabilitation system. *IEEE Transactions on Industrial Informatics, 10*(2), 1568–1577.

11. Pandey, A. K., Gelin, R., & Robot, A. M. P. S. H. (2018). Pepper: The first machine of its kind. *IEEE Robotics & Automation Magazine*, *25*(3), 40–48.

12. Fujie, H., Mabuchi, K., Woo, S. L. Y., Livesay, G. A., Arai, S., & Tsukamoto, Y. (1993). The use of robotics technology to study human joint kinematics: A new methodology. *Journal of Biomechanical Engineering*, *115*(3), 211–217.

13. Krebs, H. I., Palazzolo, J. J., Dipietro, L., Ferraro, M., Krol, J., Rannekleiv, K., ... & Hogan, N. (2003). Rehabilitation robotics: Performance-based progressive robot-assisted therapy. *Autonomous Robots*, *15*(1), 7–20.

14. Caine, R. N., & Caine, G. (1991). Making connections: Teaching and the human brain, ASCD, Virginia.

15. Olfati-Saber, R. (2001). Nonlinear control of underactuated mechanical systems with application to robotics and aerospace vehicles (Doctoral dissertation, Massachusetts Institute of Technology).

16. Lavoue, V., Zeng, X., Lau, S., Press, J. Z., Abitbol, J., Gotlieb, R., ... & Gotlieb, W. H. (2014). Impact of robotics on the outcome of elderly patients with endometrial cancer. *Gynecologic Oncology*, *133*(3), 556–562.

17. Brougham, D., & Haar, J. (2018). Smart technology, artificial intelligence, robotics, and algorithms (STARA): Employees' perceptions of our future workplace. *Journal of Management & Organization*, *24*(2), 239–257.

18. Swank, M. L., Alkire, M., Conditt, M., & Lonner, J. H. (2009). Technology and cost-effectiveness in knee arthroplasty: Computer navigation and robotics. *American Journal of Orthopedics (Belle Mead NJ)*, *38*(2 Suppl), 32–36.

19. Lafaye, J., Gouaillier, D., & Wieber, P. B. (2014, November). Linear model predictive control of the locomotion of Pepper, a humanoid robot with omnidirectional wheels. In *2014 IEEE-RAS International Conference on Humanoid Robots* (pp. 336–341). IEEE.

20. Singh, S. K., Rathore, S., & Park, J. H. (2020). BlockIoTIntelligence: A blockchain-enabled intelligent IoT architecture with artificial intelligence. *Future Generation Computer Systems*, *110*, 721–743.

21. Goebel, R., Chander, A., Holzinger, K., Lecue, F., Akata, Z., Stumpf, S., ... & Holzinger, A. (2018, August). Explainable AI: The new 42? In *International Cross-domain Conference for Machine Learning and Knowledge Extraction* (pp. 295–303).

22. Ona, E. D., Cano-de La Cuerda, R., Sánchez-Herrera, P., Balaguer, C., & Jardón, A. (2018). A review of robotics in neurorehabilitation: Towards an automated process for upper limb. *Journal of Healthcare Engineering*, *2018*, Article ID 9758939, 19 pages, https://doi.org/10.1155/2018/9758939

23. Yigitcanlar, T., Desouza, K. C., Butler, L., & Roozkhosh, F. (2020). Contributions and risks of artificial intelligence (AI) in building smarter cities: Insights from a systematic review of the literature. *Energies*, *13*(6), 1473.

24. Moritz, P., Nishihara, R., Wang, S., Tumanov, A., Liaw, R., Liang, E., ... & Stoica, I. (2018). Ray: A distributed framework for emerging {AI} applications. In *13th {USENIX} Symposium on Operating Systems Design and Implementation ({OSDI} 18)* (pp. 561–577).

5 IoT-Based Laser Trip Wire System for Safety and Security in Industries

Souvik Das, Agniv Sarkar,
O. B. Krishna, and J. Maiti
Indian Institute of Technology Kharagpur,
Kharagpur, West Bengal, India

CONTENTS

5.1 INTRODUCTION

The physical world has become hyperconnected and is now rapidly being augmented with a layer of "smartness." The fourth industrial revolution—like the ones that came before—is driven by new technologies [1]. In India, the adaptation of Industrial Revolution 4.0 is at its infancy. From the onset of the revolution, the technology started disrupting the total industrial work system structure. Along with the advancement of various industries comes the lack of advancement of the safety and security aspect of it, which has been developing at quite a slow rate [2]. The major challenges are to the smaller industries that have less resources, time, and money to secure their plants, data, and processes [3].

Today, with technological advancements, virtual reality and/or augmented reality can be used to simulate new models; their products, processes, and safety can be analyzed for better performance.

Operational technology (OT), the key driver for success in "Industry 4.0," is a broad term referring to any computer system that controls or identifies a static or kinetic action [4]. The biggest challenges to OT cyber security and safety are resources, time, and money to secure plants, processes, and data [4]. The security industry in India can be comprehensively arranged into digital security, electronic security, fire-safety location and aversion, private security, and individual safety attire and gear [5].

DOI: 10.1201/9781003181613-5

The Industry 4.0 revolutionary framework was made to be adaptable and empowers individualization with tweaked objects. Production before these days was mainly shaped by worldwide challenges of customization and quick and frequent adjustments in the constantly-evolving business sector [6]. With the advent of Industry 4.0, the industries can now be equipped for demands by radical advances in new assembling innovations. Industry 4.0 has a promising methodology that is dependent on combining the business and assembling forms among the organization's worth chain (producers and clients) [7]. Specialized and customized parts of these necessities are tended to by the use of the nonexclusive ideas of cyber physical systems (CPS) [8] and modern Internet of Things (IoT) [9] in the production frameworks.

IoT framework designs enable users to attain further automation, investigation, and combination inside a structure [10]. IoT uses existing and innovating technology for sensing, systems management, and mechanical autonomy. IoT encourages billions of gadgets, both individual and mechanical, to be empowered with organizational availability to gather and trade ongoing data for intelligent and prompt outputs [11].

With the advent of the fourth industrial revolution, new technology is playing a massive role in the transformation of the safety and security in industry as well, and one of the most promising technologies with a multitude of applications is the Industrial Internet of Things (IIoT) [12, 13]. The IIoT can create a highly-connected, unified system, which—together with machine learning and cloud technology—provides companies with unprecedented levels of insight into their operations. IoT electronics, such as sensors, are wireless, cheap, and simple to set up—they can be placed on an ad hoc basis as and when they are needed, moved around, or removed without major disruption to the system. Some current uses of IoT are:

- **Combining IoT Sensors with Computer Vision** [14]: Computer vision aims to function like the human brain and eyes do. It detects differences between environments and situations, then responds accordingly.
- **Fitting Workers with High-Tech Wearables** [15]: There are IIoT safety wearables that increase worker visibility or monitor things like posture, noise, and physiological data. The data from the sensors goes to a central database and interface for further analyses by the management.
- **Installing Integrated Building Safety Systems** [16]: Schools have smart lights that help people identify a building's safe zones and warn them of potential danger nearby. A person can use panic button during a threatening situation, which alerts authorities. Integrated IoT network systems are used in big warehouses for tracking and store management.
- **Relying on Industrial Robots That Aren't in Cages (COBOTS)** [17]: Some systems use computer vision, sensors, and artificial intelligence to help large industrial robots work around humans without cages. These machines, made to work alongside people, automatically slow/stop when detecting people in close proximity.

- **Location-Based Analytics and Real-Time Location Systems for IIoT Safety** [18]: Industries such as health care use real-time location systems (RTLS) to track assets and people. It can give warnings if unauthorized people walk into restricted areas. The technology can also connect to a panic button that lets individuals summon aid and send assistance to their exact location.

5.2 METHODOLOGY

Initially, we analyzed various mega industries and plants, and reviewed various plant operations and zones. After review, we found that infrastructure and zone surveillance is necessary with the increasing concern regarding security and safety. Conventional fence faces two major problems while protecting a building or a zone: the physical barrier of conventional fencing allows the intruder to hide before intruding into the area or the protected zone, and management fails to understand when the intruder crosses the physical barrier. Other than these, management has no knowledge about any defects of leak in the fence; large-scale fencing incurs high capital costs, and there is little or no wiggle room for future customization. In other words, one can say there is complete lack of automation either in terms of customization or intrusion detection.

Various studies have shown that people in India are quite concerned about the aesthetic value rather than safety of a site. Most available safety and security systems are fixed (hard barricading) and thus are not customizable. Regulation of the number of personnel and unknown personnel in a zone is also a major concern. Current safety and security systems don't engage in any collection of data causing infrastructural loss, intellectual property loss, and power loss.

Looking upon the focused issues as discussed we formulated our objective to be:

Development of a customizable laser trip wire soft fencing system with imbedded IoT to detect, collect data, and provide analysis for improved work systems in various industries in India.

5.2.1 COMPONENTS AND SETUP

Soft fencing systems are made from a group of small soft fencing units. The circumference of the boundary walls of the zone to be secured is divided into many straight-line segments. Every segment is protected against intrusion using a pair of soft fencing units having one light sensor unit with both laser transmitters and photo receivers, including a wireless tag to send the intrusion signal wirelessly to the control station for alert generation. A soft fence unit plays an important role in overall performance of a soft fence system using a wireless sensor network.

The main component of the soft fencing unit includes:

- **Node MCU ESP8266 (9 V, 4 MB, 450 kbps)**: Node MCU, an open-source firmware, uses the Lua scripting language and is based on the eLua project, and built on the Espressif Non-OS SDK for ESP8266 as shown in figure 5.1. It uses open-source projects, such as luacjson and SPIFFS.

- **Laser Non-Modulator Tub Sensor Receiving Module (5 V DC)**: This laser non-modulator tub sensor receiving module receives laser signals at a high-output level and does not receive any laser signal at low output level. After detecting the laser signal, output provides a high-level (5V) signal until the input laser signal is there.
- **Laser Module (5 V, 5 mW, 40 mA, 650 nm)**: This adjustable laser is easy to install, durable, and portable. It gives the best output for a 3–5 V 5 mW standard, with 6 mm outside diameter.
- **OV7670 640×480 VGA CMOS Camera Image Sensor Module**: A camera image sensor module, a low-cost image sensor, can operate at a maximum of 30 fps and 640 × 480 (VGA) resolutions, equivalent to 0.3 MP.

Some other components include N-Channel MOSFET-2N7000, N-Channel MOSFET-BS170 (200 mA, 60 V, 5 Ω), wire USB cables, Vero boards, patch wires, N-Channel MOSFET- IRFZ44N, relays, DC-DC boosters, capacitors, resistors, DC buzzers, variable potentiometer rheostats.

Arduino IDE was used to develop a program for the circuit. It is a cross-platform application used to write and upload programs to Arduino-compatible boards. To view program code, please refer to the annexure.

A trial campaign had been conducted to test the working of the soft fencing wireless network system. The soft fence was set up at a height of one foot above the ground before the entry door of a laboratory. To turn on the system, we initiated the power supply to the system. The entire system consisted of three main parts—laser activation cycle, laser trip wire system, and camera module sensor cycle as shown in figure 5.2.

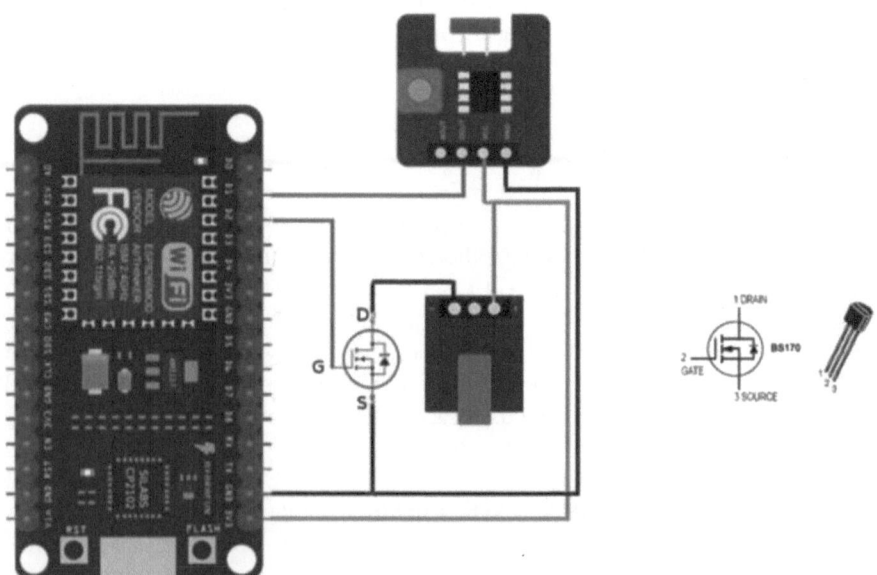

FIGURE 5.1   Basic circuit diagram.

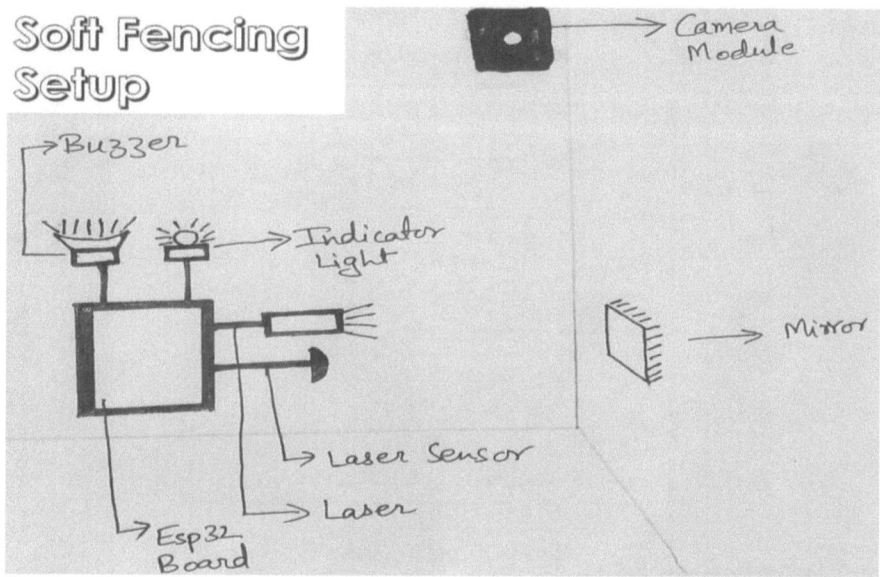

FIGURE 5.2 System flowchart.

In the laser activation cycle as shown in figure 5.3, initially as the power is supplied to the system, the whole system gets a diagnostics check. So as the power is supplied, the system automatically checks the laser alignment against the laser photo sensor. In such a process, if the system finds a fault it will immediately transmit the information to the server where the management will be notified about the error in the system. In the other case, no information would be sent to the server and the soft fencing unit would be armed and ready for intrusion detection.

The Laser trip wire system is the main functioning unit of the system. Its main purpose is to detect the tripping of the laser, i.e., when laser at any point in time doesn't get detected by the laser photo sensor. In such cases, the main computing board instantaneously get alerted and sends the information at the direct server sources with the date and time stamp of the intrusion. In case of a localized intrusion, the

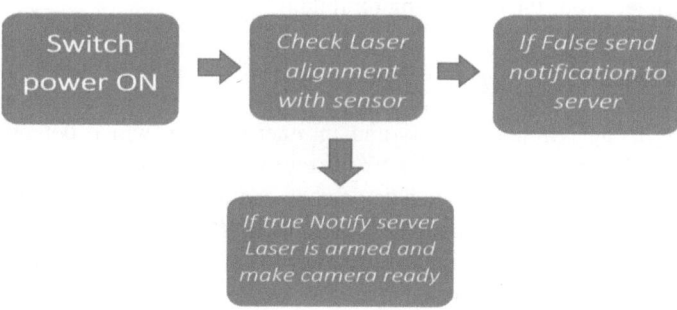

FIGURE 5.3 Laser activation cycle.

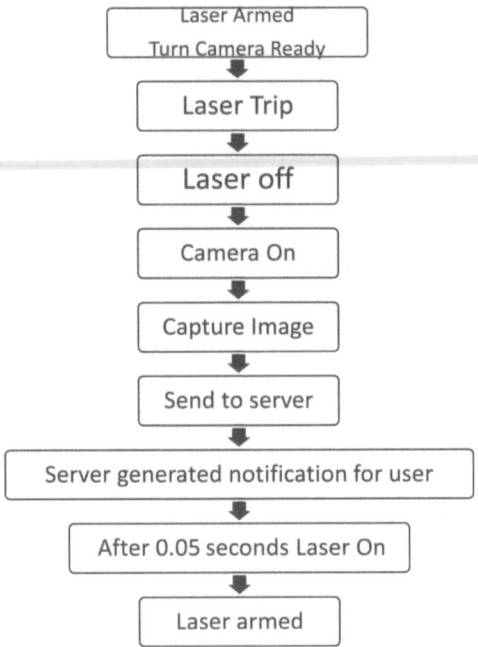

FIGURE 5.4   Steps of a working trip wire system.

computing board also activates the buzzer and flash lighting informs the immediate surroundings about the obtrusion.

The third part of the soft fencing unit consists of the camera module sensor cycle. The main component of this cycle is the camera module, which is placed at an appropriate height and distance to capture the right space of intrusion. The cycle instantly gets activated and snaps a pic of the intruder on detection of the obtrusion in the laser tripwire system as shown in figure 5.4. Once the camera module captures the image it instantly sends the image file stored in the computing module to the http local host server. Once the computing module gets a go-ahead server response, the image gets transferred to the server at a given directory with the timestamp and date. Also it gets instantly deleted from the computing module.

5.3   RESULTS AND DISCUSSION

Soft fencing using wireless sensors and intranetworking, which detects, identifies, and localizes the intrusion region can be used to solve most of the earlier-discussed problems. With the advent of such a device one can detect intrusion with precision to microseconds, collect various data such as number of intrusion, location of intrusion, and time and identity of intruder. Using such an IoT system, users can have full control at their fingertips and instant intrusion alerts. A user can also avoid huge loads of bad data while storing only the required data and further use this to customize the work by increasing or decreasing the zones and incurring very little expenses in the making.

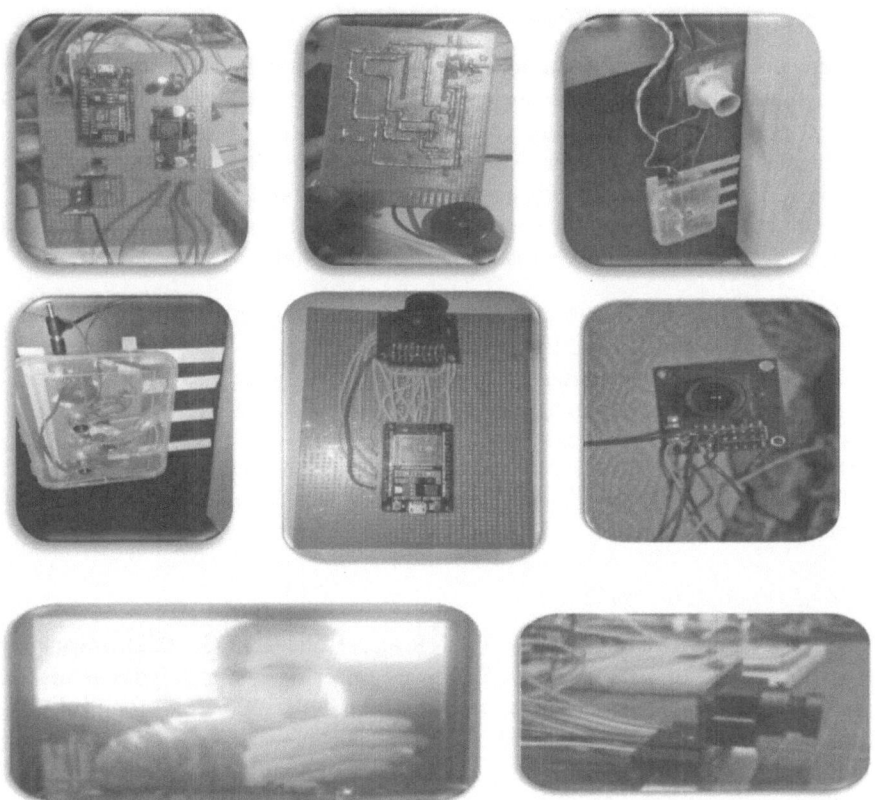

FIGURE 5.5   Demo setup.

Apart from that there are some hindrances in the unit as well. The soft fencing unit has capability functions only when the laser has a clear line of sight to the laser photo sensor as shown in figure 5.5. This in many cases, such as in topographical differences or various elevations of the perimeter, etc., may not work. Apart from that, a continuous supply of power is necessary for the system to run. In cases of power loss, the unit will fail.

Some of the major industrial purposes of the unit are perimeter fencing, entry/exit monitoring, swift counting of various moving objects (increasing productivity times), intruder detection and identification, etc. Implementation of an individual or group of such soft fencing units will be quite easy and cost-effective for various industries. Apart from that, they help industries in terms of safety and security. The addition of such small fencing units can be done at various other institutions. as well as hospitals, banks, educational institutes, personal homes, or for office security, etc.

Another important application of the device is in border fencing across countries. As known, it's a costly affair for countries to protect their borders every second. The soft fencing unit with additional sensors like vibrational, pressure plates, proximity, motion detectors, etc. can be used efficiently to give an extra hand in border protection at very economical rates.

5.4   CONCLUSION AND FUTURE SCOPE

Diverse technologies are becoming available that align with the connected nature of the IIoT and help prevent accidents or other catastrophes. Safety should be a primary concern when implementing any IIoT technology.

The performance of soft fence units may vary in different environment and orientations. The height, sensitivity, and indoor and outdoor environment influence the maximum sensing range of the soft fence. To increase the function ability of the soft fence unit, other customization that can be added include:

1. Addition of servomotors to Open or Close gates on tripping of a fence. (Tuning of soft fencing to hard fencing.)
2. Using RFID tags and readers to provide selected entry.
3. Using motion and vibrational sensors to increase security.

The same working principles can be used in border surveillance for information collection from various types of sensors and devices such as seismic, cameras, thermal cameras, and motion detectors.

ACKNOWLEDGMENTS

The work is funded by UAY project, GOI (Project Code: IITKGP_022). We acknowledge the Centre of Excellence in Safety Engineering and Analytics (CoE-SEA) (www.iitkgp.ac.in/department/SE) and Safety Analytics & Virtual Reality (SAVR) Laboratory (www.savr.iitkgp.ac.in) of the Department of Industrial & Systems Engineering, IIT Kharagpur, for experimental/computational and research facilities for this work. We would like to thank the management of the plant for providing relevant data and their support and cooperation during the study.

REFERENCES

1. S. Madakam, R. Ramaswamy, and S. Tripathi, "Internet of Things (IoT) : A Literature Review," *J. Comput. Commun.*, vol. 3, pp. 164–173, May 2015.
2. D. Lund and M. Morales, "Worldwide and Regional Internet of Things (IoT) 2014–2020 Forecast : A Virtuous Circle of Proven Value and Demand," *IDC*, vol. 3, pp. 2014–2020, 2014.
3. E. Sisinni, S. Han, U. Jennehag, and M. Gidlund, "Industrial Internet of Things: Challenges, Opportunities, and Directions," *IEEE Trans. Ind. Informatics*, vol. 65, pp. 280–290, July 2018.
4. T. R. G. Asir and H. L. Manohar, "Key challenges and success factors in IoT—A study on impact of data," in *2018 International Conference on Computer, Communication, and Signal Processing* (ICCCSP) 2018, pp. 1–5.
5. F. Wu, T. Wu, and M. R. Yuce, "An Internet-of-Things (IoT) Network System for Connected Safety and Health Monitoring Applications," *Sensors*, vol. 19, no. 21, pp. 1–21, 2018.
6. Y. Hui, K. Soundar, B. S. T. S, and T. Fugee, "The Internet of Things for Smart Manufacturing : A Review," *IISE Trans.*, vol. 0, no. 0, pp. 1–27, 2019.
7. J. Weking, M. Stocker, M. Kowalkiewicz, B. Markus, and H. Krcmar, "Leveraging Industry 4.0—A Business Model Pattern Framework," *Int. J. Prod. Econ.*, vol. 225, December pp. 107588, 2020.

8. H. I. Al-salman and M. H. Salih, "A Review: Cyber of Industry 4.0 (Cyber-Physical Systems (CPS), the Internet of Things (IoT) and the Internet of Services (IoS)): Components," *J. Phys. Conf. Ser. Pap.*, vol. 0, pp. 0–6, 2019.

9. S. Kumar, P. Tiwari, and M. Zymbler, "Internet of Things is a Revolutionary Approach for Future Technology Enhancement : A Review," *J. Big Data*, vol. 6, no. 1, pp. 1–21, 2019.

10. J. Horn, A. Koohang, and J. Paliszkiewicz, "The Internet of Things : Review and Theoretical Framework," *Expert Syst. With Appl. J.*, vol. 133, pp. 97–108, 2019.

11. S. Bera, G. S. Member, S. Misra, and S. Member, "Software-Defined Networking for Internet of Things: A Survey," *IEEE Internet Things J.*, vol. 4, no. 6, pp. 1994–2008, 2017.

12. J. Prinsloo, S. Sinha, and B. von Solms, "A Review of Industry 4.0 Manufacturing Process Security Risks," *Appl. Sci.*, vol. 9, no. 23, pp. 5105, 2019.

13. A. Bicaku, C. Schmittner, M. Tauber, and J. Delsing, "Monitoring industry 4.0 applications for security and safety standard compliance," in *2018 IEEE Ind. Cyber-Physical Syst.*, 2018, pp. 749–754.

14. L. J. Murugesan and R. S. Shanmugasundaram, "Applications of IoT Technology and Computer Vision in Higher Education—A Literature Review," *Int. J. Res. Eng. Appl. Manag.*, pp. 1120–1135, July 2019.

15. N. Izzah, M. Nor, L. Arokiasamy, and R. A. Balaraman, "The Influence of Internet of Things on Employees' Engagement among Generation Y at the Workplace," in *SHS Web of Conferences 56*, 2018, vol. 03003, pp. 1–7.

16. S. Park *et al.*, "Design and Implementation of a Smart IoT Based Building and Town Disaster Management System in Smart City Infrastructure," *Appl. Sci.*, vol. 8, pp. 1–28, 2018.

17. L. Romeo, A. Petitti, and R. Marani, "Internet of Robotic Things in Smart Domains : Applications and Challenges," *Sensors*, vol. 20, no. 3355, pp. 1–23, 2020.

18. J. Rezazadeh, K. Sandrasegaran, and X. Kong, "A location-based smart shopping system with IoT technology," in *2018 IEEE 4th World Forum on Internet of Things (WF-IoT)*, 2018, pp. 748–753.

6 Detecting Payment Fraud Using Automatic Feature Engineering with Harris Grey Wolf Deep Neural Network

*Chandra Sekhar Kolli, Mohan Kumar Ch,
Ganeshan Ramasamy, and
Gogineni Krishna Chaitanya*
Koneru Lakshmaiah Education Foundation,
Vaddeswaram, AP, India

CONTENTS

6.1 INTRODUCTION

According to Wikipedia, fraud can be referred as an intentional trick to secure one-sided or illegal gain. Examples include accounting fraud, credit card fraud, insurance fraud, online transactions fraud, etc. These days, the number of online transactions has increased drastically, and this is a golden opportunity for establishing digital trust. New digital customer bases are not entirely comfortable online, and they are

DOI: 10.1201/9781003181613-6

particularly vulnerable to fraud schemes. Web application and web security, and phishing websites are also possible ways used by the fraudsters. To effectively fight financial crime, banks must develop a comprehensive fraud strategy to provide accurate insights and enact quality decisions.

In the age of open banking and real-time payments, funds can be transferred to fraudsters' accounts immediately, and the victim does not notice suspicious movements until it is too late. It is evident that with faster payment methods, the benefits for consumers far outweigh the disadvantages, but hackers will continue to exploit this area. Fraud analysts within a bank must pivot and adapt to working in new environments to ensure that teams integrate efficiently and decrease human error that occurs through a lack of physical communication. Artificial intelligence and machine learning (ML) can also play a part in fraud and financial crime prevention, consuming disparate unstructured data and creating structured insight and conclusions.

6.2 LITERATURE REVIEW

Most of the time, financial institutions are not willing to provide the data sets or the provided data quantity is very low. The provided data set is not readily usable since the data will not have the required features to validate new methods. Diverse sets of bank transactions data presents new challenges; ML algorithms work effectively on homogeneous data, either numeric or categorical data [1]. The k-means–based methods are efficient in processing large data sets but often are restricted to numerical data [2] thus the existing outlier detection (OD) techniques are ineffective for mixed data sets. Moreover, some ML platforms, like Apache Spark, only accept numerical data. One-hot encoding (OHE) [3] technique is widely used to transform categorical features into numerical features.

Classification accuracy and efficiency of the model are very critical in fault detection (FD). The technique of ripple-down rules (RDR) is ideal among the existing rule-based techniques for fraud detecting due to its lower maintenance requirements and support for incremental learning [4]. However, the performance of RDR on distributed and dig data platforms like Spark has not been studied as RDR is not available on these platforms. Donato et al. [16] and Bolton and Hand [17] demonstrated the importance of considering combination of features of the past transactions when using the unsupervised models for analyzing the fraudulent transactional data. Goud and Mathur [5] and Kokate and Rani [6] developed and demonstrated a model by considering descriptive statistics as the feature. Kolli and Devi [15] have demonstrated the importance of classification algorithm in combination with the feature engineering. The combination of Hidden Markov Model (HMM)-based features improves and generates an efficiency in the accuracy of the model in classifying the fraudulent transaction from legitimate ones.

During the model construction, various weaknesses that are affecting the model performance and the type of features are to be considered are analyzed [9, 10]. Abhishek and colleagues [8] have clearly explained the modern tools and techniques of promoting banking financial services and insurance and improved techniques for ranking fraud discovery for mobile apps, since these days mobile and Internet are so common. Selvaraj et al. [11] have mentioned the ontology-based recommender systems. These can also be considered because various types of facilities are available to perform transactions on mobile, tablet, laptop, or any other handheld device. Depending on the type of need, the application can recommend what type of facility is most suitable based on location, time, and any other attributes similarly to perform secure transactions. Sajana and Narasingarao [12] have mentioned the Ensemble framework for classification of imbalanced data handling.

6.3 PROPOSED HARRIS WATER OPTIMIZATION–BASED FEATURE ENGINEERING WITH DEEP RECURRENT NEURAL NETWORK FOR FRAUD DETECTION

In the literature, several fraud detection methods are explored, but none of the methods was able to produce effective solution. Hence, a real and effective fraud discovery algorithm named the Howard Water Optimization (HWO) with feature engineering with deep recurrent neural network (RNN) is proposed. The proposed method consists of three phases: pre-processing phase, feature extraction phase using HMM, and fraud detection phase. Figure 6.1 illustrates the schematic representation of the proposed model. First, we give the input data set to the preprocessing phase, where the data is analyzed and all the unnecessary and redundant, missing values, noise data are removed. In the next step, the refined data will pass through the HMM-based feature engineering model.

This model will generate the groups of features and then the features that are chosen will be fed as input to the fraud detection model. The deep recurring neuronic classifier will undergo the training process using the HWO to detect fraudulent transactions.

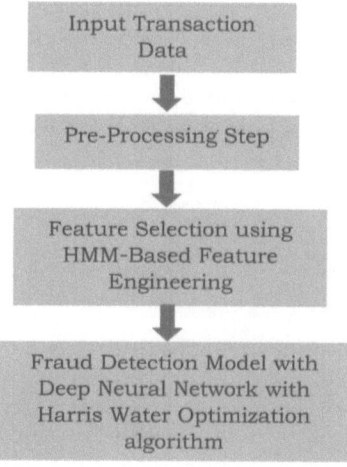

FIGURE 6.1   Schematic representation of the proposed fraud detection model.

6.3.1 Input Data Representation

The data set is fed as input to the pre-processing step. We consider the database as K with n number of data points; T is represented as follows:

$$K = \{T1, T2, T3, Ti, Tn\}; 1 \leq i \leq n$$

where K denotes the whole input data, $T1$, $T2$, Tn are the transactions data items; n represents the number of data points. The input data Ti is selected and passed to pre-processing phase.

6.3.2 Pre-Processing Phase

The input data T_i is selected and passed through the pre-processing phase, and the data points are analyzed with the help of the Box-Cox conversion technique. The main objective is to remove the redundant values. It is a kind of power transformation method. It will convert the non-normal data to normal data using some optimal parameters. The transformation can be represented as:

$$PD = \left\{ \frac{T_{ij}}{j}; \ j \neq 0 \text{ and } l_n(T_i) = 0 \right.$$

where T_i is the managed data, T_i is the input data, and j specifies the undecided constraint value. The transformation process adopts a complex value of T_j ($T_j > 0$), then only the transformation process will occur. Input data is altered as normal distribution created on the charge of j. Finally, the data PD is passed to the feature-selecting phase.

6.3.3 Feature Selection Using HMM and Feature Engineering Model

The pre-processed data PD is input to the feature collection stage, where most appropriate features are effectually and carefully chosen. It is done with a Hidden Markov Model. The feature engineering technique is applied to get descriptive features from past transactions.

The selected features are represented as F as $[u \times v]$ and passed as input to the deep RNN. Figure 6.2 illustrates the architecture of the model.

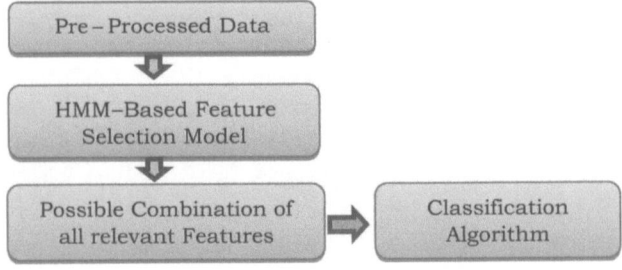

FIGURE 6.2 Improving the quality of transactions with HMM-based feature engineering.

6.3.4 FRAUD DISCOVERY BY MEANS OF HARRIS WATER OPTIMIZATION WITH DEEP RECURRENT NEURAL NETWORK

The process is carried out by taking the input from the previous phase. The selected features are now fed into the HWO classifier. This is an integration of Harris hawks optimization [20] and water wave optimization [21]. The fraud detection model is worked with the use of a deep neural network classifier. It takes the feature F as input then processes in the recurrent hidden layers. It uses fitness function (FF) to compute the solution, and it is represented as:

$$FF = \frac{1}{n} \sum_{i=0}^{n} Di^{(s,z)} - \vartheta i$$

where FF is fitness function, $D_i^{(s,z)}$ denotes output of the classifier, and ϑi is estimated output.

6.3.4.1 Architecture of the Neural Network and Hidden Layers

The features generated from HMM are voted for classification. It will use the earlier results and then feed as input to the next iterations. Then, the recurrent nature improves the classifier's detection results. Figure 6.3 explains the construction of the RNN classifier, input vector ith layer at jth time represented as: $X(i, j) = \{X_1^{(i, j)}, X_2^{(i, j)}, X_3^{(i, j)}, \dots X_n^{(i, j)}\}$, and the similarly, output vector of ith layer at jth time embodied as: $Y(i, j) = \{Y_1^{(i, j)}, Y_2^{(i, j)}, Y_3^{(i, j)}, \dots Y_n^{(i, j)}\}$. The components of the input vector are scientifically denoted as in Figure 6.2.

Assume p specifies the at a particular position, that is an arbitrary unit at position sth layers, and specifies the quantity of units at the sth layers. Along with the input parameter and output parameter, there is a random unit value of $(s - 1)$th layers,

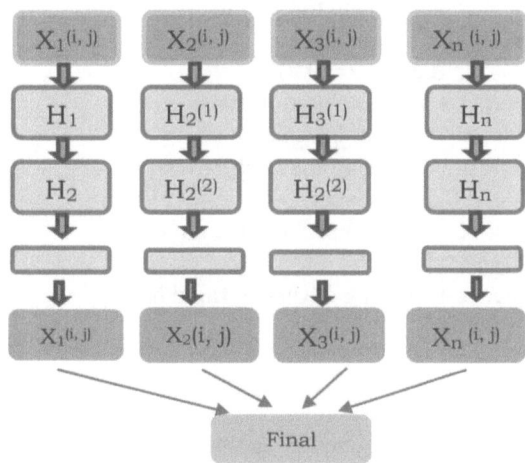

FIGURE 6.3 Deep recurrent neural network structure.

which is stated as g, and the total number of units of $(s-1)$th layers are denoted. The input vectors are logically represented as:

$$X_l^{(i,j)} = \sum_{b=1}^{T} \alpha_{lb}^s Y_b^{(i-1, j)} + \sum_{l'}^{a} \beta_{ll'}^{(s)} Y_{l'}^{(i, j-1)}, \text{ where } \alpha_{lb}^s \text{ and } \alpha_{lb}^s \text{ are the elements of } \omega(i), \text{ and}$$

$\omega(i)$. l' indicates the arbitrary unit number of the ith layer. The output vector of ith layers is mathematically represented as:

$$Y_1^{(i,j)} = \delta^i \left(X_l^{(i,j)} \right)$$

where δ^i is activation function.

6.3.4.2   Proposed Harris Water Optimization Algorithm

The classifier recurrent neural network is achieved with the help of the Harris Water Optimization method. The steps involved in this method are population initialization, calculating the fitness value of all the hawks, and then it will calculate the best possible solution. After that, update the original energy of the rabbit. Update the position direction during the exploration phase. The Harris hawks keep track for the prey, and once it detects it will move. By joining the parametric value from optimization and the proposed algorithm the model outperforms in detecting the fraudulent transactions.

The proposed algorithm is given below:

Step 1: Input: Assume the population size N, and maximum iterations are X
Step 2: Output: $R(x+1)$
Step 3: Initialize the population
Step 4: While (stops when all items are over)
Step 5: Compute the value of F
Step 6: Set Rabbit is the top position of rabbit
Step 7: For (every individual hawk (Rj))
Step 8: Update the value of P0 and E and P
Step 9: If $(|p| >= 1)$
Step 10: Update $R(x+1)$ during the investigation stage
Step 11: if $(|p| < 1)$ then
Step 12: if $(m >= 0.5 \ \&\& \ |p| >= 0.5)$
Step 13: Update $R(x+1)$ at exploration stage
Step 14: Else if $(m >= 0.5 \ \&\& \ |p| < 0.5)$
Step 15: Update $R(x+1)$ during the exploration phase
Step 16: else if $(m < 0.5 \ \&\& \ |p| >= 0.5)$
Step 17: Update $R(x+1)$ using progressive rapid dives
Step 18: Else if $(m < 0.5 \ \&\& \ |p| < 0.5)$
Step 19: Update $R(x+1)$ using advanced rapid bars
Step 20: Return Rabbit

6.4   RESULTS

This section explains the experimentation and results attained using the proposed model for fraudulent transaction detection. The data set contains various transactions data, which is used for fraud detection. The performance of the created model is assessed with the help of the performance metrics of accuracy, specificity, and sensitivity.

Accuracy measure calculates the proportion of true positives and true negatives in all the cases of the confusion matrix. Sensitivity is the measure used to detect the fraudulent activities correctly, and specificity is the measure that detects the legitimate activities correctly. These metrics are represented as follows:

$$A = \frac{Tp + Tn}{Tp + Tn + Fp + Fn}$$

where, Tp is True Positives, Tn is True Negatives, Fp is False Positives, and Fn is False Negatives. A denotes Accuracy of the proposed model.

Sensitivity is the measure used to sense the fake actions, which is signified as follows:

$$Sn = \frac{Tp}{Tp + Fn}$$

where Sn specifies sensitivity calculated by the model.

Specificity is the measure that senses the genuine actions properly, which is signified as follows:

$$Sp = \frac{Tp}{Tp + Fp}$$

where Sp specifies specificity calculated by the model.

The values of sensitivity, specificity, and accuracy calculated by the proposed model are 0.7923, 0.9102, 0.9845, respectively.

6.5 CONCLUSION

In this investigation, an effective and real fraud discovery approach named automatic feature engineering with Harris Grey Wolf Deep Neural Network is proposed for detecting fraudulent transactions. The anticipated method accomplishes the discovery in three phases: pre-processing, feature selection phase, and a fraud-detection phase. The input data is used to perform the fake transactions discovery process. First, the input data is subjected to input in the pre-processing phase to remove redundant values; we have used the Box-Cox transformation method in this phase.

After that, the pre-processed data is sent as input to the feature collection stage to choose appropriate features by means of the feature engineering along with HMM. In the last phase, the selected features can enter the detection phase; we used the deep RNN classifier to spot the fake transactions. For precise fake transaction discovery, the deep RNN is trained with the planned HWO procedure, and the performance attained by the proposed HWO-based deep RNN is assessed using the metrics. In our example, sensitivity, specificity, and accuracy values are 0.7923, 0.9102, and 0.9845, respectively.

The results reveal that the proposed method is better than current approaches. In the future, we can integrate any semi-supervised approach in combination with a hybrid classifier to get an advanced approach and better performance in fraud detection.

REFERENCES

1. Shih, M.-Y., Jheng, J.-W., & Lai, L.-F. (2010), "A two-step method for clustering mixed categorical and numeric data", *Tamkang Journal of Science and Engineering*, Vol. 13, No. 1, pp. 11–19.
2. Chen, C.-M., Guan, D. J., Huang, Y.-Z., & Ou, Y.-H. (2012). "Attack sequence detection in cloud using hidden Markov model". Paper presented at the 2012 Seventh Asia Joint Conference on Information Security, Tokyo.
3. Harris, D. M., & Harris, S. L. (2012). *Digital Design and Computer Architecture* (N. McFadden Ed., 2nd ed.). New York, NY: Morgan Kaufmann.
4. Compton, P., & Jansen, R. (1988). "Knowledge in context: A strategy for expert system maintenance". Paper presented at the Australian Joint Conference on Artificial Intelligence, Adelaide, Australia.
5. Swapna Goud, N., & Mathur, A. (2019), "A certain investigations on web security threats and phishing website detection techniques", *International Journal of Advanced Science and Technology*, Vol. 28, No. 16, pp. 871–879.
6. Kokate, S., & Dr Rani, C. M. S. (2020), "Fraudulent detection in credit card transactions using radial basis function kernel method based support vector machine", *International Journal of Advanced Science and Technology*, ISSN: 2005-4238, Vol. 29, No. 12s, pp. 2557–2565.
7. Adinarayana, P. J., & Kishore Babu, B. (2019), "Modern techniques of promoting the banking financial services and insurance (BFSI)", *International Journal of Innovative Technology and Exploring Engineering*, Vol. 8, No. 10, pp. 1715–1719.
8. Abhishek, P., Kumar, K. R., & Naga Bhavya, B. (2018), "Enhanced technique for ranking fraud discovery for mobile apps", *International Journal of Engineering and Technology (UAE)*, Vol. 7, No. 2, pp. 172–174.
9. Anila, M., & Pradeepini, G. (2017), "Study of prediction algorithms for selecting appropriate classifier in machine learning", *Journal of Advanced Research in Dynamical and Control Systems*, Vol. 9, No. special issue 18, pp. 257–268.
10. Sundari, M. S., & Nayak, R. K. (2020), "Master card anomaly detection using random forest and support vector machine algorithms", *Journal of Critical Reviews*, Vol. 7, No. 09, pp. 2384–2390, Innovare Academics Sciences Pvt. Ltd, ISSN: 2394-5125, Cite Score-0.6 (Scopus indexed publication), doi: 10.31838/jcr.07.09.387.
11. Selvaraj, P., Burugari, V. K., Sumathi, D., Nayak, R. K., & Tripathy, R. (2019). "Ontology-based recommendation system for domain specific seekers", In 2019 Third International Conference on I-SMAC (IOT in Social, Mobile, Analytics and Cloud)(I-SMAC), ISBN: 978-1-7281-4365-1(E), ISBN: 978-1-7281-4366-8(P), DOI: 10.1109/I-SMAC47947.2019.9032634, pp. 341-345, IEEE Xplore Publication (Scopus Indexed publication).
12. Sajana, T., & Narasingarao, M. (2017), "Ensemble framework for classification of imbalanced malaria disease data", *Journal of Advanced Research in Dynamical and Control Systems*, Vol. 9, No. Special Issue 18, pp. 473–482.
13. Arshad, M., & Hussain, M. A. (2018), "A real-time LAN/WAN and web attack prediction framework using hybrid machine learning model", *International Journal of Engineering and Technology (UAE)*, Vol. 7, No. 3.12 Special Issue 12, pp. 1128–1136.

14. Chaitanya, G. K., Amarendra, K., Aslam, S., Soundharya, U. L., & Saikushwanth, V. (2019), "Prevention of data theft attacks in infrastructure as a service cloud through trusted computing", *International Journal of Innovative Technology and Exploring Engineering*, Vol. 8, No. 6 Special Issue 4, pp. 1278–1283.
15. Kolli, C. S. et al. (2020), "Fraud detection in bank transaction with wrapper model and Harris water optimization-based deep recurrent neural network". *Kybernetes*, Vol. 1, pp. 1–10.
16. Donato, J. M., Schryver, J. C., Hinkel, G. C., Schmoyer, R. L., Leuze, M. R., & Grandy, N. W. (1999), "Mining multi-dimensional data for decision support", *Future Generation Computer Systems*, Vol. 15, pp. 10–20.
17. Bolton, R., & Hand, D.J. (2001). Unsupervised profiling methods for fraud detection, Credit Scoring and Credit Control VII.
18. Kolli, C., & Uma Devi, T. (2019), "Isolation forest and XG boosting for classifying credit card fraudulent transactions", *International Journal of Innovative Technology and Exploring Engineering*, Vol. 8, No. 8, pp. 41–47.
19. Kolli, C., & Uma Devi, T. (2020). "Potential of automated feature engineering to classify credit card fraudulent transactions of e-commerce and business in the mobile age," *Kybernetes*, Vol. 13, No. 4, pp. 347–355.
20. Heidari, A. A., Mirjalili, S., Faris, H., Aljarah, I., Mafarja, M., & Chen, H. (2019), "Harris hawks' optimization: algorithm and applications", *Future Generation Computer Systems*, Vol. 97, pp. 849–872.
21. Zheng, Y. J. (2015), "Water wave optimization: A new nature-inspired metaheuristic", *Computers and Operations Research*, Vol. 55, pp. 1–11.

7 Real-Time Agent-Based Load-Balancing Algorithm for Internet of Drone (IoD) in Cloud Computing

Savita Saini, Ajay Jangra, and Gurpreet Singh
Department of Computer Science & Engineering,
University Institute of Engineering & Technology (U.I.E.T),
Kurukshetra University, Kurukshetra, Haryana, India

CONTENTS

7.1   INTRODUCTION

Late improvement in the territory of unmanned aerial vehicles (UAVs) has encouraged different open doors at a successful expense. Because of the capacity of dynamic reconfigurability, quick reaction, and simplicity of sending, UAVs can be considered a foremost arrangement in numerous regions of reconnaissance, three-dimensional administrations, clinical, farming, and transportation. Despite being beneficial, high versatility is a worry for organizations that require adequate power over these elevated vehicles. It is additionally one reason for interface contortion in UAV organizing. With the modernization of Internet of Things (IoT), UAVs' organizing has been given another wording, yet with a comparable working Internet of Drones (IoD), which upholds coordination of UAVs in the sky. With this, UAVs-helped networks

DOI: 10.1201/9781003181613-7

work to encourage remote availability in the territories where sending of actual foundation is troublesome or costly.

7.1.1 INTERNET OF DRONES (IoD)

IoD is instituted from IoT by supplanting "Things" with "Robots" while contributing comparable properties. IoD is foreseen to turn into a fundamental achievement in the improvement of UAVs. Gharibi et al. [7] characterized IoD as a "layered organization control engineering," which upholds UAVs in planning. In the IoD climate, numerous robots consolidate and structure an organization while communicating and accepting information from one another. IoD offers an arrangement for working distantly or through the Internet by means of IP addresses. UAVs clear a path for a ton of utilizations, yet their utilization case faces various difficulties

regarding usage, designing, and organization. The building design of robot correspondence needs standard and unification. Additionally, UAV-empowered correspondence networks experience the ill effects of the issue of committed range-sharing. UAV arrangement and way-arranging are different issues to be considered during range portions, as these have a strong effect on energy effectiveness. Additionally, UAV correspondences require consistency with large, solid organizations alongside their security. IoD can be considered as a critical answer as it upholds coordination of UAVs and different issues identified with UAV interchanges. IoD upgrades putting-away abilities and builds on the help of algorithmic arrangements, which thus permit arrangements enlivened by machine learning and artificial intelligence to be utilized for agreeable developments [5]. Regarding applications, IoD can be utilized as airspace designation interface, as a supplier of extremely reliable dependability, routes, and precision to drones in different regions as introduced in Figure 7.1. Different innovations like long-range wide-zone organization (LoRaWAN), low-power wide-region organization (LPWAN), and narrow-band IoT (NB-IoT) are reasonable arrangements in the sending of IoD in combination with IoT.

The appropriation of IoD with the properties of low cost, little size, high reconfigurability, operational usefulness, and ongoing responsibility is the need of the hour [9]. The market patterns are coasting toward the appropriation of IoD with following qualities:

- IoD upholds gadget to gadget just as gadget to multi-gadget interchanges.
- IoD encourages availability to the logical organizations.
- IoD assists with exploring and upgrading the survivability of vehicles through V2X or V2V mode (drone-to-drone mode).

Independent of the headways and plenty of answers for drone interchanges, security is as yet an essential worry in IoD as transmissions include touchy and basic data. UAVs are typically worked distantly while getting control and order messages from ground stations composed in IoD; in any case, a few cases may include self-ruling UAVs too. These order and control messages are communicated over various channels and variable transmission rates, which require impressive exertion for the board

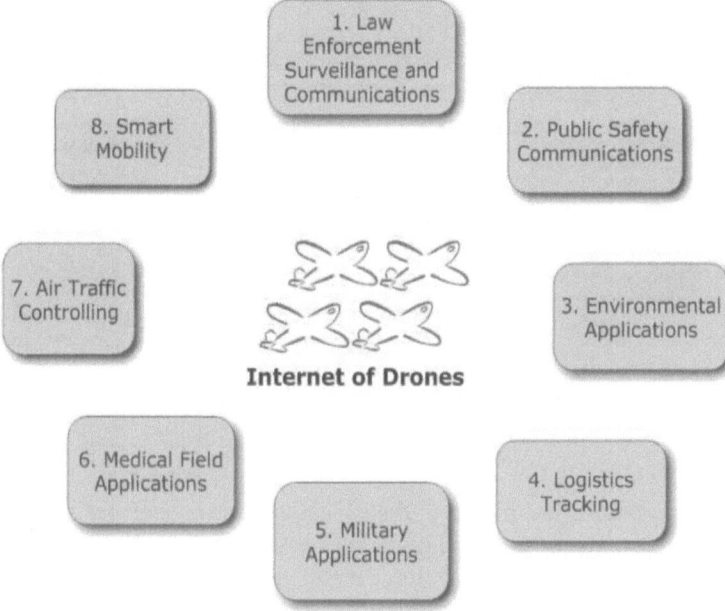

FIGURE 7.1   An overview of IoD application areas.

and control. Consequently, the security of IoD is one of the main prerequisites in the genuine usage of various UAVs.

7.1.2   IoD: Application Scenario

Quickly deployable and dependable strategic correspondence applications are the essential prerequisites for an effective mission. IoD assumes a critical part in these uses. IoD can encourage different administrations for the military applications and observation. IoD gadgets/UAVs fill in as a transitional stage for different sensors and gadgets, which have been sent to gather information and data in mission-situated assignments. One of the most appropriate application situations of IoD, UAVs-empowered combat zones, is portrayed in Figure 7.2. In this application situation, IoD can be utilized as an airspace assignment interface that assists with exploring the neighboring military UAVs. IoD gathers the information from the sent sensors alongside explicit kinds of military hardware. Thinking about the area of use and substance, the gathered data in such a situation is extremely basic and ought to be shielded from interlopers.

7.1.3   IoD as a Service

IoD is an essential piece of things to come Internet, which might be utilized for cutting-edge applications like savvy following, keen stopping, aviation authority, and cell organizations. The ongoing improvement in the correspondences' organization

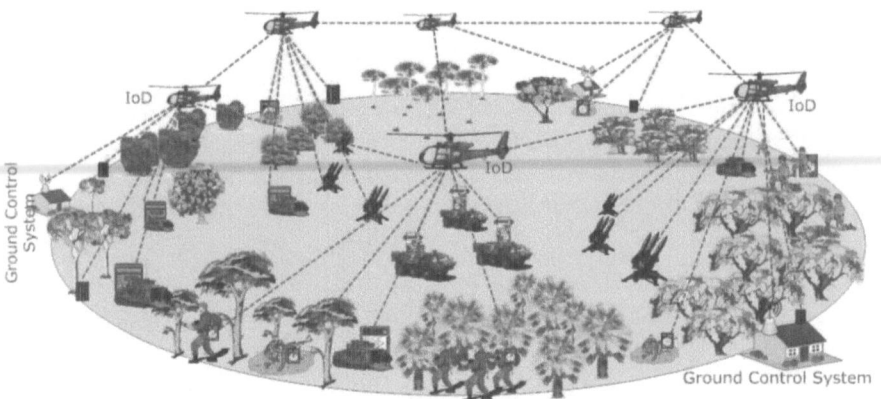

FIGURE 7.2   An exemplary illustration of IoD-assisted battlefield scenario.

and robot applications has pulled in the specialists and market for the headway of IoD. The correspondence joining IoD in various applications is another impression for the cutting-edge advancement of these organizations. IoD has attributes like more modest armadas with restricted batteries that send and get a lot of information [3].

IoD can be utilized to permit clients to build up the associated gadgets' work process by means of joining different incorporated robots and information bases with a legitimate asset arranging. For these components, different machine learning and information-mining procedures can be utilized for preparing enhanced data [1].

The cutting edge IoD is qualified for be coordinated with loads of outsiders' organizations and object-oriented administration over the sky. The future IoD market consolidates with the assortment of undertakings in the coordinated effort mode. UAVs in IoD have capacities for performing various tasks and work in metropolitan and country conditions. What's more, this cutting-edge IoD should adapt to security answers to relieve the dangers related with their missions.

7.1.4 Challenges and Future Research

IoD is dramatically developing innovation encouraging networking to the Internet. This developing innovation is confronting heaps of security dangers and issues. In this manner, the target of IoD security is the improvement of insurance as far as security for IoD and its foundation. Some future advancements are needed to improve IoD for additional applications. Initially, the IoD is confronting issues in the turn of events, arrangement, correspondence, and coordination stages. The absolute most critical difficulties and examination to target incorporate the following.

1. **Low Overheads and Dynamic Burden Adjusting**: The overheads of IoD in memory, energy, and deferments lead to different issues like asset debasements and low effectiveness. Dynamic burden adjusting in IoD engages the operational productivity. The constant catching and handling of information requires a great deal of energy and memory. Accordingly, supporting

different assignments with low overheads is an open test. Dynamic burden adjusting in IoD can be considered a viable arrangement and a point to follow for additional exploration.

2. **Survivability and Lifetime**: IoD includes UAVs that are battery-worked and have restricted force; subsequently, the coordination ought to be encouraged at low force. The survivability is influenced by the distorted signs and starvation conditions. These issues increment the consumption pace of assets like memory, energy, and so on, improving IoD survivability and upgrading the lifetime with these restricted assets, which requires an impressive consideration from the exploration community [18]. It ought to be mandatory to oblige better survivability that requires new and imaginative advancements, with powerful and more secure abilities in the mechanization and enhancement of mission arranging in unstructured conditions of IoD.

3. **Low Expense of Activity and High Throughput**: The expense is likewise a significant factor in the turn of events and sending of IoD. The assembling cost is straightforwardly relative to the necessities of IoD. Non-checked arrangement of IoD prompts enormous capital expenditures (CAPEX)/ operating Expenses (OPEX) issues which influence the exhibition regarding cost. Moreover, utilizing IoD for improving the administrations to bigger gatherings of detached clients is completely reliant on the expense of activity. In this way, minimal effort arrangements ought to be created for improving the utilities of IoD. As proposed in the current investigations, throughput is significantly influenced through the expense of activity, and there exists a tradeoff between the expense of activity and normal throughput accomplished in IoT, which must be adjusted for compelling correspondences.

4. **Performance and Unwavering Quality**: The exhibition of IoD is refined with the mission finish time and productive asset usage. Execution and dependability rely upon great assembling by thinking about the essential requirements while conveying end answers for IoD. To comprehend asset productivity and dependability, utilization-centered markers ought to be fused in IoD gadgets. The checking of such assets is assessed as far as execution. The productive utilization of assets prompts fruitful missions [15]. Accordingly, the asset usage is considered as a huge boundary for execution and unwavering quality in IoD.

5. **Dynamic Geographies and Versatile Directing Instruments**: The dynamic geographies and versatile steering systems encourage on-request and practical conditions for IoD. The dynamic steering geographies uphold waypoint coordination in IoD networks. The deferrals displayed through the ordinary steering calculations don't proficiently oblige IoD coordination. Subsequently, versatile directing systems ought to be considered for progressively-changing geographies in IoD.

6. **Low Disappointment Pace of IoD**: IoD has been designed, created, and utilized for explicit applications like ecological checking, agribusiness, and so on. To help these administrations, IoD developments ought to be considered with asset limitations, dynamic geography, flexibility, versatility,

security, and QoS backing to the clients. The disappointment pace of the organization corrupts the general presentation of IoD. Thus, the disappointment pace of the organization ought to be limited to improve the general exhibition of IoD.

7. **Emphasized Security Answers for IoD**: The ongoing advancement in IoD stresses the information rate and security. The security is a consolidated part of classification, trustworthiness, confirmation, and non-renouncement of communicated information in IoD networks [10]. The snooping, network sticking, feeble confirmation, and portability of the board are significant open issues in IoD correspondences.

8. **Efficient Organization of Countermeasures and Support**: The security arrangement fuses an intrusion detection system, confirmation instruments, and cryptographic calculations. Customary systems are delayed in speed, huge in estimate, and burn through more force and may neglect to give essential assurance to information in the organization [13]. In this manner, the productive arrangement of countermeasures and upkeep can spare the extra computational cost and can forestall extreme force utilizations.

9. **End-to-end Network and Participation**: The start-to-finish availability mirrors the coordination effectiveness. In the IoD framework, availability between the source and the objective ought to consistently be held. The start-to-finish network upgrades the participation among neighboring IoD and different sensors. Some application regions, similar to ongoing observing, require a broad start to finish availability for better inclusion and collaboration [21]. Load balancing is a significant issue in IoD and cloud computing (CC) [1]. It is an instrument that appropriates the dynamic nearby workload uniformly over all the hubs in the whole cloud. This will maintain a strategic distance from the situation where a small number of hubs are vigorously loaded while others are inert or accomplishing little work. It assists with accomplishing high client satisfaction and asset usage proportion. Henceforth, this will improve the general implementation and asset usefulness of the structure. It likewise assures that each computing asset is dispersed effectively and decently [2]. It further forestalls bottle-necks of the structure, which may happen because of load lopsidedness. It additionally guarantees that each computing asset is appropriated produc-tively and reasonably [3, 4]. The utilization of assets and preservation of vitality is not generally a prime focal point of conversation in CC. Be that as it may, asset utilization can be kept to a base with appropriate load bal-ancing, which helps in diminishing expenses as well as making endeavors greener. Adaptability, which is one of the significant highlights of CC, is additionally empowered by load balancing [5].

7.2 LOAD BALANCING

Load balancing is a technique to disseminate tasks over numerous PC groups or PCs; it can arrange joins and focal handling units to accomplish ideal asset use, or differ-ent assets, plate drives, augment throughput, limit reaction time, and stay away from overload. Utilizing dissimilar segments with load balancing, rather than a solitary division, may enlarge reliability through recurrence. The load-balancing administration is generally given by dedicated equipment or programming, e.g., a domain name system worker or a multilayer switch.

Load balancing is the focal issue in CC [6]. It is a method that circulates the dynamic nearby workload uniformly over all the hubs in the whole cloud to keep away from a situation where a small number of hubs are vigorously loaded while others are inactive or accomplishing little work. It assists with accomplishing high client satisfaction and asset utilization proportion, subsequently improving the general implementation and asset usefulness of the framework. It likewise assurance that each computing asset is circulated proficiently and decently [7]. It further forestalls bottlenecks of the structure, which may occur because of load awkwardness.

The objective of load balancing is improving the presentation by balancing the load amid these different assets (plate drives, focal preparing units, arrange joins) to accomplish greatest throughput, ideal asset usage, most extreme response time, and to sustain a strategic gap from overload. To disperse the load on various frameworks, distinctive load balancing calculations are used [8].

7.3 LOAD BALANCING CONFRONT IN THE CLOUD COMPUTING

Even though CC has been generally embraced, examination in CC is still in its beginning phases, and some logical difficulties remain unanswered by established researchers, especially load-balancing difficulties [9].

- **Automated Administration Provisioning**: A key component of CC is flexibility; assets can be distributed or delivered naturally.
- **Virtual Machines Migration**: With virtualization, a whole machine can be viewed as a record or set of documents. To unload a physical machine vigorously loaded, it is conceivable to move a virtual machine between physical machines.
- **Energy Management**: The advantages that advocate the reception of the cloud is the economy of scale. Vitality sparing is a key point that permits a worldwide economy where a lot of worldwide assets will be bolstered by diminished suppliers rather that everyone has its assets.
- **Stored Information for Executives**: In the current scenario, data put away over the structure has an exponential increment in any event, for organizations by re-appropriating their information stockpiling or for people, the organization of information stockpiling or for people, the administration of information stockpiling turns into a significant test for CC.
- **Emergence of Little Server Farms for CC**: Little suppliers can convey CC administrations prompting geo-assorted variety computing. Load balancing will turn into a problem on a worldwide scale to guarantee a sufficient reaction time with an ideal appropriation of assets [10].

7.4 LITERATURE SURVEY

Ogden et al. [11] proposed a neuro-fluffy operator–based load-balancing instrument. By using the neuro-fluffy technique, undesirable data was disposed of, and expected data with better precision was put into the cloud node.

Ogden and Guo [11] talked about the advantages and disadvantages of existing methodologies and pinpoint the likely regions for development. They proposed a comprehensive methodology that thinks about versatile explicit variables when structuring portable surmising systems.

Moradi et al. [12] proposed uPredct, a client-level prescient system for multitenant framework Software-as-a-Service (SaaS) clouds.

Belgaum et al. [13] audited the conversation with a scientific categorization of current developing load-balancing methods and Software-defined networking (SDN) deliberately by arranging the strategies as ordinary and man-made consciousness-based procedures to improve the administration quality.

Balasubramanan et al. [14] proposed a novel yet basic model that takes help from the RL worldview. Creators planned an operator that takes motivation from the fretful scoundrel file strategy.

Patel et al. [15] concentrated on disseminating the workload similarly among all the compartments and lessening the makespan.

Junad et al. [16] proposed a crossover calculation Additive Criterion with μ determined off-line (ACOFTF), which considers significant QoS measurements, for example, relocation time, throughput time, overhead time, and advancement time. The proposed model has demonstrated a strategic distance from untimely intermingling, which was one of the destinations of crossbreed metaheuristics even within the sight of differing data sets.

Babou et al. [17] proposed a strategy called HEC clustering balance method that fundamentally diminishes the preparing season of solicitations sent by clients while effectively using assets of HEC node, MEC node, and the focal cloud on the three-level engineering of HEC.

Shahd et al. [18] intentional on one of CC man problems, i.e., load balancing (LB). The objective of LB was equilibrating the calculation on the cloud nodes with the end goal that no host is under/overloaded. A few LB calculations have been executed in writing to give compelling organization and fulfill client demands for fitting cloud nodes, to improve the general effectiveness of cloud administrations, and to furnish the end client with more fulfillment.

Devraj et al. [19] proposed load-balancing calculation as a half and half of the Firefly and improved multi-objective particle swarm optimization (MPSO) procedure, condensed as FMPSO. This procedure sends Firefly (FF) calculation to limit the pursuit space, whereas the MPSO method was actualized to recognize the improved reaction.

7.5 PROPOSED WORK

In Agent Walk1 as shown in Figure 7.3, a specialist is actuated at any arbitrary worker and secures several positions in line at that worker. The specialist will rehash this procedure for all workers of that common group After that it will figure AVERAGE. Based on AVERAGE, it will detect the worker's status regarding underloaded and overloaded.

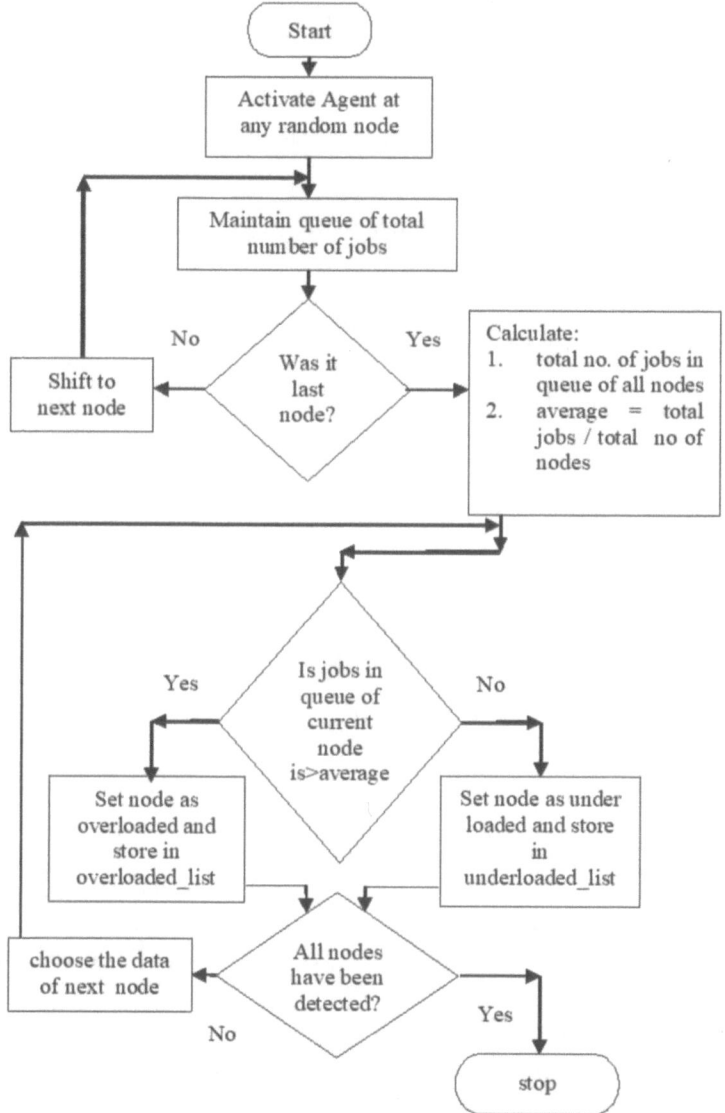

FIGURE 7.3 First walk of the agent from first to the last node for gathering load information.

In Figure 7.4, the operator will begin backtracking from the last worker to the first worker for balancing the load of workers.

7.6 RESULTS

In this section, results are presented; the proposed algorithm is implemented using MATLAB.

1. In the existing unified plan, CPU time utilization is 10 units as node-to-node correspondence takes bunches of CPU time in passing load data to the central node and the other way around.

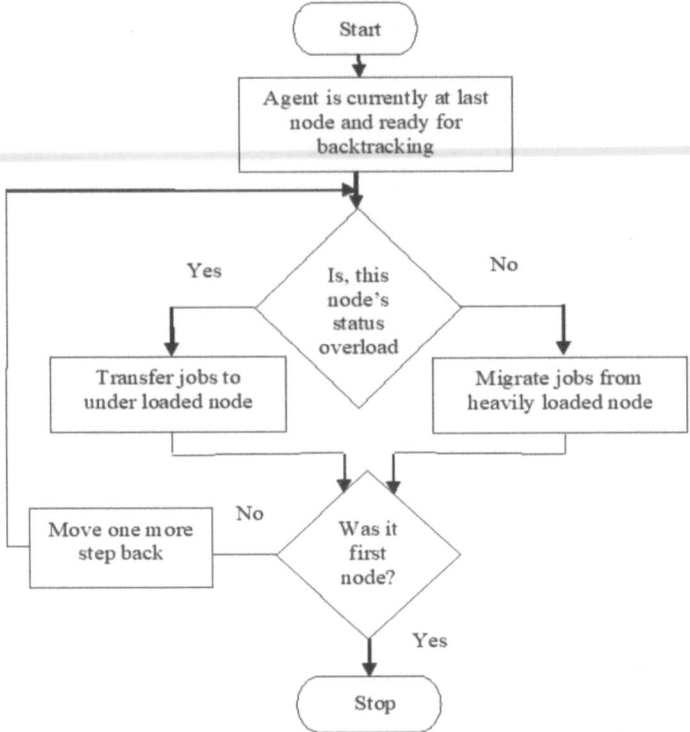

FIGURE 7.4   Second walk of agent from last node to first node for balancing load.

2. In the proposed plot, CPU time utilization is 1 unit as agent-node correspondence costs less in passing load data to agent and the other way around.
3. Time utilization by both the plans during the time spent balancing load
after the assortment of load data from all the hubs is taken as equivalent to
50 units of time.

Execution is done based on adaptability for 12, 24, 48, 96, and 192 hubs having an
irregular number of employments for every hub in a cloud (Table 7.1). In Figure 7.5,

TABLE 7.1
Simulation Parameters

Factors	Values
No. of Nodes	12, 24, 48, 96, 192
CPU Time Consumption (Process of balancing load)	50 Units
CPU Time Consumption (Proposed)	1 Unit
CPU Time Consumption (Existing centralized scheme)	10 Units
Platform	MATLAB 2013 v2
Operating System	Windows 10

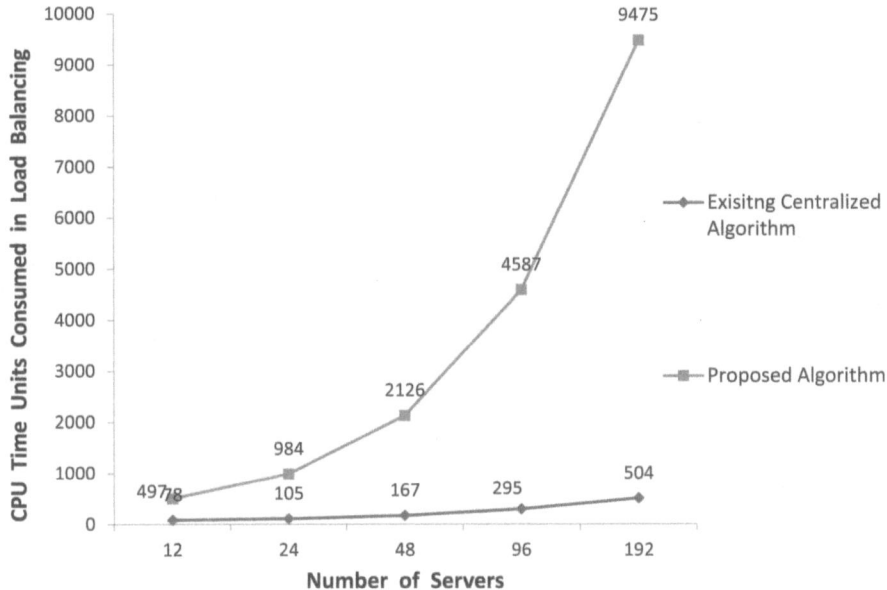

FIGURE 7.5   Comparison of performance w.r.t. CPU time unit consumed and load balancing.

the CPU costs for existing and proposed load-balancing plans are portrayed based on adaptability. The diagram shows that the CPU time unit devoured in the load-balancing is projected as much less than the current brought together to plan.

Throughput examination of proposed and existing concentrated calculation is shown in Figure 7.6. The chart shows better throughput results proposed when contrasted with the existing one.

Correlation of average waiting season of projected and centralized calculation is shown in Figure 7.7. The diagram illustrates that the proposed framework conspire

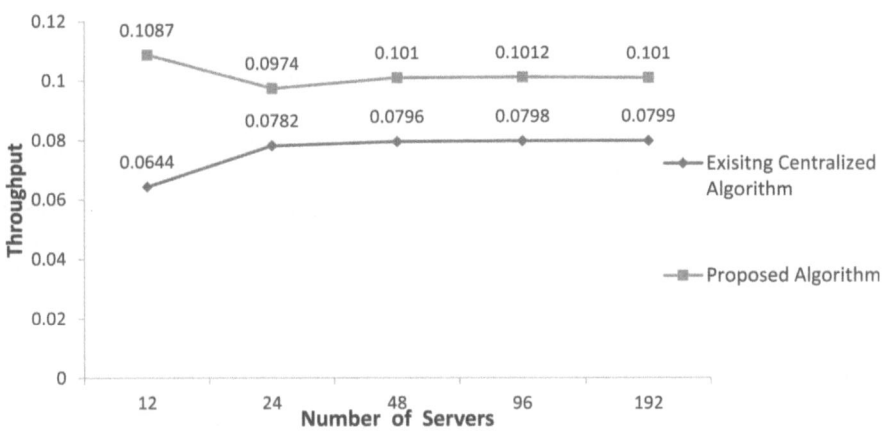

FIGURE 7.6   Performance comparison in terms of throughput.

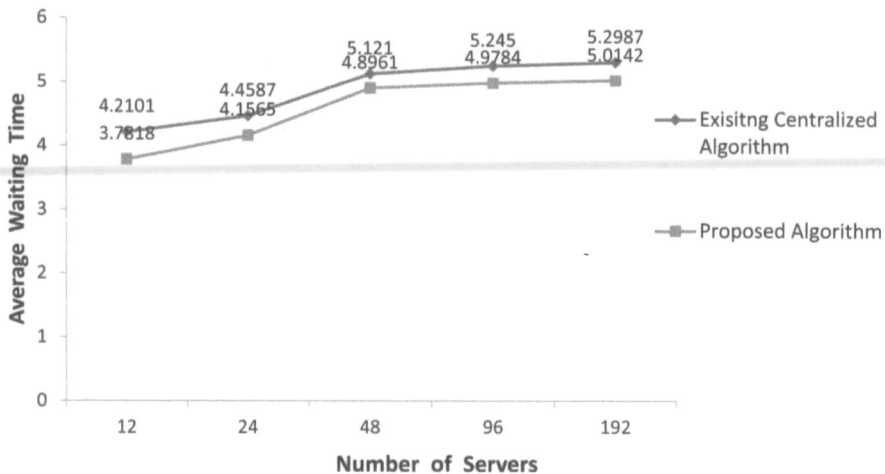

FIGURE 7.7 Performance comparison in terms of average waiting time.

performs better as far as average waiting time. The throughput and average holding-up time delineated in Figures 7.6 and 7.7 individually likewise affirm the pitiful actuality that the CPU can use spared time (outlined in Figure 7.5) in engaging different employments.

In the proposed framework, CPU cost diminishes because, in the proposed plot, the agent moves from node to node, expends insignificant CPU time, and spares the node's season of correspondence that it needs to utilize while speaking with the central node sent for load balancing in the existing framework. This CPU time can be utilized to engage different employments in line by the node, and it naturally will expand throughput and reduction response time of the cloud.

7.7 CONCLUSION

This work presents a proposed operator-based plan for CC. In the wake of contrasting it and a customary load-balancing plan, that is the necessity of existing unique load-balancing calculations in CC, and we likewise saw that this technique enormously decreases the correspondence cost of nodes, quickens the pace of load balancing, and gets better response time and throughput of the cloud. Subsisting load-balancing plans rely on the data-transfer capacity or traffic condition to move the load data starting with one node then onto the next. However, in a purposed load-balancing plan, a lightweight operator can move effectively starting with one node then onto the next without influencing the system's load excessively. It is expected to keep taking a shot at the proposed component and attempting to accomplish much better throughput and start-to-finish delay. Additionally, it will attempt to include security instruments in the proposed system to make sure about the transmission of messages in CC.

REFERENCES

1. Jadeja, Y., & Modi, K. (2012). Cloud computing-concepts, architecture and challenges. In International Conference on Computing, Electronics and Electrical Technologies (ICCEET) (pp. 877–880). IEEE.
2. Fernando, N., Loke, S. W., & Rahayu, W. (2013). Mobile cloud computing: A survey. Future Generation Computer Systems, 29(1), 84–106.
3. Le, N., Hieu, T. N. T. H., Xuan, P. N., & Cong, H. T. (2020). MCCVA: A new approach using SVM and K-means for load balancing on cloud. International Journal on Cloud Computing: Services and Architecture (IJCCSA), 10(3), 1–14.
4. Kumar, P., & Kumar, R. (2019). Issues and challenges of load balancing techniques in cloud computing: A survey. ACM Computing Surveys (CSUR), 51(6), 1–35.
5. Ahmad, M. O., & Khan, R. Z. (2018). Load balancing tools and techniques in cloud computing: A systematic review. In Advances in Computer and Computational Sciences (pp. 181–195). Springer, Singapore.
6. Raghava, N. S., & Singh, D. (2014). Comparative study on load balancing techniques in cloud computing. Open Journal of Mobile Computing and Cloud Computing, 1(1), 18–25.
7. Narwal, A., & Dhingra, S. (2018). Analytical review of load balancing techniques in cloud computing. International Journal of Advanced Research in Computer Science, 9(2), 550–553.
8. Patel, S., Patel, H., & Patel, N. (2016). Dynamic load balancing techniques for improving performance in cloud computing. International Journal of Computer Applications, 138(3), 1–5.
9. Ghomi, E. J., Rahmani, A. M., & Qader, N. N. (2017). Load-balancing algorithms in cloud computing: A survey. Journal of Network and Computer Applications, 88, 50–71.
10. Shakir, M. S., & Razzaque, A. (2017). Performance comparison of load balancing algorithms using cloud analyst in cloud computing. IEEE 8th Annual Ubiquitous Computing, Electronics and Mobile Communication Conference (UEMCON) (pp. 509–513). IEEE.
11. Ogden, S. S., & Tian G. (2020). MDInference: Balancing inference accuracy and latency for mobile applications. IEEE International Conference on Cloud Engineering (IC2E) (pp. 28–39). IEEE.
12. Moradi, H., Wei, W., Amanda, F., & Dakai, Z. (2020). uPredict: A user-level profiler-based predictive framework in multi-tenant Clouds. IEEE International Conference on Cloud Engineering (IC2E) (pp. 73–82). IEEE.
13. Belgaum, M. R., S. Musa, Alam, M. M., & Mazliham M. S. (2020). A systematic review of load balancing techniques in software-defined networking. IEEE Access, 5, 1–15.
14. Venkatraman, B., Aloqaily, M., Tunde-Onadele, O., Yang, Z., & Reisslein, M. (2020). Reinforcing cloud environments via index policy for bursty workloads. NOMS 2020-2020 IEEE/IFIP Network Operations and Management Symposium (pp. 1–7). IEEE.
15. Patel, D., Patra, M. K., & Sahoo, B. (2020). GWO based task allocation for load balancing in containerized cloud. International Conference on Inventive Computation Technologies (ICICT) (pp. 655–659). IEEE.
16. Junad, M., Sohal, A., Ahmed, A., Baz, A., Khan, I. A., & Alhakam, H. (2020). A hybrid model for load balancing in cloud using file type formatting. IEEE Access, 5, 8118135–118155.
17. Babou, C. S. M., Fall, D., Kashihara, S., Taenaka, Y., Bhuyan, M. H., Niang, I., & Kadobayashi, Y. (2020). Hierarchical load balancing and clustering technique for home edge computing. IEEE Access, 8, 127593–127607.
18. Shahid, M. A., Islam, N., Alam, M. M., Su'ud, M. M., & Musa, S. (2020). A comprehensive study of load balancing approaches in the cloud computing environment and a novel fault tolerance approach. IEEE Access, 8, 130500–130526.

19. Devaraj, A. F. S., Elhoseny, M., Dhanasekaran, S., Lydia, E. L., & Shankar, K. (2020). Hybridization of firefly and improved multi-objective particle swarm optimization algorithm for energy efficient load balancing in cloud computing environments. Journal of Parallel and Distributed Computing, 142, 36–45.
20. Siddiqui, S., Darbari, M., & Yagyasen, D. (2020). An QPSL queuing model for load balancing in cloud computing. International Journal of e-Collaboration (IJeC), 16(3), 33–48.
21. Naz, I., Naaz, S., & Biswas, R. (2020). A parametric study of load balancing techniques in cloud environment. In Inventive Communication and Computational Technologies (pp. 881–890). Springer, Singapore.

8 Ultrawide Band Antenna for Wireless Communications
Applications and Challenges

Sapna Arora, Sharad Sharma*,
and Arun Kumar Rana#*
*Maharishi Markandeshwar (Deemed
to be University), Mullana, India
#Panipat Institute of Engineering and
Technology, Samalkha, Panipat, India

CONTENTS

8.1 INTRODUCTION

The health sector is one of the most favorable places for the use of Internet of Things (IoT) technologies. In the field of health, a series of studies have been carried out in order to minimize the use of resources and increase productivity. Along with other developments, the use of IoT has taken quality production in the health sector to a minimum. One such technology is the use of wireless body area networks (WBANs), which in the future will significantly help patients and make their treatment more efficient because they will not have long wait times at home or in a hospital. WBANs and IoT have an integrated future, as a collection of heterogeneous sensor-based systems are WBANs, like any IoT application. To improve the combination of IoT and WBANs with UWB antennas, it is important to resolve several obstacles that block their integration.

With the advancement of wireless communication, a number of new technologies have arisen. These technologies have significantly impacted human life. Broadband wireless technology employs signals transmission between one point to another or one point to many points. Broadband technologies offer fast-speed communications. There are numerous advantages of using higher frequencies for communications,

DOI: 10.1201/9781003181613-8

e.g., large bandwidth, reduction in antenna size, and large distance coverage. Wireless technologies can be categorized by different features such as power, data rate, and distance covered [1]. Wireless personal area networks (WPANs) are getting very popular these days. Maximum distance coverage is 10 m and is used for wireless interconnection of devices, such as mobiles, laptops, and personal digital assistants (PDAs). WPAN has many applications in both homes and offices to transfer audio and video data. This requires larger data rates than existing technologies. Maximum data rates for a band-limited ideal additive white Gaussian noise (AWGN) channel is given by the Shannon channel capacity Equation (8.1):

$$C = B \log2(1 + SNR) \tag{8.1}$$

where:
  B = bandwidth of channel
  C = channel capacity
  SNR = signal-to-noise ratio

So channel capacity can be enhanced by increasing bandwidth or signal power. But signal power cannot be increased beyond a limit due to battery usage by all concerned devices, and interference will also increase with radio channels. So the only solution to this problem will be high bandwidth.

UWB frequency ranges from 3.1 GHz–10.6 GHz employs bandwidth of 7.5 GHz; this spectrum was declared as unlicensed in 2002 by the Federal Communications Commission (FCC) [2]. UWB technology offers data rates of up to 1 Gb/s and distance coverage is up to 30 m. The main advantage of UWB technology is low power requirements with immunity to noise and Also UWB signals can penetrate in large variety of materials' technology. The main component of UWB technology is UWB antennas because they influence the overall performance of the system. For effective communication in all directions, UWB antenna should have omnidirectional radiation patterns and should be compact in size for indoor applications. Broad bandwidth can be achieved by existing frequency-independent log periodic [3] and log spiral antennas [4]. But these existing structures are large, so they cannot be incorporated in small electronic devices. Other options to achieve wide bandwidth are biconical and Vivaldi antennas [5]. These antennas have bidirectional radiation patterns that are not suited to UWB applications.

For UWB applications, monopole antennas with planar structures [6–8] are proposed with various shapes such as polygonal, circular, and elliptical. Printed monopole antennas are small in size and provide omnidirectional radiation patterns. UWB printed antennas are attracting researchers due to their capability to be fabricated with other circuits on the same printed board.

In the UWB frequency range from 3.1–10.6 GHz, the IEEE 802.16 standard for WLAN and IEEE 802.11a standard for WIMAX exist at 5.15–5.825 GHz and 3.3–10.6 GHz, respectively. These bands introduce interference to the UWB band. So it becomes necessary to filter out the signals in the UWB range to improve communication in this band. Therefore, band stop filters are required to avoid interference from these narrow band signals. Using an extra filter circuit will increase the size and complexity of the system. So UWB antennas with inbuilt band stop filters are solutions to minimize the interference. In last few years, several methods have been investigated to integrate the band stop filters in UWB antennas. The effective method to get notch bands is to incorporate slots and stubs in patch or ground plane [9–12]. More recently, in response to consumer demands, we have witnessed increased research activities focused on the production of flexible electronics, reporting a rising interest in lightweight, portable, and wearable devices. Innovations in layered semiconductors and novel manufacturing techniques have greatly allowed the success of flexible electronics. Consequently, for advanced wireless communication, the integration of versatile UWB and MIMO antennas with such devices is ultimately necessary. Due to performance degradation caused by potential structural deformation, implementation of diversity antenna schemes for IoT applications that are subject to bending and flexing or have curved surfaces is a daunting task. Furthermore, in applications with restricted space constraints, the close proximity of the radiating elements contributes to an increased mutual coupling that compromises the efficiency of the device.

8.2 LITERATURE SURVEY

Today, several groups of researchers are focused on designing of UWB antennas due to their inherent advantages in indoor applications. Some of the prominent studies are listed in the following.

Wu and Li [12] designed an antenna with dual U-shape slots in radiation patches having band rejection capability in the 3.20–3.52 GHz and 5.13–6.09 GHz frequency bands covering 2.45–11.0 GHz with VSWR < 2 with dimension 38×40 mm². The patch is fabricated on Teflon substrate with a thickness of 1.5 mm and relative $\epsilon_r = 2.65$, and it provides an omnidirectional radiation pattern. By varying the length and width of slot, desired notch bands are obtained [12].

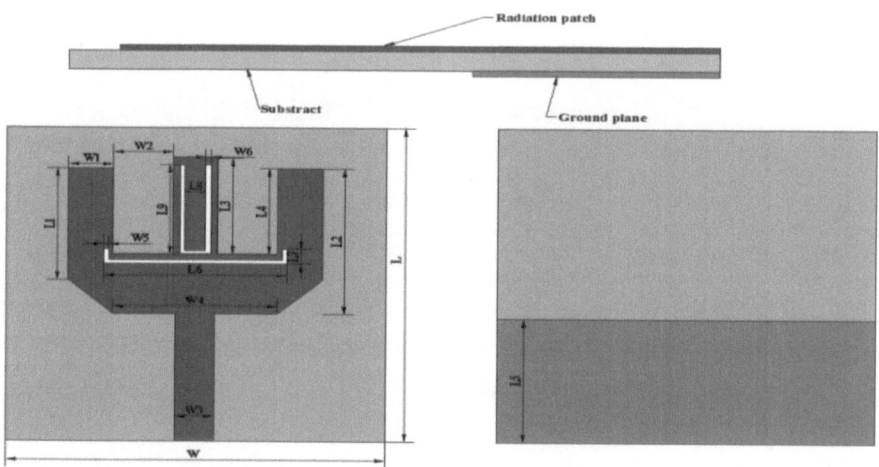

FIGURE 8.1 Geometry and configuration of the proposed antenna [12].

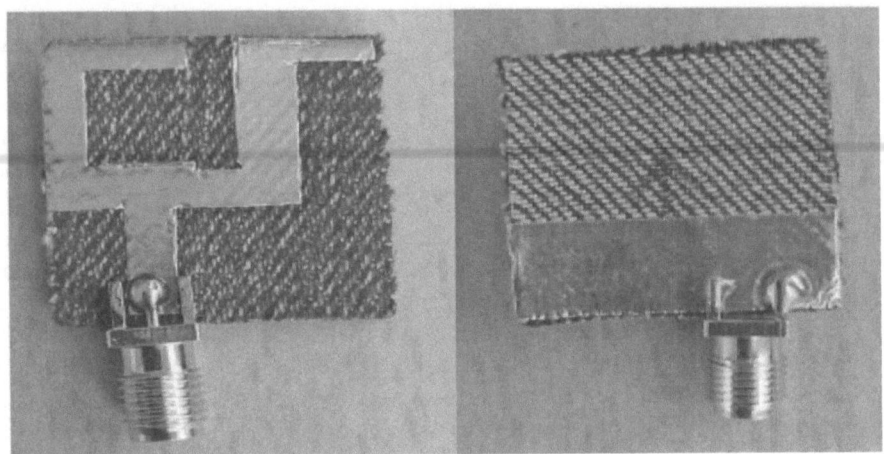

FIGURE 8.2   Optimized textile UWB fabricated antenna [14].

Yadav et al. designed a wearable antenna with an impedance bandwidth of 2.96–11.6 GHz and which has application for a UWB range of 3.1–10.6 GHz. This structure provides a maximum gain at 7.3 GHz, which is 5.47 dBi. Due to cutting, notch and slot on the patch antenna bandwidth have been increased. Jeans fabric has been used as a dielectric substrate having a low dielectric constant (ε_r = 1.7). See Figure 8.2 [13].

Sopa and Rakluea fabricated an antenna having a patch with hexagonal geometry with a C-shaped slot having capability with band-notched characteristics from 2.5–3.1 GHz. To improve antenna performance, the ground plane is etched with rectangular slots (Figure 8.3). The substrate of antenna is made from military cotton with dielectric constant 2.25 with dimensions 58 × 42 × 0.77 mm. The bandwidth obtained was 1.76–17.78 GHz. Omnidirectional radiation pattern was obtained [32].

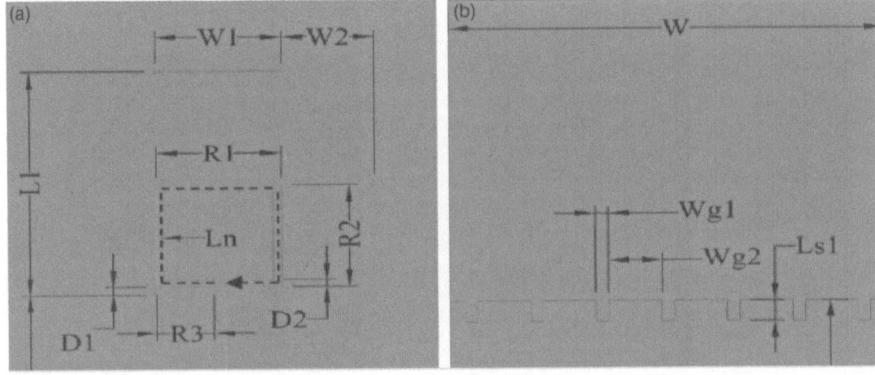

FIGURE 8.3   The hexagonal-shaped antenna with C-shaped slot: (a) front view and (b) back view [14].

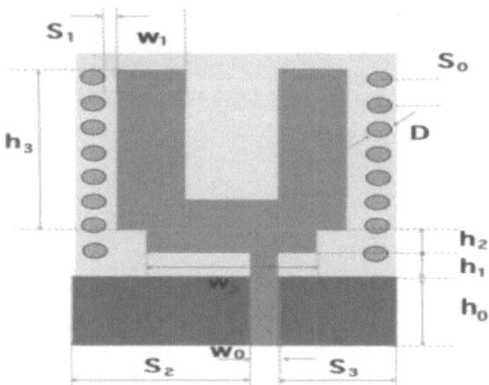

FIGURE 8.4   UWB antenna with SIW structure [33].

Donelli designed an UWB antenna using an substrate-integrated waveguide (SIW) resonator. SIW structure tends to decrease antenna size and leads to enhanced bandwidth. The bandwidth obtained is 3–10 GHz. But the radiation pattern obtained is not omnidirectional so it cannot be used for mobile phone applications but can be used for wireless sensor network applications to connect sensor nodes [33].

Samal designed an UWB antenna with full ground plane for a WBAN application. A small–size antenna with dimensions 39 mm × 42 mm × 34 mm was designed with a full ground plane, which makes the antenna suitable to be used near the vicinity of human body because radiation in backward direction has been reduced. SAR (specific absorption rate) is under the limit for WBAN applications [34].

Chandel, et al. has proposed UWB antenna with one inverted T-stub and two C-shaped slots to incorporate notch bands in WiMAX, WLAN, and X band provides bandwidth 2.49–19.41 GHz with dimension 20×20 mm^2 as shown in Figure 8.5 [17]. Design provides good efficiency, which is greater than 85% in the complete radiating band except in band stop bands [16].

Doddipalli et al. presented another monopole antenna with slotted substrate for WBAN applications. This antenna has an elliptical-shaped patch, and slots are etched on the patch surface. This modification leads to return loss improvement and

FIGURE 8.5   Antenna with full ground plane [16].

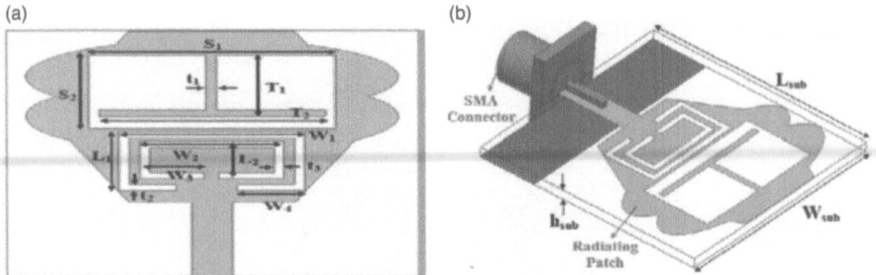

FIGURE 8.6   Antenna with stub and notches: (a) front view and (b) side view [17].

improves return loss performance 2.94–17.6 GHz. The antenna area of 648.88 mm^2 provides overall bandwidth of 14.66 GHz. Through parametric analysis, the design has been optimized as shown in Figure 8.6 [18].

Doddipalli et al. proposed a hexagonal-shaped patch with a slotted ground plane for UWB applications as shown in Figure 8.8. Defected ground plane made of FR4 substrate having square slots with dimensions 15 mm × 25 mm are able to achieve UWB bandwidth in range of 2.9–11GHz in free space and human proximity as shown in Figure 8.7. SAR is achieved according to FCC guidelines so the antenna can be used for WBAN applications as well [19].

Ray presented design aspects of monopole antennas with various shapes for UWB applications. The author has described the design equations for printed monopole antennas for lower band-edge frequency in terms of dimensions of antennas. Dimensions for planar monopole antennas were derived by equating the area to that of cylindrical monopole antennas. This paper describes the starting point for antenna designing [20].

Zhu et al. described an antenna having a half-circle–shaped patch having an open rectangular slot and a half-circle–shaped ground plane. This antenna is fed with a CPW transmission line [39].

In Figure 8.9a, quarter-wavelength strips are connected to the transmission line horizontally and act as band stop filter wide-notched bands in 4.6–6.4 GHz with center frequency $f_r = 5.5$ GHz. In Figure 8.9b, a narrow notched band at 5.725–5.825 GHz is obtained using two slots with quarter wavelength in the feed line. In Figure 8.9c,

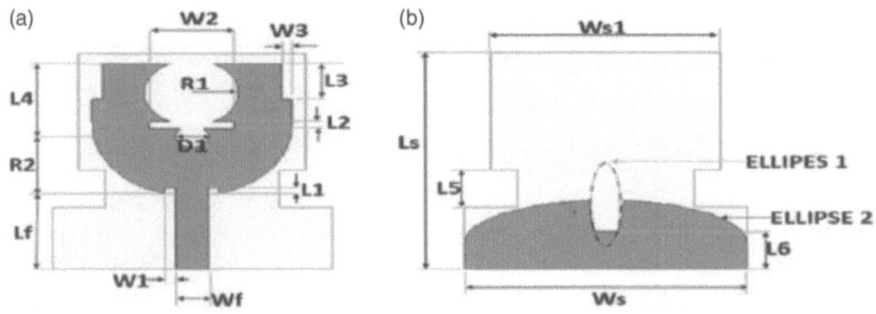

FIGURE 8.7   Structure of antenna: (a) Front view (b) Back view [18].

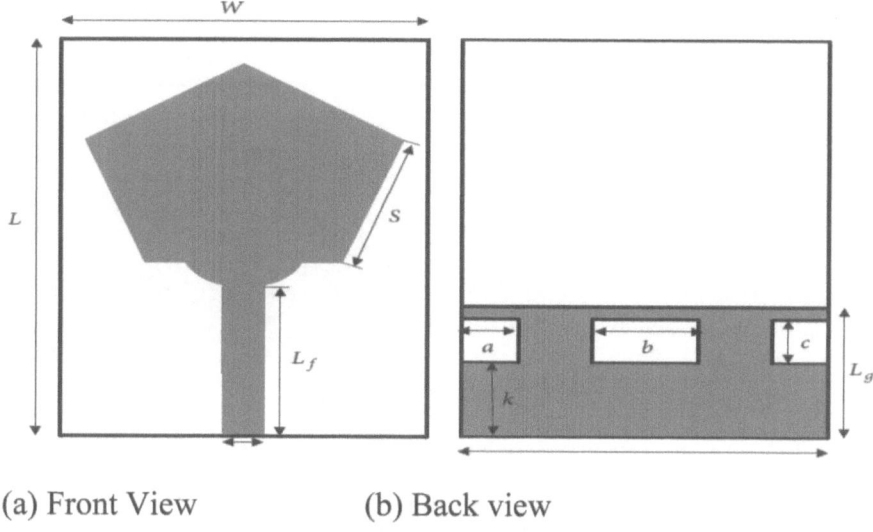

(a) Front View   (b) Back view

FIGURE 8.8   Antenna with hexagonal geometry [19].

dual notch bands are obtained at 5–5.35 GHz and 5.7–6.1 GHz using two stubs in rectangular slots and one quarter-wavelength stub in the transmission line [21–22].

Tripathi et al. presented elliptical planar antenna with a size of 26 × 30 mm² on dielectric substrate FR4 having permittivity $\varepsilon_r = 4.4$ with thickness $h = 0.8$ mm. The bandwidth of antenna is enhanced by making slots under the feedline. Further, to enhance bandwidth, ground plane is truncated from corners that control upper and lower frequencies. The bandwidth covered in range of UWB application from 2.94–12.57 GHz. Effect of ground plane modifications on bandwidth has been discussed [23].

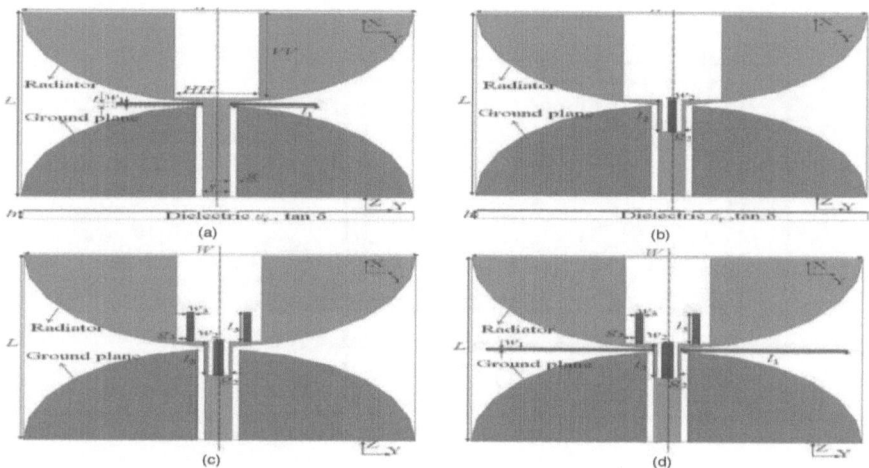

FIGURE 8.9   Geometry of antenna [39].

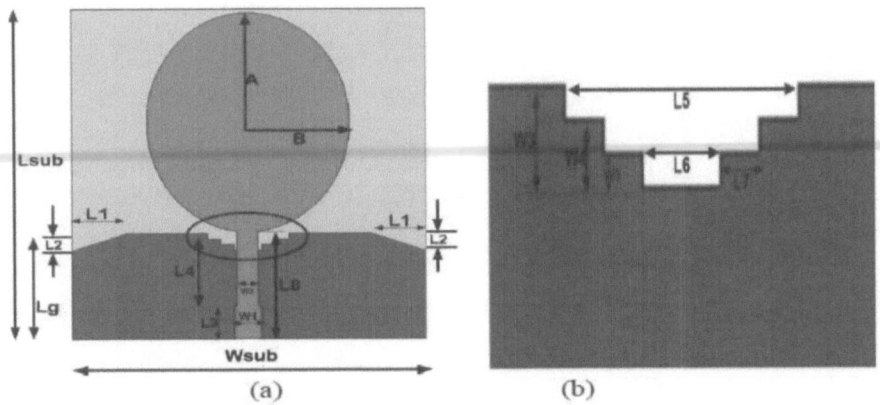

FIGURE 8.10   Antenna geometry: (a) top view and (b) matching slot in ground [23].

Shakib described a ring-shaped antenna that covers tri-bands for wireless applications. In its patch, two rings are connected by a rectangular strip, and stubs are extended from the coplanar ground plane to both sides of the radiating patch, and a CPW-based feeding structure has been used. The dimensions of antenna are 35 mm × 26.6 mm and fabricated on FR4 epoxy substrate having a thickness of 1.6 mm and a relative permittivity of 4.4. Reflection coefficients less than −10 dB at 3.10–4.34 GHz, 5.14–5.60 GHz, and 7.33–10.43 GHz observed in free space as shown in Figure 8.10, 8.11 [24].

8.3   APPLICATIONS AND CHALLENGES

The user's demand for wearable technology has risen quickly with the launch of current monitoring systems [25–30]. Specifically, the healthcare sector is increasingly using monitoring systems, which will have a positive impact on society. Key

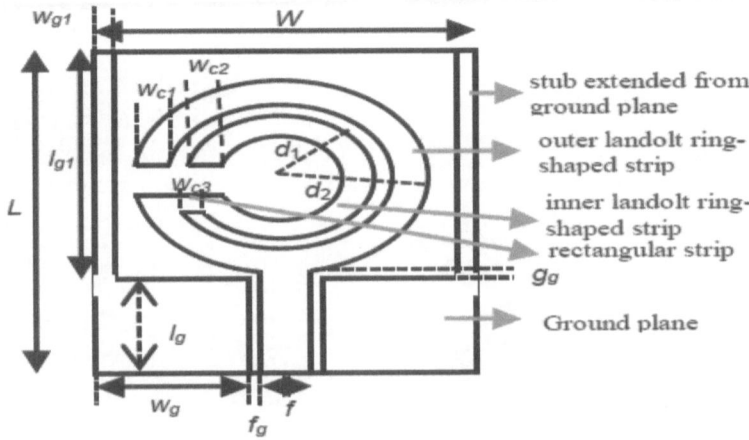

FIGURE 8.11   Geometry of antenna [24].

health measures are physiological problems and their monitoring could be useful in initially diagnosing any type of disease [31–37]. Significant health issues, such as disease prevention and costs, can be reduced by smart wearable devices. Typically, clinical devices are nonwearable for measurement; therefore, researchers have focused on unique smart shapes that would make wearable devices that are low-profile, robust, and high-performance [38–42]. Normal implementations have been approved by IEEE 802.15 for body-centric communications. Different types of wearable antennas, including reconfigurable antennas and dielectric resonant antennas, can be used for WBAN applications. Demand for smaller wearable devices is increasing day by day, particularly in everyday appliances. In areas such as patient-health monitoring, security safety, and public entertainment, WBAN-centric connectivity has a number of applications [43–46]. Cloth-integrated antennas are more suitable for personal wireless devices in close proximity to the body, sufficient to ensure a quality of service independent of body movements. Wearable electronic devices do not have any size and shape restrictions due to the large body space available. In comparison, conventional antennas are designed to be flexible and ideally suited for integration with wireless networking devices into the IoT network [47–50]. There are several applications for which UWB antenna is used. Since UWB technology can provide 500 Mb/s, this can be used for personal area networks to transfer files among mobile, mobiles to laptops, etc. WUSB application is designed to achieve 480 Mb/s for distance up to 3 m. Due to large data rates, UWB antenna are used for radio frequency identifications (RFIDs). UWB antennas are also used for location tracking and positioning. Realtime location systems can be designed with UWB antennas [51, 52].

Several challenges impact implementing UWB antennas; bandwidth enhancement on microstrip antennas with limited dimensions is a main challenge. Various methods are investigated for bandwidth enhancements, such as use of partial ground planes, different-shaped slots, use of stubs, etc.

8.4 CONCLUSION

The portable antenna is a key component and must be portable, low-profile, low-power, versatile, and lightweight in WBAN systems in IoT networks. This chapter has addressed a comprehensive review on UWB antennas with notch characteristics for wireless communication. Different structures, materials, and techniques for creating notches, and tools for design and analysis have been explored. C-slots, stubs, U-shaped slots and ring-shaped slots have been identified for creating notches in UWB frequency band. Also the bandwidth enhancement techniques have been taken into account.

REFERENCES

1. Sengaliappan, M., & Kumaravel, K. (2017). Analysis study of wireless technology and its communication standards using IEEE 802.11. *2017 International Journal of Advanced Research in Science, Engineering and Technology (IJARSET)*, 4(6), 220–230.

2. Federal Communications Commission, Washington, D.C. (2002). First report and order revision of Part 15 of the commission's rule regarding ultra-wideband transmission system FCC 02-48. *Federal Communications Commission.*

3. Licul, S., Noronha, J. A., Davis, W. A., Sweeney, D. G., Anderson, C. R., & Bielawa, T. M. (2003, October). A parametric study of time-domain characteristics of possible UWB antenna architectures. In 2003 IEEE 58th Vehicular Technology Conference. VTC 2003-Fall (IEEE Cat. No. 03CH37484) (Vol. 5, pp. 3110–3114). IEEE.

4. Sego, D. J. (1994, May). Ultrawide band active radar array antenna for unmanned air vehicles. In Proceedings of IEEE National Telesystems Conference-NTC'94 (pp. 13–17). IEEE.

5. Linardou, I., Misliaccio, C., Laheurte, J. M., & Papiernik, A. (1997). Twin Vivaldi antenna fed by coplanar waveguide. *Electronics Letters, 33*(22), 1835–1837.

6. Kim, S. G., & Chang, K. (2004, June). Ultra wideband exponentially-tapered antipodal Vivaldi antennas. In IEEE Antennas and Propagation Society Symposium, 2004 (Vol. 3, pp. 2273–2276). IEEE.

7. Su, S. W., Wong, K. L., & Tang, C. L. (2004). Ultra-wideband square planar monopole antenna for IEEE 802.16a operation in the 2–11-GHz band. *Microwave and Optical Technology Letters, 42*(6), 463–466.

8. Chen, Z. N., Ammann, M. J., & Chia, M. Y. W. (2003). Broadband square annular planar monopoles. *Microwave and Optical Technology Letters, 36*(6), 449–454.

9. Chen, Z. N., Ammann, M. J., Chia, M. Y. W., & See, T. S. P. (2002). Annular planar monopole antennas. *IEE Proceedings-Microwaves, Antennas and Propagation, 149*(4), 200–203.

10. Chen, Z. N., Wu, X. H., Yang, N., & Chia, M. Y. W. (2003, June). Design considerations for antennas in UWB wireless communication systems. In IEEE Antennas and Propagation Society International Symposium. Digest. Held in conjunction with: USNC/CNC/URSI North American Radio Sci. Meeting (Cat. No. 03CH37450) (Vol. 1, pp. 822–825). IEEE.

11. Chen, Z. N., Wu, X. H., Li, H. F., Yang, N., & Chia, M. Y. W. (2004). Considerations for source pulses and antennas in UWB radio systems. *IEEE Transactions on Antennas and Propagation, 52*(7), 1739–1748.

12. Wu, J. and Li, J. (2013, October). Compact ultra-wideband antenna with 3.5/5.5 GHz dual band-notched characteristic. In 2013 5th IEEE International Symposium on Microwave, Antenna, Propagation and EMC Technologies for Wireless Communications (pp. 446–450). IEEE.

13. Ojaroudi, M., Ojaroudi, N., & Ghadimi, N. (2013). Dual band-notched small monopole antenna with novel coupled inverted U-ring strip and novel fork-shaped slit for UWB applications, *IEEE Antennas and Wireless Propagation Letters, 12,* 182–185.

14. Ojaroudi, N., & Ojaroudi, M. (2013). Novel design of dual band-notched monopole antenna with bandwidth enhancement for UWB applications, *IEEE Antennas and Wireless Propagation Letters , 12* (), 698–701.

15. Abbas, S. M., Ranga, Y., Verma, A. K., & Esselle, K. P. (2014). A simple ultra wideband monopole antenna with high band rejection and wide radiation patterns, *IEEE Transactions on Antennas and Propagation, 62*(9), 4816–4820.

16. Chandel, R., Gautam, A. K., & Kanaujia, B. K. (2014). Microstrip-line fed beak-shaped monopole-like slot UWB antenna with enhanced band width, *Microwave and Optical Technology Letters, 56*(11), 2624–2628.

17. Foudazi, A., Hassani, H. R., Mohammad, S., & Nehzad, A. (2012). Small UWB planar monopole antenna with added GPS/GSM/WLAN bands, *IEEE Transactions on Antennas and Propagation, 60*(6), 2987–2992.

18. Mohammad, S., Nehzad, A., Hassani, H. R., & Foudazi, A. (2013). A dual-band WLAN/UWB printed wide slot antenna for MIMO/diversity applications, *Microwave and Optical Technology Letters, 5*(6), 461–465.

19. Dikmen, C. M., Cimen, S., & Cakir, G. (2014). Planar octagonal-shaped UWB antenna with reduced radar cross section, *IEEE Transactions on Antennas and Propagation, 62*(6), () 2945–2953.

20. Unnikrishnan, D., Kaddour, D., Tedjini, S., Bihar, E., & Saddaoui, M., 2015. CPW-fed inkjet printed UWB antenna on ABS-PC for integration in molded interconnect devices technology, *IEEE Antennas and Wireless Propagation Letters, 14* 1125–1128.

21. Mobashsher, A. T., & Abbosh, A. (2015). Utilizing symmetry of planar ultrawide band antennas for size reduction and enhanced performance, *IEEE Antennas and Wireless Propagation Letters, 57*(2), 153–166.

22. Natarajan, R., George, J. V., Kanagasabai, M., & Shrivastava, A. K. (2015). A compact antipodal Vivaldi antenna for UWB applications, *IEEE Antennas and Wireless Propagation Letters., 14*, 1557–1560.

23. Alnahwi, F. M., Abdulhasan, K. M., & Islam, N. E. (2015). An ultrawide band to dual-band switchable antenna design for wireless communication applications, *IEEE Antennas and Propagation Magazine, 57*(14), 1685–1688.

24. Moghadasi, M. N., Sadeghzadeh, R. A., Sedghi, T., Aribi, T., & Virdee, B. S., 2013. UWB CPWfed fractal patch antenna with band-notched function employing folded T-shaped element, *IEEE Antennas and Wireless Propagation Letters, 12*, 504–507.

25. Beigi, P., Nourinia, J., Mohammadi, B., & Valizade, A., 2015. Bandwidth enhancement of small square monopole antenna with dual band notch characteristics using U-shaped slot and butterfly shape parasitic element on backplane for UWB applications, *ACES, 30*(1), 78–85.

26. Yeoh, W. S., & Rowe, W. S. T. (2015). An UWB conical monopole antenna for multiservice wireless applications, *IEEE Antennas and Wireless Propagation Letters, 14*, 1085–1088.

27. Wu, A., & Guan, B. (2013). A compact CPW-fed UWB antenna with dual band-notched characteristics, *IEEE Antennas and Wireless Propagation Letters, 12*, 151–154.

28. Gautam, A. K., Yadav, S., & Kanaujia, B. K. (2013). A CPW-fed compact UWB microstrip antenna, *IEEE Antennas and Wireless Propagation Letters, 12*, 151–154.

29. Verma, A. K., Awasthi, Y. K., & Singh, H. (2009). Equivalent isotropic relative permittivity of microstrip on multilayer anisotropic substrate, *International Journal of Electronics, 96*(8), 865–875.

30. Zarrabi, F. B., Mansouri, Z., Gandji, N. P., & Kuhestani, H. (2016). Triple-notch UWB monopole antenna with fractal Koch and T-shaped stub, *International Journal of Electronics and Communications, 70*, 64–69.

31. Yadav, A., Kumar Singh, V., Kumar Bhoi, A., Marques, G., Garcia-Zapirain, B., & de la Torre Díez, I. (2020). Wireless body area networks: UWB wearable textile antenna for telemedicine and mobile health systems. *Micromachines, 11*(6), 558.

32. Sopa, P., & Rakluea, P. (2020, March). The hexagonal-shaped UWB wearable textile antenna with band-otched Characteristics. In 2020 8th International Electrical Engineering Congress (iEECON) (pp. 1–4). IEEE.

33. Donelli, M., Menon, S. K., Marchi, G., Mulloni, V., & Manekiya, M. (2020). Design of an ultra wide band antenna based on a SIW resonator. *Progress in Electromagnetics Research, 103*, 187–197.

34. Samal, P. B., Soh, P. J., & Zakaria, Z. (2019). Compact microstrip-based textile antenna for 802.15. 6 WBAN-UWB with full ground plane. *International Journal of Antennas and Propagation, 1*, 1–10.

35. Sharma, M., Awasthi, Y. K., Singh, H., Kumar, R., & Kumari, S. (2016). Compact printed high rejection triple band-notch UWB antenna with multiple wireless applications. *Engineering science and technology, an international journal, 19*(3), 1626–1634.

36. Doddipalli, S., Kulkarni, A., Kaole, P., & Kothari, A. (2018, July). Slotted substrate miniaturized ultra wide band antenna for WBAN applications. In 2018 9th International Conference on Computing, Communication and Networking Technologies (ICCCNT) (pp. 1–4). IEEE.

37. Doddipalli, S., Kothari, A., & Peshwe, P. (2017). A low profile ultrawide band monopole antenna for wearable applications. *International Journal of Antennas and Propagation*, *5*, 80–89.

38. Ray, K.P. (2008). Design aspects of printed monopole antennas for ultra-wide band applications. *International Journal of Antennas and Propagation*, *15*, 1–8.

39. Zhu, F., Gao, S., Ho, A. T., Abd-Alhameed, R. A., See, C. H., Brown, T. W., Li, J., Wei, G., & Xu, J. (2013, April 25). Multiple band-notched UWB antenna with band-rejected elements integrated in the feed line. *IEEE Transactions on Antennas and Propagation*, *61*(8), 3952–3960.

40. Tripathi, H. N., Shukla, S., Aggarwal, R., Jha, K. R., & Tripathi, N. (2014, May). Effects of planar ground plane structure on elliptical-shaped patch antenna for wireless UWB applications. In 2014 Students Conference on Engineering and Systems (pp. 1–4). IEEE.

41. Shakib, M. N., Moghavvemi, M., & Mahadi, W. N. L. B. W. (2016). Design of a tri-band off-body antenna for WBAN communication. *IEEE Antennas and Wireless Propagation Letters*, *16*, 210–213.

42. Doddipalli, S., & Kothari, A. (2018). Compact UWB antenna with integrated triple notch bands for WBAN applications. *IEEE Access*, *7*, 183–190.

43. Khan, M. A., &Salah, K. (2018). IoT security: Review, blockchain solutions, and open challenges. *Future Generation Computer Systems*, *82*, 395–411.

44. Rana, A. K., Krishna, R., Dhwan, S., Sharma, S., & Gupta, R. (2019, October). Review on artificial intelligence with Internet of Things—Problems, challenges and opportunities. In 2019 2nd International Conference on Power Energy, Environment and Intelligent Control (PEEIC) (pp. 383–387). IEEE.

45. Rana, A. K., & Sharma, S. (2021). Contiki Cooja Security Solution (CCSS) with IPv6 routing protocol for low-power and lossy networks (RPL) in Internet of Things applications. In *Mobile Radio Communications and 5G Networks* (pp. 251–259). Springer, Singapore.

46. Kumar, A., Salau, A. O., Gupta, S., & Paliwal, K. (2019). Recent trends in IoT and its requisition with IoT built engineering: A review. In *Advances in Signal Processing and Communication* (pp. 15–25). Springer, Singapore.

47. Dalal, P., Aggarwal, G., & Tejasvee, S. (2020). Internet of Things (IoT) in healthcare system: IA3 (Idea, Architecture, Advantages and Applications). *Available at SSRN 3566282*.

48. Rana, A. K., & Sharma, S. (2020). Industry 4.0 manufacturing based on IoT, cloud computing, and big data: Manufacturing purpose scenario. In *Advances in Communication and Computational Technology* (pp. 1109–1119). Springer, Singapore.

49. Wang, Q., Zhu, X., Ni, Y., Gu, L., & Zhu, H., 2020. Blockchain for the IoT and industrial IoT: A review. *Internet of Things*, *10*, 100081.

50. Rana, A. K., Salau, A., Gupta, S., & Arora, S., 2018. A Survey of Machine Learning Methods for IoT and their Future Applications.

51. Kumar, K., Gupta, E. S., & Rana, E. A. K. (2018). Wireless sensor networks: A review on Challenges and Opportunities for the Future World-LTE.

52. Sachdev, R. (2020, April). Towards security and privacy for edge AI in IoT/IoE-based digital marketing environments. In 2020 Fifth International Conference on Fog and Mobile Edge Computing (FMEC) (pp. 341–346). IEEE.

9 Internet of Underwater Things
Applications and Challenges

Sanyam Jain*, Nitin Yadav*, Mubarak Husain*,
Mohit Aggarwal*, Priyesh Arya*, and
Shobhit Aggarwal#
*Panipat Institute of Engineering & Technology, Samalkha
Kurukshetra University, Kurukshetra, Haryana, India
#Department of Electrical and Computer
Engineering, University of North Carolina
at Charlotte, North Carolina, US

CONTENTS

9.1   INTRODUCTION

Due to urbanization, there has been a rushed developing pattern in creating keen coasting urban areas. Over 70% of the Earth's surface is enclosed by water of which over 90% of the submerged regions are as yet unexplored. Apparently, the idea of IoT was concocted in earlier 1985 while the term Internet of Underwater Things was first talked about in 2012. IoUT is characterized as "the organization of savvy interconnected submerged objects."

Atmospheric conditions and patterns of winds that affect life ashore are dictated by the sea temperature. Also, 1% of fresh water in lakes and streams is debased, truly harming the ecosystem [1–4]. This reality has drawn the consideration of numerous analysts toward the submerged correspondences and organizing and a need has arisen to investigate what is covered up in the unfathomed world that

DOI: 10.1201/9781003181613-9

lies under the oceans by network correspondence on remote connections. It is considered an organization of interconnected smart submerged items, for example, submerged sensors, autonomous underwater vehicles (AUVs), autonomous surface vehicles (ASVs), remotely operated vehicles (ROVs), surface floats, ships, and so forth. There has been an enormous interest for rapid ongoing submerged remote connects to stack a wide scope of submerged applications, for example, natural checking and contamination control, submerged investigation, seaward oilfield investigation, sea paleontology, observing logical applications (live aquariums), modern applications (fish ranches and pipeline checking). Likewise, it is utilized for seaward oil and gas extraction, oil slicks, military observation, observation, mine derivation, contamination checking, characteristic catastrophes like tsunamis and typhoons, coral reefs, acclimating observation of marine life and fish farming, scientific information assortment, port security, and strategic reconnaissance among others.

IoUT can assume a significant part in data transfer from submerged sensor frameworks to server farms ashore [5]. The organization has links in the ocean, empowering individuals to interface their gadgets, which can then send information to and from the shore [6–7]. The seawater cools the water circuit from 30°C to 17°C. The networking inside the sea is a propelling innovation that has been increasingly considered over the last 20 years. An innovation that permits correspondence among various acoustic clients managing various applications that goes from the profundities of the sea to the ocean surface is called an underwater acoustic network (UAN). Thus correspondence by methods for acoustic signs was discovered, which performed submerged remote communications [8–12]. In actuality, acoustic signs can proliferate with lower weakening and assimilation for any long distance in submerged conditions relying upon the plan of correspondence protocol. To defeat this test, the self-administration measure needs to work without human intercession. For this reason, self-arrangement, self-recuperating, self-enhancement, and self-protection capabilities and are required. Underwater communication is genuinely restricted by the brutal states of the submerged path (channel) bringing in high error rates because of refraction of the signal based on the depth of water. Another significant challenge is the improvement of the different types of tracking methodologies applied to marine creatures for their security. Energy efficiency is also a great challenge for acoustic communication, which is power-hungry by reaping energy with supercapacitors that can supplant batteries when they run out of power.

IoUT is likewise an incredible asset for security reasons also. A modern seaside security framework comprises cameras, and an acoustic sensor network is needed for harbor security to maintain exchange and development and also to alert regarding potential security challenges. An underwater acoustic sensor network (UWASN) is an organized arrangement of various versatile sensor hubs remotely associated with acoustic correspondence modules set up underwater. UWASN, with the assistance of automated robotic vehicles. can notice and anticipate the nature of the sea climate. It can observe contamination, such as synthetic, natural, and atomic, and perform undersea investigation for identifying submerged oil fields, laying of undersea links, and investigation of significant minerals. Additionally, it can give tidal wave and seismic tremor admonitions. The sensors can likewise be utilized to recognize perils on the seabed, identify risky rocks in shallow water, and distinguish mine-like items.

Different epic procedures and guidelines have been proposed that assume a pivotal part in building next-generation submerged organizations. Underwater wireless sensor networks (UWSNs) have been viewed as a promising organization framework and comprise a few segments. The fundamental segments are the underwater sensors. These sensors are the hubs with acoustic modems and are circulated in one or the other shallow or deep water. Every sensor hub can detect (various sensors can detect distinctive natural data, for example, the temperature, pressure, water quality, metal, and compounds, as well as organic components), hand-off, and forward information. This information is supposed to be moved to the fundamental components on the outside surface of the sea, known as a sink. Sinks are the hubs with acoustic as well as radio modems. It should be noted that these sinks can be floats, ships, or even autonomous surface vehicles (ASVs).

Radio waves multiply at huge distances through conductive seawater exactly at lower frequencies in between 30–300 Hz, which requires an extremely large receiving wire and higher transmission power. So, the radio waves are not sensible for lowered correspondence. Acoustic correspondences experience the negative impacts of a small open exchange speed under 100 kHz, very low celerity, and gigantic latencies as a result of the low multiplication speed. This chapter gives a careful examination of the IoUT applications, troubles, and channel models.

9.2 LITERATURE SURVEY

The IoT was developed in 1985 but it was not until 2012 that the IoUT was first analyzed. IoUT is portrayed as "the firm of splendid interconnected submerged things." The arrangement of encounters dates back to the Cold War when hydrophones were used to control the advancement of Soviet submarines. With the sit back, most complex acoustic associations were made for undermining boats and distinguishing disguised submarines [13]. In the continuous past, ease sensors were made, which deal with comparability equipment to examine oceans. Submarine signs were at the time genuine with the rough maritime radio navigation organization. The later Fessenden oscillator allowed correspondence with submarines.

IoUT is a three-layered plan, first depicted in 2012, that fulfills the solicitations of usage customers. The wisdom layer joins sensors, vehicles, and stations that are subject to social affair data [14]. The middle layer, i.e., the association layer joins web, wired/far-off associations, conveyed figuring stages, and so forth for sending information from the acknowledgment layer, and the application layer satisfies expanded solicitations of different customers. King Abdullah University of Science and Technology KAUST, on the Red Sea coast, has been related to another zone of specific examination for an extended period, recollecting for making some early, pivotal lowered data trades [15–20]. In 2015, it ran a 16-QAM-OFDM transmission with a 450-nm laser, 4.8 Gb consistently. Orthogonal frequency division multiplexing (OFDM) disengages single data streams into different channels to reduce obstacles. There is moreover a necessity to measure cooling, something that ocean water can give. As a procedure for driving laborers, even wave energy suggests that ocean and data are getting bound. In the wake of building an undersea water-cooling worker ranch a hundred and seventeen feet underneath the ocean's surface in 2018, even Microsoft ricocheted in a mixed bag of data. Likewise, while many acknowledge that our correspondences are being passed on via satellites, basically the whole of our intercontinental trades are truly happening by methods for undersea fiber-optic connections. These nursery hose–sized connections hold all open, public Internet traffic. However, these headways are not immaculate, helpful energy. Alongside the eco-monitoring PC drivers, one of the most powerful and tremendous reasons ocean-based handling is broadly examined is that on the high seas, there is no rent payable and no jurisdictional belonging [21–25].

The impact of climate change on our web accessibility and web-based prosperity is subverted by a collection of malicious performers that have provoked the use of an immense number of checks from our side. One of these, and it's one we sadly can't do anything especially about, is ecological change. A progressing report has overlaid sea-level assault figures on top of US basic web-establishment advisers. This has revealed that ecological change by 2033 would douse around 4000 miles (6400 km) of fiber joins: wires that were hindered numerous years and never anticipated lowered undertakings. Plainly, by a long shot, the majority of web interfaces that pass on ordinary data fixes are by and by lowered, winding the ocean bottom over an enormous overall association that ranges around 550,000 miles (885,000 km) of submarine information transmission. These interfaces are good to move data at 2000 Mb/s consistently, and no more. The truth remains that undersea connections were unequivocally proposed to be waterproof. Regardless, when these submarine connections show up at coastline towns, the genuine cabling ordinarily changes from waterproof wires to ones that are just water-safe. Affiliations began their association system plans at the time the commercialization of the web increased. Everyone clung to their game plans and strategies and passed, subject to the current necessities of the time. Unfortunately disregarding the way that this cycle happened quite a while back, before the current degree of an overall temperature adjustment care, the future-fixing of the associations against an unforeseen flooding event never formed a bit of the masterminding. Miami, Seattle, and New York are the feeblest metropolitan networks in the United States, yet considering how the web works, the results would not be pressed in those regions alone as all data joins brought through influenced regions may be impacted. Solidify this with the way that researchers only viewed American networks; a comparative issue may be a test in countries around the world demonstrating we're facing a significant issue. A bit of the accessibility goliaths like AT&T and Verizon are as of now aware of the issue and are presenting structures that can withstand the rising tide. However, the truth remains that a huge piece of the web advancement did not consider the aftereffects of natural change, and all future associations must see and consider its effects [26–28].

9.3 IoUT ARCHITECTURE

IoUT functionalities are subdivided into three noticeable layers:

1. **Perception Layer:** It accumulates data by recognizing objects with the assistance of static submerged sensors, portable submerged vehicles, surface and observing stations like PCs, cell phones, information stockpiling radios, and beneficiary labels.
2. **Network Layer:** With the assistance of an exclusive organization, web, network managerial framework, and distributed computing stages, the information from the discernment layer is handled and communicated to the next layer.
3. **Application Layer:** It is a set comprising canny arrangements attained through IoUT innovation to adequately address the client's issue [29].

9.3.1 ONE-DIMENSIONAL ARCHITECTURE

In one-dimensional IoUT design, the sensors are set up and they work separately. The sensor hub is planned as an independent organization performing detecting and sending the prepared signs to the surface station (SS). Autonomous Underwater Vehicles (AUVs) are likewise conveyed and submerged, and the AUVs accumulate the detected information from the hubs and send it to the SS. Hubs communication and sign transmission can be acoustic, optical, or radio recurrence. The sign exchange from the hubs to the SS occurs in a solitary bounce. One-dimensional engineering (Figure 9.1) is otherwise called static design, where the situation of the hubs is static and solitary organization geography is followed all through, in contrast to the two- and three-dimensional structures.

9.3.2 TWO-DIMENSIONAL ARCHITECTURE

Two-dimensional engineering (Figure 9.2) alludes to an organization game plan of a group of sensor hubs sent on the ocean bed with an anchor hub. The submerged properties accumulated by the sensor hubs are shipped off anchor hubs, and from that point, it is communicated to the surface floats. The correspondence here is two-staged: (i) Sensed information transmission from sensor hubs to moor hub using an even connection, and (ii) Signal hand-off as acoustic or optical structures anchor the hub to the surface floats as a vertical connection. Best detecting results and immense correspondence inclusion is accomplished in ref. [12]. In ref. [13], the exhibition of the hubs is improved utilizing virtual sinks. Acoustic, optical, and radio recurrence joins are utilized for handing-off signs. However, acoustics is most favored when the distance between the anchor hub and the surface float is dominatingly high. The hub

FIGURE 9.1   One-dimensional architecture [20].

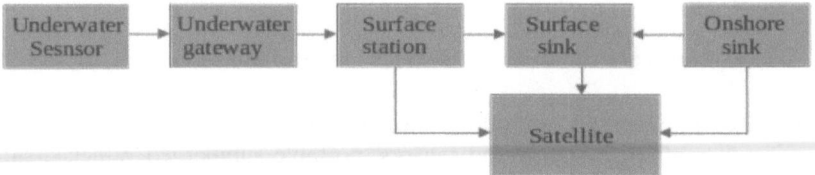

FIGURE 9.2   Two-dimensional architecture.

group can follow any organization's geography like a star, ring, or network regarding the plan necessity. Two-dimensional UWASN is for the most part favored for their deference resilience and time lack.

9.3.3   THREE-DIMENSIONAL ARCHITECTURE

Mostly in the organization, the sensor bunches are sent and secured at different profundities of the seabed. The situating of sensors decides the idea of the correspondence, which happens in three levels: (i) Communication between the sensors at the differing profundities' intercluster correspondence; (ii) Sensor groups to secure hub correspondence—intracluster correspondence; and (iii) Anchor hub to surface float correspondence [14]. The data/signal transferring utilizes acoustic, optical and radio recurrence joins are utilized in a mix. The three-dimensional organization (Figure 9.3) favors guaranteed network inclusion with the utilization of a basic directing convention, yet the expense of the watery sensors is a little costly and the sensor thickness is low when contrasted with earthbound correspondence.

9.4   APPLICATIONS

In the most recent decade, scientists have introduced various reasonable and possible IoUT applications (Figure 9.4). We group the applications into five kinds:

1. Monitoring environment
2. Submerged exploration
3. Prevention of disasters
4. Military
5. Others

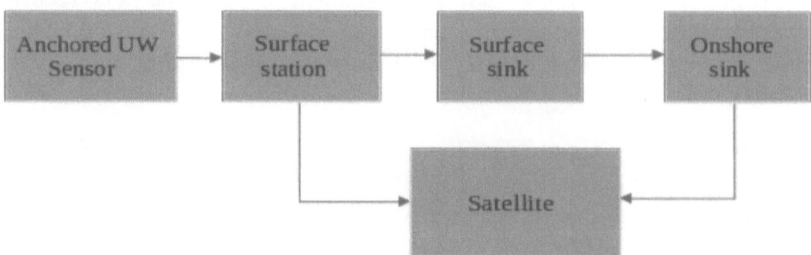

FIGURE 9.3   Three-dimensional architecture.

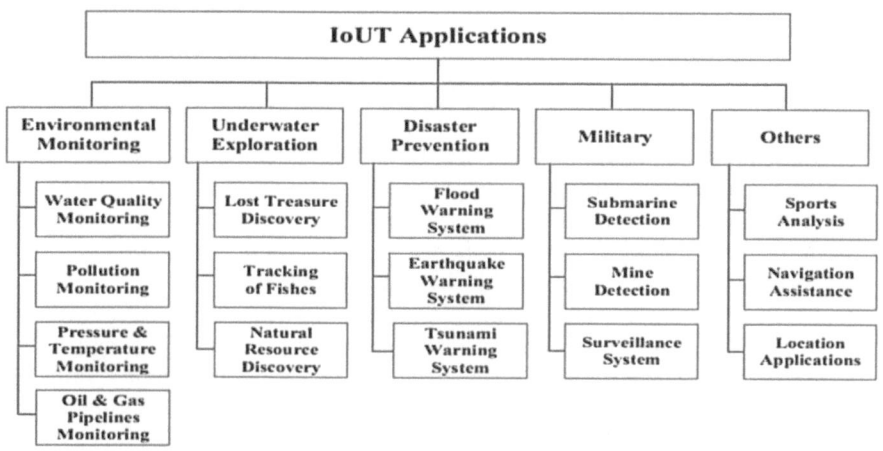

FIGURE 9.4   The IoUT applications.

9.5   MONITORING ENVIRONMENT

One of the most-used types of IoUT applications is regular noticing, checking the water quality, substance, and natural tainting checking, warm defilement checking, pressure checking, and temperature checking. Also, oil and gas pipelines' noticing can similarly be refined by using the concept of UWSNs. It should be noted that the common checking applications/systems are increasingly renowned and are as of now receiving unprecedented interest for overall superior metropolitan zones [30].

- **Monitoring Pollution:** Focusing on ocean defilement neutralization, a far-off sensor association can be used to screen the ocean tainting. Differentiated and the standard long-range lowered correspondence, this work centers around the short-range multi-bob correspondence. Short-range correspondence recommends that the sensor association can reuse a part of the acoustic bandwidth and besides sidestep an extensive parcel of the troubles of long-range correspondence.
- **Pressure and Temperature Monitoring:** A checking system in Queensland, Australia, notices the lowered temperature and brightness or luminosity, that is, the information essential to induce the prosperity status of the coralline impediment. Considering distant sensor associations, the makers present a certifiable reasonable framework.
- **Oil and Gas Pipelines Monitoring:** The quick progress in far-off correspondence developments has improved each part of current practices related to oil and gas pipeline monitoring.

Submerged Exploration: A significant investigation of different UWSN models shows how the plans empower robotized examinations. The objective is to make advanced correspondence techniques for capable, consistent assessment of the oceans.

- **Lost Treasure Discovery:** The possibility of the Internet of Underwater Things can be applied to the deepwater treasure explorations. For example, the exploration of the *Titanic* in 1985 benefitted from the use of autonomous lowered vehicles. As referred to in this chapter, different compelling recovered fortune discoveries were made with the help of UWSNs.

- **Natural Resource Discovery:** Moreover, submerged regular resource disclosure, for instance, minerals, metals, corals, and coral reefs, could all be helped by the establishment of UWSNs too. Bainbridge et al. finished an applied assessment in Australia. The assessment covers the activities picked up from two years of action of UWSNs passed on at seven coral reefs in northeastern Australia along the Great Barrier Reef.

- **Flood Warning System: Flood Warning System** provide the arrangement of a rational nonstop assessment system for flood checking. They passed on a real checking network with a couple of steady assessment contraptions in Spain. Different testing results illustrate that the suggested checking structure is energy-capable, and has a solid correspondence limit with regards to the long-stretch variety of water-level data in various zones.

- **Earthquake Warning System:** Professionals highlight that the headway of the initial notification age structure reliant on the UWSNs can unmistakably help save lives. The exhortation age system can help save lives when a disastrous occasion, for example, a tremor, occurs. Mainly the standard responsibilities of this chapter is to highlight the real layer, which is used in developing a UWSN structure for notice age. The makers assume that strong correspondence, low-power plan, and beneficial resources will remain the critical obstacles for UWSN-based advice-age systems.

- **Tsunami Warning System:** Professionals talk of developing different strategies for tsunami-warning systems and offer a beneficial design for different applications. The suggested configuration uses seismic compel sensors to foresee waves, and to propel/hand-off the recognized statistics by using the organized scattering directing system. The professionals in like manner analyze and portray the examination and response instruments in this task.

Military: The military as often as possible mirrors the limit of a nation to shield themselves against an attack, including submerged attack. The IoUT is compulsory for assurance purposes, and can be used for submarine acknowledgment, submerged mine disclosure, and lowered observation structures. These operations have exceptional possibility for future sea powers.

- **Submarine revelation:** For submarine acknowledgment, the makers research the arrangement of the sensor course of action and suggest prearrangement to update the noticing consideration.

Others: Because of the progression in UWSNs, there is a growing number of different IoUT applications, for instance, in sports, course, and constraint applications. It should be noted that the limitation applications are being developed, yet they will be significant. Perceiving the submerged zone is difficult, considering that global positioning system (GPS) can't work in aquatic conditions. Visualizing, for instance, under water (UW) sensors can be used as zone reference centers, and thus can give swimmer delivery vehicles with critical territory information.

- **Navigation Assistance:** For a combination of uses, the identified data should be translated genuinely when alluded to the region of the sensors. The makers suggest a multi-stage AUV-upheld limitation scheme for UWSNs. They guided reenactments to survey the introduction of the proposed scheme similar to the control incorporation, precision, and correspondence costs. The results exhibit that this area can be developed and cultivated by properly picking the correspondence range.
- **Location Applications:** The makers suggest an anchor-free localization algorithm (AFLA) for aquatic conditions. By far most of the standard lowered constraint calculations anticipate that one-of-a kind, lowered center points or AUVs can be used as "anchor center points" to help recognize the submerged territory. The proposed AFLA needn't mess with information from the unprecedented anchor center points. Taking everything into account, AFLA uses the neighboring relationship of standard sensor center points to recognize the zone. The reenactment results show that we can use AFLA in both the dynamic and static UWSNs.

9.6   CHALLENGES

The stream development of explorations using IoUT is moderate owing to the troubles arising out of the distinctiveness of UWSNs. By the day's end, the obstacles for IoUT are the variances in UWSNs and Territorial Wireless Sensor Networks (TWSNs). We examine these troubles for IoUT-reliant systems based on the ten perspectives:

1. Transmission media
2. Engendering speed
3. Transmission range
4. Transmission rate
5. Recharge difficulties
6. Versatility
7. Dependability
8. Threats from marine life
   Different epic procedures and guidelines have been proposed that assume a pivotal part in building next-generation submerged systems. UWSNs have been viewed as a capable organizational framework. UWSNs comprise a few segments. The fundamental segments include underwater sensors. In underwater sensor networks, the sensors are the hubs with acoustic-based modems and are circulated in one or the other shallow or deep water. Every sensor hub can detect (various sensors can detect distinctive natural data, for example, the temperature, pressure, water quality, metal, and compound and organic components), hand-off, and forward information. The information should be moved to the fundamental component on the outside surface of the sea, known

as a sink. Sinks are the hubs with acoustic as well as radio-based modems. Note that sinks can be floats, ships, or ASVs.
9. Security/issues
10. Potential harm to biodiversity

We have described the different challenges as follows:

1. **Transmission Media:** In TWSNs, the radio waves are frequently utilized for interchanges. Nonetheless, UWSNs normally depend on acoustic correspondences instead of radio interchanges. That's because radio signs would be rapidly consumed by water. Lamentably, the acoustic wave's properties are altogether unlike that of the radio waves; consequently, a considerable number of correspondence conventions applied to TWSNs can't be straightforwardly applied to UWSNs. As needs are, the transmission media is one of the primary difficulties for IoUT.
2. **Engendering Speed:** The spread speed in UWSNs is multiple times more sluggish than that in TWSNs. In particular, the engendering rate of TWSN radio channels is nearly 300,000 km/s, while the spread speed of UWSN acoustic channels is just near to 1.5 km/s. Consequently, ensuring a limited start to finish postponement would be a difficult issue for IoUT.
3. **Transmission Range:** The UWSN's transmission range can be almost 10 times greater than that in TWSNs. The signals in underwater environments need to be transmitted by using a low frequency to prevent it getting absorbed by water. The lower frequency allows longer transmission range, and the longer transmission range allows much more possibilities that cause interferences, thus collisions occur during data transmission. So, the prevention of collisions and interferences is viewed as one of the major problems in IoUT.
4. **Transmission Rate:** When compared with radio interchanges in TWSNs, acoustic correspondences in UWSNs utilize a thin data transfer capacity. Attributable to the limited transfer speed, the rate of transmission in UWSNs is by and large low (roughly 10 kbps). Henceforth, data transmission usage is a significant worry for IoUT.
5. **Recharging Difficulties:** In UWSNs, submerged sensors are hard to revive because the sensors are sent in the submerged regions. At the point when we consider the possible expense of reviving the batteries of the submerged sensors, energy proficiency would be another significant worry for IoUT.
6. **Versatility:** UWSNs are portable WSNs essentially. When there are water flows, the UWSN sensors may move and experience the ill effects of dynamic organization geography changes. It is a provoking undertaking to manage the dynamic changes for IoUT.
7. **Dependability:** The connection of unwavering quality in UWSNs is actually regularly insecure and low. The unwavering quality of a connection implies the effective conveyance proportion (fruitful conveyance proportion is characterized as the proportion of the quantity of information that has been effectively conveyed to a recipient contrasted with the

quantity of information that has been conveyed by the sender) between a couple of sensor hubs. In UWSNs, the fruitful conveyance proportion would be seriously influenced by transmission misfortune (transmission misfortune is characterized as the collected diminishing in force of waveform energy when a wave engenders outward from a source) and natural commotions. Since signs would be consumed by water in submerged conditions, the transmission misfortune is one of the serious issues in UWSNs. Moreover, the natural clamors in UWSNs are made out of a few confounded components, including choppiness, transporting, waves, and so on Thus, the dependability subject is one of the most testing subjects for IoUT.

8. **Threat from Marine Life:** Marine life and potential noise from ships can damage the whole network setup and infrastructure, and tracing and reviving would be more difficult.
9. **Security Issue:** Acoustic networks can be used for long distances, but they lack stealth and can be used by third-party easily.
10. **Potential Harm to Biodiversity:** This high-energy wave can cause harm to coral reefs, marine biodiversity, fragile marine ecosystem, etc.

The solutions to these problem include:

1. A framework that can simultaneously send light and energy to submerged energy gadgets and gear is under development. A self-fueled IoUT, it will have to collect energy and translate data transferred by light fault and can enhance detecting and correspondence in oceans and seas.
2. Simultaneous Lightwave Information and Power Transfer (SLIPT) can help charge gadgets in blocked off area where nonstop driving is costly or unrealistic.
3. By and by submerged correspondence framework uses electromagnetic, optics, and acoustic information transmission procedures to send information among various positions. Electromagnetic correspondence procedure is influenced by the leading nature of ocean-water while optic waves are material on extremely short distance since optic waves are prematurely ended via seawater. Acoustic correspondence is just a single method that has better execution regarding submerged correspondence because of less weakening in seawater. Acoustic correspondence likewise has less weakening in deep and thermally stable seas.
4. UWSNs are powerless against different dangers and assaults. To accomplish the goals of the security necessities, a bunch of components and security advancements should be proposed to forestall UWSN from assaults. As per the OSI networks, the security issues of UWSNs are legitimately partitioned into isolated segments. The security design of USWNs can be separated into four layers, actual layers, connect layers, transport layers, and application layers. The security issues principally include key administration, interruption discovery, trust the executives, secure administration, secure synchronization, and directing security.

9.7   CONCLUSION

In this chapter, we have discussed the diverse class of IoT, called IoUT. We have offered knowledge regarding the IoUT: (i) applications; (ii) troubles; and (iii) solutions. Precisely, when the transmitter's power is higher, the contrasting SNR is marked higher; when the basic error rate (BER) is lower, the productive transport extent is higher. Also, when the transmission distance constricts, the relating SNR decreases, and when the BER extends, the movement extent reduces. Automated underwater vehicles (AUVs) are emerging as a capable and strong response for various submerged complexities and will continue to play an enormous part of the examination of the seas. As of now, the acoustic waves are used with the ultimate objective of lowered correspondence. As of now, we are executing Li-Fi as an IoUT. When the photodetector gets the signals, it ships off the sink. Later on, the results can be conveyed to the cloud. We have gotten the nuances through the flexible application. The review of this chapter has outlined the flow works, and they are open for expansive future assessment to uncover the captivating bit of the oceans. The paper hopes to give a significant root stray pieces to future novel movements in underwater sensor associations.

REFERENCES

1. Fang, S., Xu, L.D., Zhu, Y., Ahati, J., Pei, H., Yan, J. and Liu, Z., 2014. An integrated system for regional environmental monitoring and management based on internet of things. *IEEE Transactions on Industrial Informatics, 10*, pp. 1596–1605.
2. Petrioli, C., Petroccia, R., Potter, J.R. and Spaccini, D., 2015. The SUNSET framework for simulation, emulation and at-sea testing of underwater wireless sensor networks. *Ad Hoc Networks, 34*, pp. 224–238.
3. Martins, R., Sousa, J.B., Caldas, R., Petrioli, C. and Potter, J., 2014, 3–5 September. SUNRISE project: Porto university testbed. In *Proceedings of the IEEE Underwater Communications and Networking (UComms)*, Sestri Levante, Italy.
4. Petrioli, C. and Petroccia, R., 2012, 12–14 September. SUNSET: Simulation, emulation and real-life testing of underwater wireless sensor networks. In *Proceedings of the IEEE Underwater Communications and Networking (UComms)*, Sestri Levante, Italy.
5. Petrioli, C., Petroccia, R. and Spaccini, D., 2013, 11–13 November. SUNSET version 2.0: Enhanced framework for simulation, emulation and real-life testing of underwater wireless sensor networks. In *Proceedings of the ACM International Conference on Underwater Networks and Systems (WUWNet)*, Kaohsiung, Taiwan.
6. Felemban, E., Shaikh, F.K., Qureshi, U.M., Sheikh, A.A. and Qaisar, S.B., 2015. Underwater sensor network applications: A comprehensive survey. *International Journal of Distributed Sensor Networks, 11*, p. 896832.
7. Davis, A. and Chang, H., 2012, 14–19 October. Underwater wireless sensor networks. In *Proceedings of the IEEE OCEANS*, Virginia Beach, VA.
8. Lloret, J., 2013. Underwater sensor nodes and networks. *Sensors, 13*, pp. 11782–11796.
9. Menon, K.A.U., Divya, P. and Ramesh, M.V., 2012, 26–28 July. Wireless sensor network for river water quality monitoring in India. In *Proceedings of the IEEE International Conference on Computing, Communication and Networking Technologies (ICCCNT)*, Tamil Nadu, India.
10. Saeed, H., Ali, S., Rashid, S., Qaisar, S. and Felemban, E., 2014, 9–13 June. Reliable monitoring of oil and gas pipelines using wireless sensor network (WSN): REMONG. In *Proceedings of the IEEE International Conference on System of Systems Engineering (SOSE)*, Adelaide, Australia.

11. Gupta, O., Goyal, N., Anand, D., Kadry, S., Nam, Y. and Singh, A., 2020. Underwater networked wireless sensor data collection for computational intelligence techniques: Issues, challenges, and approaches. *IEEE Access, 8*, pp. 122959–122974.

12. Gupta, O., Kumar, M., Mushtaq, A. and Goyal, N., 2020. Localization schemes and its challenges in underwater wireless sensor networks. *Journal of Computational and Theoretical Nanoscience, 17*(6), pp. 2750–2754.

13. Goyal, N., 2020. Architectural analysis of wireless sensor network and underwater wireless sensor network with issues and challenges. *Journal of Computational and Theoretical Nanoscience, 17*(6), pp. 2706–2712.

14. Choudhary, M. and Goyal, N., 2020. routing protocol design issues and challenges in underwater wireless sensor network. In *Energy-Efficient Underwater Wireless Communications and Networking* (pp. 1–15). IGI Global, PA.

15. Goyal, N., Sandhu, J. K. and Verma, L., 2020 CDMA-based security against wormhole attack in underwater wireless sensor networks. In *Advances in Communication and Computational Technology* (pp. 829–835). Springer, Singapore.

16. Ghassemlooy, Z., Hayes, A., Seed, N. and Kaluarachchi, E., 1998. Digital pulse interval modulation for optical communications, *IEEE Communications Magazine, 36*(12), pp. 95–99.

17. Rana, A.K., Krishna, R., Dhwan, S., Sharma, S. and Gupta, R., 2019, October. Review on artificial intelligence with Internet of Things—Problems, Challenges and Opportunities. In *2019 2nd International Conference on Power Energy, Environment and Intelligent Control (PEEIC)* (pp. 383–387). IEEE.

18. Rana, A.K. and Sharma, S., 2021. Contiki Cooja Security Solution (CCSS) with IPv6 routing protocol for low-power and lossy networks (RPL) in Internet of Things applications. In *Mobile Radio Communications and 5G Networks* (pp. 251–259). Springer, Singapore.

19. Rana, A.K. and Sharma, S., 2019. Enhanced Energy-Efficient Heterogeneous Routing Protocols in WSNs for IoT Application.

20. Ahmed, E., Islam, A., Ashraf, M., Chowdhury, A.I. and Rahman, M.M., Internet of Things (IoT): Vulnerabilities, Security Concerns and Things to Consider.

21. Veeramanickam, M.R.M. and Mohanapriya, M., 2016. IoT enabled futurus smart campus with effective e-learning: i-campus. *GSTF Journal of Engineering Technology (JET), 3*(4), pp. 8–87.

22. Cho, S.P. and Kim, J.G., 2016. E-learning based on Internet of Things. *Advanced Science Letters, 22*(11), pp. 3294–3298.

22. Kumar, A. and Sharma, S., Demur and routing protocols with application in underwater wireless sensor networks for smart city. In *Energy-Efficient Underwater Wireless Communications and Networking* (pp. 262–278). IGI Global, PA.

23. Abbasy, M.B. and Quesada, E.V., 2017. Predictable influence of IoT (Internet of Things) in the higher education. *International Journal of Information and Education Technology, 7*(12), pp. 914–920.

24. Kumar, A., Salau, A.O., Gupta, S. and Paliwal, K., 2019. Recent trends in IoT and its requisition with IoT built engineering: A review. In *Advances in Signal Processing and Communication* (pp. 15–25). Springer, Singapore.

25. Dalal, P., Aggarwal, G. and Tejasvee, S., 2020. Internet of Things (IoT) in Healthcare System: IA3 (Idea, Architecture, Advantages and Applications). Available at SSRN 3566282.

26. Rana, A.K. and Sharma, S., Industry 4.0 manufacturing based on IoT, cloud computing, and big data: Manufacturing purpose scenario. In *Advances in Communication and Computational Technology* (pp. 1109–1119). Springer, Singapore.

27. Wang, Q., Zhu, X., Ni, Y., Gu, L. and Zhu, H., 2020. Blockchain for the IoT and industrial IoT: A review. *Internet of Things, 10*, p. 100081.

28. Rana, A.K., Salau, A., Gupta, S. and Arora, S., 2018. A Survey of Machine Learning Methods for IoT and their Future Applications.

29. Kumar, K., Gupta, E.S. and Rana, E.A.K., Wireless Sensor Networks: A review on "Challenges and Opportunities for the Future world-LTE".

30. Sachdev, R., 2020, April. Towards security and privacy for edge AI in IoT/IoE-based digital marketing environments. In *2020 Fifth International Conference on Fog and Mobile Edge Computing (FMEC)* (pp. 341–346). IEEE.

10 IoT-Enabled Wireless Mobile Ad-Hoc Networks

Introduction, Challenges, Applications: Review Chapter

Manwinder Singh*, Navdeep Kaur Jhajj#, and Anudeep Goraya†

*School of Electronics and Electrical Engineering, Lovely Professional University, Phagwara, Jalandhar, Punjab, India
#Assistant Professor, SBSSTC-Ferozepur, Punjab, India
†IKG-PTU Jalandhar, Punjab, India

CONTENTS

10.1 INTRODUCTION

Few decades back, wired connection was the only means of communication between different connections over the Internet. Since 2010, however, wireless connectivity has changed the way of traditional connectivity over the Internet by enabling different wireless technologies to exchange information between different computers, devices, mobiles, and grids, as well as cloud computing. This impacts every area of human life. The IoT provides facilities to identify and communicate with physical objects (smart devices). The smart devices can transfer data in mobile ad hoc networks (MANETs) across all active devices without the need for a centralized approach. The sensor network is a backbone of IoT. Since their appearance in 1970 in the form of ALOHAnet, wireless packet radio networks have come an extended way in terms of several applications, as well as the feature set, among several other things. The two major attractions of wireless communication have been easing of deployment as well as mobility—arranging cables are not only time-consuming and laborious, their maintenance is likewise troublesome. Wireless communication nowadays surrounds us in numerous forms, each one of them with its distinctive coverage, range of applications, and frequency band. It has developed to a great extent; in addition, standards have been developed for local area network (LANs), broadband wireless access, as well as personal area networks (PANs) [1–10].

When two or more than two devices are connected together, and they provide networking and communication capabilities, an ad hoc network is formed. In other words, it is a network that is made up of, or we can say carries, individual devices which communicate directly to each other as shown in figure 10.1 and 10.2.

Many ad hoc networks are LAN-structured. Ad hoc is a Latin phrase meaning "for this purpose," and so as name suggests these networks are developed for specific functions [11–15].

The nature of these networks is

- Temporary
- Decentralized
- Spontaneous
- Adaptive
- Self-organizing
- Infrastructure-less [2]

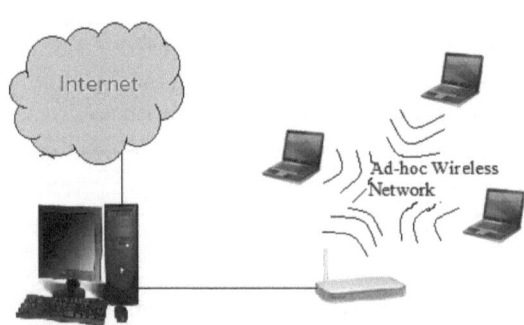

FIGURE 10.1   Basic ad hoc network [3].

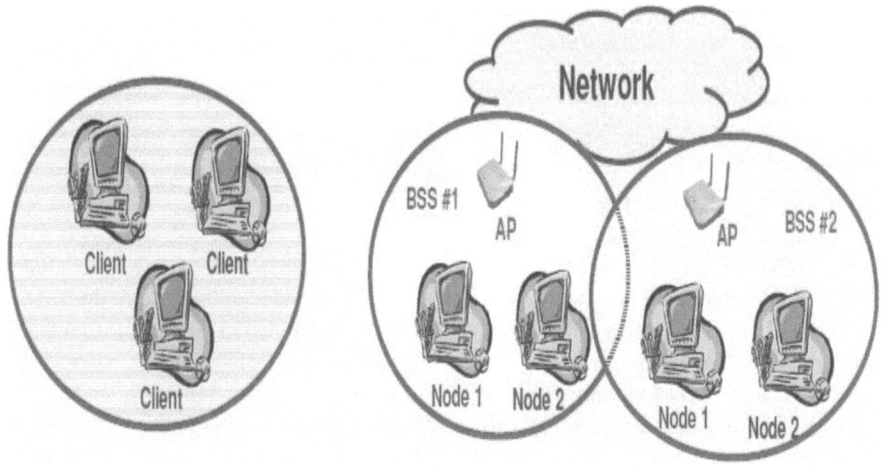

FIGURE 10.2   Ad hoc and infrastructure network topology [4].

10.1.1   ASPECTS OF AD HOC NETWORKS

- Independent in nature
- Multi-hop routing
- Distributed architecture
- Active in nature
- High nodal connectivity

10.1.2   ARRANGEMENT OF AD HOC NETWORK

There are two main types of arrangements as shown in Figure 10.3.

10.1.3   MAIN TYPES

- **MANET**
  - Stands for mobile ad hoc networks
  - Temporary structure
  - Decentralized

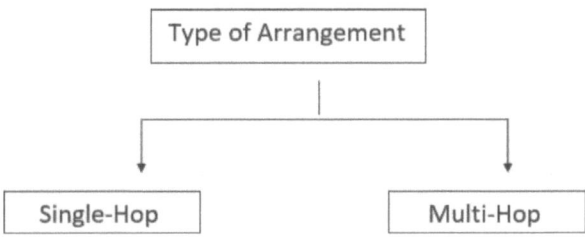

FIGURE 10.3   Types of arrangement.

- Short-lived, without the aid of any infrastructure
- Self-organized wireless
- Dynamic topology
- **WSN**
  - Wireless sensor network—non-wired network
  - Follows open systems interconnection (OSI) model
  - Independent sensor devices
  - Uses environmental, physical areas
- **VANET**
  - Vehicular ad hoc network use (VANET) vehicles as mobile node
  - Provides communication between vehicles and road side equipment
  - Sub form of MANET
  - Fast topology change
  - Highly challenging aspects

10.2 MANET

MANET is a self-organized structure of mobile devices attached by wireless link (Figure 10.4). This is an infrastructure-less temporary design. Mobility of each device is high and independent, and therefore links to other devices change

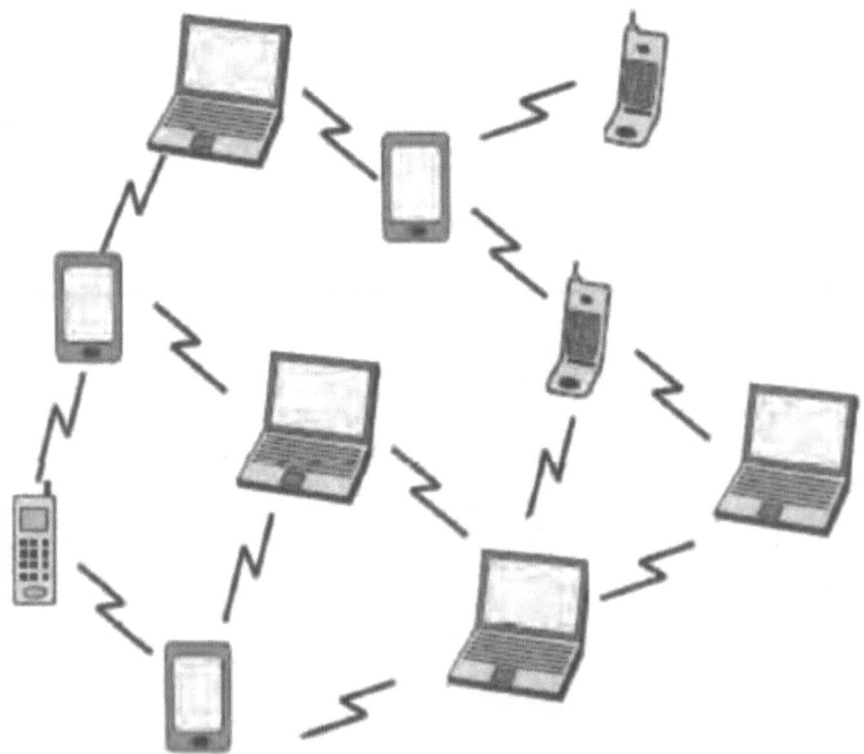

FIGURE 10.4   Basic MANET.

FIGURE 10.5 MANET structure [13].

frequently. In network, every mobile device is independent in nature. Wireless medium is shared in between nodes, and topology changes dynamically. In MANET, as nodes are free to move to anywhere, communication links break frequently [8–12].

Over a past decade, there has been an excessive increase in the popularity of mobile devices and wireless networks. Due to this, MANET has become a vibrant and active field of communication. Interconnected mobile network is another term frequently used for MANET. For research and development of wireless networks, it's a promising field [13–18] as shown in Figure 10.5.

10.2.1 MANET Architecture

The architecture as shown in Figure 10.6 is grouped into three main classifications:

- Enabling part
- Middleware and application
- Networking

10.2.2 MANET Characteristics

The main characteristics of MANET are as follows:

- Distributed operation
- Autonomous terminal
- Wide coverage area
- Changing/driving topology
- Wireless medium
- Multi-hop routing
- Infrastructure-less network

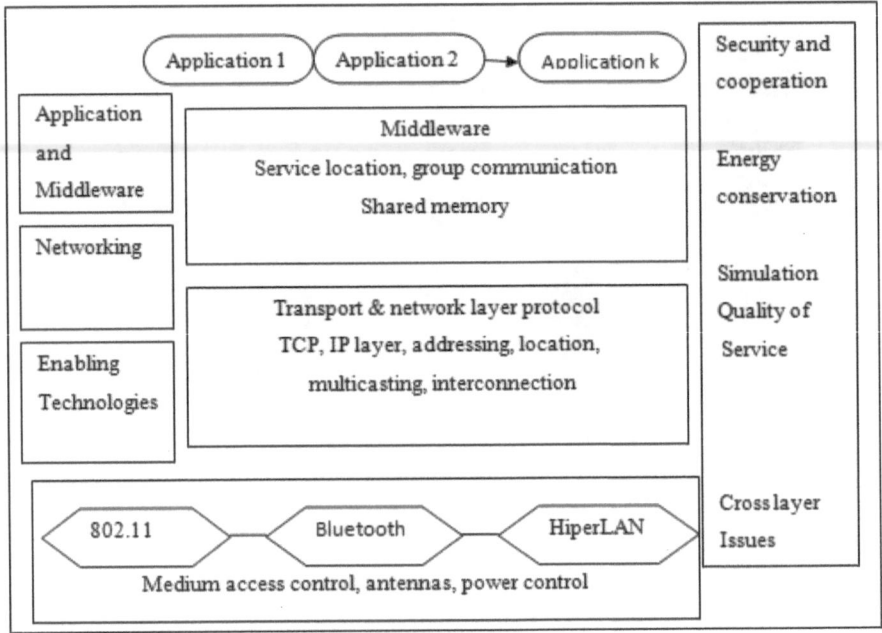

FIGURE 10.6   MANET architecture [14–19].

- Power management
- Peer-to-peer nature
- Limited bandwidth

10.2.3   APPLICATION IN MANET

The applications of ad hoc networks include:

- **Emergency services**
  - Search operation
  - Disaster recovery
- **Home and workplace networking**
  - Network at construction sites
  - Office chat rooms
  - Conference halls
  - Home safety networks
- **Educational sector**
  - Campus shaping
  - Smart rooms
  - Seminar halls
- **Military networks**
  - Communication between military base camps
  - Automated battlefields

- **Entertainment**
  - P2P networking
  - Games multiport
  - Amusement parks
- **Sensor networks**
  - Data tracking
  - Home applications
  - Body area networks (BANs)
- **Commercial and civilian**
  - E-commerce
  - Business sector
  - Vehicular services
  - Networks at airports [20–24].

10.2.4 MANET Advantages

The main advantages of MANET are as follows:

- Good coverage
- Cost estimation is less
- Easy to set up
- Multi-hop network with autonomous terminal and dynamic network topology
- Use as temporary network
- Less time to set up
- Self-management
- Mobility facts [25]

10.2.5 Overview of Challenges in Manet

Despite the facts that MANET has a large number of applications, there are many challenges, which must be thoroughly studied before its deployment. These are shown in Table 10.1.

10.2.6 Energy-related Issues in MANET

Along with network connectivity maintenance and period enhancement of the network, energy consumption is also an important issue in network, as nodes are battery-powered. Energy directly relates with lifetime and connectivity of the network. An energy-efficient route selection and network approach should be proposed [21–25].

10.2.6.1 Energy Consumption in MANET

For energy consumption in MANET, there are three steps:

- First, transmission of individual packet → energy consumed.
- Second, packet forwarded → energy consumed.
- Third and final, not transmitting or forwarding or idle nodes → energy consumed.

TABLE 10.1

Challenges in MANET [1–25]

Routing	Since the topology of network is constant, the issue of routing packet between any pair of nodes becomes a challenging task. Most protocols should be based on reactive routing instead of proactive. Multicast routing is another challenge because the multicast tree is no longer static due to the random movement of nodes within the network. Routes between nodes may potentially contain multiple hops, which makes it more complex than the single hop communication.
Security and reliability	In addition to the common vulnerabilities of wireless connection, an ad-hoc network has its particular security problem due to, for example, nasty neighbor relaying packets. Further, wireless link characteristics also introduce reliability problem because of the limited wireless transmission range, broadcast nature of the wireless medium, mobility-induced packet losses, and data transmission error.
Quality of service	Providing different quality of service level in a constantly changing environment will be a challenge. The inherent stochastic feature of communication quality in a MANET makes it difficult to offer fixed guarantees over the traditional resource reservation to support the multimedia services.
Inter-networking	In addition to the communication within an ad-hoc network, inter-networking between MANET and fixed network is often expected in many cases. The coexistence of routing protocol in such a mobile device is a challenge for the harmonious mobility management
Power consumption	For most of the lightweight terminal, communication-related function should be optimized for lean power consumption. Conservation of power and power-aware routing must be taken into consideration.
Location aided routing	Location-aided routing uses positioning information to define associated region so that the routing is spatially oriented and restricted broadcast in ABR.
Restricted wireless transmission range	The radio group will be restricted in the wireless network, and as a result, data amounts it can provide much slighter than what a bound network can provide. This involves routing procedures of wireless network must be use bandwidth in ideal way. This can be achieved through protection the overhead as minimum as conceivable. The restricted transmission range also enforces restraint on routing procedure for sustaining the topographical information. Particularly, in MANETS, because of regular variation in topology, preserving the topological data for every node includes more controller overhead, which results in additional bandwidth depletion.
Topology maintenance	Updating information of dynamic links among node in MANET is a major challenge.
Packet loss due to transmission error	Ad-hoc wireless network practices very advanced packet damage due to reasons such as extraordinary bit error rate (BER) in the wireless channel, enlarged crashes because of the existence of unseen terminals, occurrences of intervention, position reliant controversy, single directional associated, regular pathway breakage due to device movement, and the integral declining characteristics of the wireless passage.
Mobility induced route change	The system topography in ad-hoc wireless network is extremely active because of node movement; as a result, a constant meeting undergoes numerous pathway breakages. Such positions often result in regular path alteration. So, flexibility administration is a massive investigation theme in ad-hoc 99,420 network.

(Continued)

TABLE 10.1 *(Continued)*
Challenges in MANET [1–25]

Mobility induced packet loss	Communication contact in an ad-hoc network is insecure such that consecutively conservation procedure for MANETs over a great damage frequency will suffer from performance deprivation. Though, with large frequency of inaccuracy, it is problematic to supply a data packet to its target.
Battery constraints	It is due to restricted resources that arrange main limitation on the mobile devices in an ad-hoc network. Nodes which are contained in such a network have restriction on the supremacy foundation in order to preserve immovability, dimension, and capacity of node. The accumulation of power and the processing capacity make the nodes heavyweight and less portable. Consequently, only MANET device has to use this resource
Limited bandwidth	Wireless link continues to have significantly lower capacity than infrastructure network. In addition, the realized throughput of wireless communication after accounting for the effect of multiple access, fading, noise, interference condition, etc. is often much less than a radio's maximum transmission rate.
Dynamic topology	Dynamic topology membership may disturb the trust relationship among nodes. The trust may also be disturbed if some nodes are detected as compromised.
Routing overhead	In wireless ad-hoc network, nodes often hang their location within network. So, some stale routes are generated in the routing table, which lead to unnecessary routing.
Hidden terminal problem	The hidden terminal problem refers to collision of packet at a receiving node due to simultaneous transmission of those nodes that are not within the direct transmission range of sender, but are within the transmission range of the receiver.
Routing in dynamic topology	In MANET, the presence of node mobility changes the link of connectivity between the nodes very frequently. The existing conventional Bellman Ford algorithms of classic Link State algorithm are not applicable for such dynamic network, where the topology changes with the free movement of the node.
Autonomous	No centralized administration entity is available to manage the operation of the different mobile node in MANETs.
Poor transmission quality	This is an inherent problem of wireless communication caused by several error sources that result in degradation of the received signal.
Scalability	Scalability can be broadly defined as whether the network is able to provide an acceptable level of service even in the presence of a large number of nodes.

10.2.6.2   Type of Communication in MANET

- D-2-D communication
  - Direct communication between two nodes/elements without any other node/element in between.
  - Basically layer 2 link communication, energy consumption between two end devices during data exchange.
- End-to-end communication
  - Communication between two nodes via multiple nodes in between.
  - Indicate communication happening between two applications [25–30]

Energy consumption–related issues and challenges in MANET include:

- **Overhead:** A flexible configuration is offered in a wireless network. Wireless medium is used to comprise nodes. Due to high-mobility factor, the connection or link may break and the result is path failure and route discovery. Overhead can't be ignored. An effective and fundamental data dissemination technique of route discovery is broadcasting. But due to again and again blindly rebroadcasting of packets, broadcast storm problem generates, which is the main reason of overhead issue. Also, sometimes due to encryption decryption of packets, overhead increases and energy consumption increases [28–32].
- **Packet Loss:** When a packet fails to reach destination during communication, packet loss problems occur. Because wireless links are subjected to transmission error and dynamic change in topology, the packet loss problem is much more complicated. A packet loss may be due to
- Transmission error
- No route to destination
- Broken link
- Congestion

Packet loss is a serious problem in wireless ad hoc networks. There are several classifications for packet dropping and packet-dropping detection techniques [33–35].
Major classifications for packet dropping include

- Legitimate packet drop-ping
- Stealthy packet dropping
- Malicious packet dropping

Packet-dropping detection techniques are mainly classified into

- Watch dog technique
- Side channel monitoring
- Monitoring agent techniques
- Two Ack and Path Ratter
- **Energy Consumption:** Energy consumption is one of the most concerning points in MANET.
  - Limited battery resources are used to operate mobile nodes
  - As network is non-wired, the rate of energy consumption is high
  - Energy is consumed
    1. When node is transmitting
    2. When node is receiving
    3. Idle node
    4. Sleep mode
- Ad hoc network also means there is a shared environment and neighborhood communication also consumes some energy. So the battery is not used only by sending packets but also by just overhearing packets

from other nodes. While forwarding packets for other nodes energy is also spent because of multi-hop communication process in MANET [25–34].

- **Network Link Failure:** MANET is one of the most promising wireless network architectures. It is a wireless network of mobile devices that configures itself. The major issue created due to node mobility is link failure. Factors like
  - Dynamic topology due to node mobility
  - Interface traffic congestion
  - Link stability
  - Incorrect serial bandwidth setting
  - Route flags
  - Unidirectional traffic flow
  - Router interface down
  - Fault, error, or discard in network
  - Power outage
  - Security attacks
  - Partitions in between networks leads to interruption in communication and network link failure

Routing techniques help in path establishment for transmission. Overhead suffering routing protocol releases energy and may lead to link failure. As we have to establish the link again with the increase in energy consumption [35].

- **Collision:** A number of nodes are present in MANET. When a node wants to send a packet, it first checks the signal level of the line to find out if it is free or taken. If the route is free, it sends the packet, but sometime two nodes send the packet at the exact same time, which leads to collision. And due to this, both nodes need to wait again for their turn to transmit; the result is energy loss and loss of quality in communication.
- **Network Failure:** Due to all the above-mentioned problems, network failure may occur.

10.2.7 ENERGY CONSERVATION ROUTING APPROACHES

- Power-aware routing protocol
- Efficient power and lifetime-aware protocol
- Ad hoc on-demand distance vector routing (AODV) sleep
- Maximum energy-level ad hoc distance vector
- Mobility-based minimum network coding
- Triangular energy-saving cache scheme
- Energy-enhanced AODV routing
- Maximum lifetime ad hoc routing
- E-AOMDV energy-based multipath
- Predictive energy-efficient multicast algorithm
- Location-based power scheme [30–35]

10.3 CONCLUSION

This chapter gives a detailed survey of all the aspects of MANET, its types, architecture, characteristics, applications, advantages, and various challenges in MANET-enabled IoT realization. In this chapter, we further discuss various types of energy-related issues in MANET along with various energy conservation routing protocols. We discussed MANET, its radio frequency identification, routing protocols, and its applicability in the realization of IoT. While Internet of Things as a component of the future Internet has been depicted as a worldview that predominantly incorporates and empowers a few advancements and correspondence arrangements, a striking interest is to characterize how current standard correspondence conventions could uphold the acknowledgment of the vision. Inside this specific situation, we offer a state-of-the-workmanship audit on spontaneous and remote-sensor organizations, close to handle correspondences, radio recurrence recognizable proof, and steering conventions as intended to portray their pertinence toward an Internet of Things acknowledgment.

REFERENCES

1. X.M. Zhang, E.B. Wang, J. J. Xia, D.K. Sung, "A neighbor coverage-based probabilistic rebroadcast for reducing routing overhead in mobile ad hoc network". IEEE Transactions on Mobile Computing, volume 12, March 2013, Page no. 424–433.
2. R. Ragul Ravi, V. Jayanthi, "Energy efficient neighbor coverage protocol for reducing rebroadcast in MANET". Procedia Computer Science 2015, Page no. 417–423, www.sciencedirect.com.
3. V.D. Chakravarthy, V.D. Renga, "Neighbor coverage based probabilistic rebroadcast for reducing routing overhead in mobile ad-hoc network". International Journal of Emerging Technology and Advanced Engineering, volume 3, special issue 1, January 2013, Page no. 302–308, ISSN-2250- 2459 (online), ICISC-2013, www.ijetae.com.
4. M. Vijay, M. Srinivasa Rao, A.V.N Chandrasekhar, "Efficient rebroadcast for reducing routing overhead in MANETs using continuous neighbour discovery". International Journal of Academic Research, volume 2, issue 2, July–September 2014, ISSN:2348–7666, www.ijar,org.in.
5. A.P. Reddy, Satyanarayana, "A neighbour knowledge rebroadcast method for reducing routing overhead in MANET". International Journal of Scientific Engineering and Technology Research, volume 3, issue 29, October 2014, Page no. 5773–5777, ISSN-2319-8885, www.ijsetr.com.
6. S. Shivashankar, H.N. Varaprasad Golla, Jayanthi, "Design energy routing protocol with power consumption optimization in MANET, IEEE, volume 3, issue 4, 2014, DOI:10.1109/TETC.2013.2287177.
7. R. Naeem, M.U. Aftab, Q.A. Mohammad, Q. Akbar, "Mobile ad-hoc network applications and its challenges". Scientific Research Publishing Communication and Network, volume 8, 2016, Page no. 131–136, http://dx.doi.org/10.4236/cn.2016.83013.
8. L. Raja, S.S. Baboo, "An overview of MANET, application, attack and challenges". International Journal of Computer Science and Mobile Computing (IJCSMC), volume 3, issue 1, January 2014, Page no. 408–4017, ISSN:2320-088X, www.ijcsms.com.
9. M. Singh, M. Kumar, J. Malhotra, "Energy efficient cognitive body area network (CBAN) using look-up table and energy harvesting". Journal of Intelligent Fuzzy Systems, volume 35, 2018, Page no. 1253–1265.

10. K.S. Dinesh, "Routing overhead reduction and selection of stable paths in MANET". International Journal of Inventions in Computer Science and Engineering (IJICSE), volume 1, issue 9, 2014, Page no. 15–19.

11. R.S. Paranjape, N.N. Mail, D.V. Bhosale, "A neighbor converge based probabilistic rebroadcast for reducing routing overhead in mobile ad-hoc network using cluster scheme". International Research Journal of Multidisciplinary Studies, volume 2, 2016), Page no. 424–433.

12. M. Singh, M. Kumar, J. Malhotra, "An energy efficient spectrum sensing, access and handoff concept using look up table for cognitive radios networks". International Journal of Advanced Research in Computer Science, volume 8, issue 5, 2017, Page no. 229–238.

13. M. Renuka Devi, N. Stevenraj, "Reducing routing overhead in MANET using NCPR". International Journal of Engineering and Computer Science, volume 3, issue 6, June 2014, Page no. 6702–6705, ISSN:2319-7242, www.ijecs.in.

14. P. Nagrare, V. Sahare, "A survey on reducing routing overhead in MANET by using various techniques". International Journal of Advance Research in Computer Science and Software Engineering, volume 5, issue 1, January 2015, ISSN: 2277-128X, www.ijarcsse.com.

15. K.A. Ogudo, D.M.J. Nestor, O.I. Khalaf, H. Kasmaei, "A device performance and data analytics concept for smartphones' IoT services and machine-type communication in cellular networks". Symmetry, volume 11, 2019, Page no. 593.

16. R. Sharma, M. S. Nehra, "Improving the performance of routing using neighbor coverage based probabilistic rebroadcast in mobile ad hoc network". International Journal on Recent and Innovation Trends in Computing and Communication, volume 1, issue 12, December 2013, Page no. 939–943, ISSN:2321-8169, http://www.ijritcc.org.

17. O.I. Khalaf, G.M. Abdulsahib, B.M. Sabbar, "Optimization of wireless sensor network coverage using the bee algorithm". Journal of Information and Science Engineering, volume 36, 2020, Page no. 377–386.

18. C. Rajan, C. Dharanya, Dr. N. Shanthi, "A probabilistic rebroadcast for reducing routing overhead in real time MANET environment". Journal of Global Research in Computer Science, volume 5, issue 1, January 2014, ISSN:2229-371K, www.jgrcs.info.

19. M. Singh, M. Kumar, J. Malhotra, "An energy efficient spectrum sensing, access and handoff concept using look up table for cognitive radios networks". International Journal of Advanced Research in Computer Science, volume 8, 2017, Page no. 228–238.

20. C.P. Soumya, G. Girija, N. Gowdhami, "Comparative study of node energy constraints for MANET routing protocol". International Journal of Computer Science and Mobile Computing, volume 2, issue 11, November 2013, Page no. 364–370, ISSN 2320-088X.

21. O.I. Khalaf, B.M. Sabbar, "An overview on wireless sensor networks and finding optimal location of nodes". Periodicals of Engineering and Natural Sciences (PEN), volume 7, 2019, Page no. 1096–1101.

22. A. Singh, D. Chadha, "A study on energy efficient routing protocols in MANETs with effect on selfish behaviour". International Journal of Innovative Research in Computer and Communication Engineering, volume 1, issue 7, September 2013, ISSN (online):2320-9801 ISSN (print):2320-9798.

23. A. Pandey, "Introduction to mobile ad hoc network". International Journal of Science and Research Publication, volume 5, issue 5, May 2015, ISSN:2250-3153.

24. R. Bruzgiene, L. Narbutaite, T. Adomkus, "MANET Network in Internet of Things System". In Ad Hoc Networks. InTech, Luton, 2017.

25. Q. Ye, W. Zhuang, "Distributed and adaptive medium access control for Internet-of-Things-enabled mobile networks". IEEE Internet of Things Journal, volume 4, issue 2, 2017, Page no. 446–460.

26. M.U. R. Salfi, "A study mobile ad-hoc network-issue and challenges". International Journal of Advanced Research in Computer Science, volume 6, issue 7, September–October 2015, ISSN:0976-5697.

27. G. Abdulsahib, O. Khalaf, "Comparison and evaluation of cloud processing models in cloud-based networks". International Journal of Simulation Systems Science And Technology, volume 19, 2019, Page no. 26.1–26.6.

28. D. S. Patil, P. Rewagad, "NCPR using paillier cryptosystem to reduced routing overhead and secure data transmission in MANET". International Journal of Advance research in Computer Science and Management Studies, volume 4, issue 3, March 2016, ISSN:2321-7782.

29. S. Türk, M. Deveci, E. Özcan, F. Canıtez, R. John, "Interval type-2 fuzzy sets improved by simulated annealing for locating the electric charging stations". Information Sciences, volume 547, 2020, Page no. 641–666.

30. M. Izharul Hasan Ansari, S.P. Singh, M.N. Doja, "Energy efficient routing protocol for mobile ad hoc network: A review". International Journal of Computer Application, volume 3, issue 2, December 2015, ISSN:0975-8887.

31. D. Pamucar, M. Deveci, D. Schitea, L. Erişkin, M. Iordache, I. Iordache, "Developing a novel fuzzy neutrosophic numbers based decision making analysis for prioritizing the energy storage technologies". International Journal of Hydrogen Energy, volume 45, 2020, Page no. 23027–23047.

32. Yi Lu, Yuhui Zhong, Bharat Bhargav, "Packet loss In mobile ad hoc network", Center of Education and Research in Information Assurance and Security and Department of Computer Sciences, Purdue University, West Lafayette, Indiana.

33. D. Rajasekaran, S. Saravanan, "Enhanced routing performance and overhead in mobile ad hoc network for big data Transmission in Telemedicine using computer communication network". International Journal of Advance Research in Computer and Communication Engineering, volume 3, issue 11, November 2014, ISSN online: 2278-1021, ISSN print:2319-5940 www.ijarcce.com.

34. S. Dinakar, J. Shanthini, "Study on broadcasting techniques and characteristics in mobile ad hoc networks". International Journal of Computer Science Engineering, volume 3, issue 5, September 2014, ISSN: 2319-7323, www.ijcse.com.

35. Q. Ye, W. Zhuang. "Token-based adaptive mac for a two-hop internet-of-things enabled manet". IEEE Internet of Things Journal, volume 4, issue 5, 2017, Page no. 1739–1753.

11 Internet of Things (IOT)
Hands-On Sensing, Actuating, and Output Modules in Robotics

Shivam Gupta
MRIIRS, Faridabad, Haryana, India

CONTENTS

11.1 INTRODUCTION

In IOT, the systems are made up of different blocks, which perform different functions in order to accomplish some tasks. For example, let us consider an IOT-powered temperature- and humidity- monitoring system, which can be used by people to

DOI: 10.1201/9781003181613-11

check the temperature and humidity of different areas in a city in real-time. To perform this task, different sub-tasks are accomplished, which are a sequence of operations that needs to be performed in a sequential manner. They include:

- Sensing the change in temperature and humidity in the environment using humidity and temperature sensor and sending the data to microcontroller
- Performing the arithmetic and logical operations by the microcontroller on the received data and taking necessary actions
- Displaying and visualizing the information with the help of different output devices and actuators
- Communicating with the other devices and cloud platforms to exchange the information so that users can see the data in the real-time from anywhere

As can be seen, the primary task of an IOT system includes the sensing of changes occurring in the environment and converting the sensed physical changes into the electrical output, which can be further processed and analyzed by the electronic devices to do the decision-making process regarding what to do the next. This requires the use of sensors, which are so designed and configured that they can sense the physical changes and convert the non-electrical quantity to electrical quantity. These physical quantities include temperature, humidity, gases present in the atmosphere as well as their composition, water vapor content, fire, sound, light, position, vibrations, magnetic fields, chemicals, force, pressure, and many more [1–5].

After gathering certain relevant information of the user's interest, it becomes beneficial for an IOT system to produce an output based on which the user can control the input provided to the IOT system and control the future actions and outputs. This task of actuating, displaying, and visualizing the information plays important role by producing the human sensible, readable, and understandable output. These outputs are produced with the help of different actuating and output devices like motors, hydraulic pumps, LCDs, LEDs, audio devices, display devices, etc. [6–9].

In the field of robotics and drones, too, the sensing and actuating are the two major roles without which the robots and drones can neither commute to another point nor can perform any mechanical or non-mechanical operation. They are required to perform certain tasks such as sensing different things like obstacles around them, distances of obstacles from them, physical conditions around them, etc., and based on that they produce different outputs like movements through different motors, audio through buzzers, and speakers, etc. [10–16].

Now let us practically understand different sensors, actuators, and other output devices and how they are interfaced in the real world.

11.2 SENSING MODULES

As discussed previously, the primary subsystem of an IOT system are the sensing modules, which provide the input to an IOT system by sensing the physical changes occurring in the surrounding environment. The sensors sense the physical changes like temperature, humidity, flame, gases, sound, vibration, radiations, etc., in the environment in which they are placed and then converting the non-electrical parameters into the electrical signals. These electrical signals are readily fetched by the microcontrollers and different actions are performed based on the different inputs received by the microcontrollers and information processed from that input [17–22].

The sensors can be digital or analog depending on the type of output they produce. Some sensors produce data in the form of digital signal, i.e., 0 and 1, and are called digital sensors while some can output data in the analog form ranging the value from 0 to 1023.

Sensors have been found having a wide variety of applications in various fields like production and manufacturing, vehicles and automobiles, robotics and drones, everyday life, defense and military, security and surveillance and others [23–26]. Sensors have different parameters, which are directly related to their output values. They are:

- **Sensitivity:** It is defined as the quantity by which the output changes when the input quantity is changed.
- **Accuracy**: It is defined as the ratio of the number of times the sensor produces correct output to the number of total outputs produced.
- **Precision:** It is defined as how close the predicted output is to actual output.
- **Resolution**: It is defined as the smallest change in the input quantity, which produces a change in the output quantity.

Sensors are also differentiated broadly based on the physical conditions they can sense. The different categories of sensors are shown in Figure 11.1.

Let us discuss a few sensors along with their interfacing with the Arduino, schematic diagrams, C codes, and results obtained.

11.2.1 DHT11 Temperature and Humidity Sensor

There are many temperature and humidity sensors available in the market that have their own technical specifications and applications. The commonly-used temperature

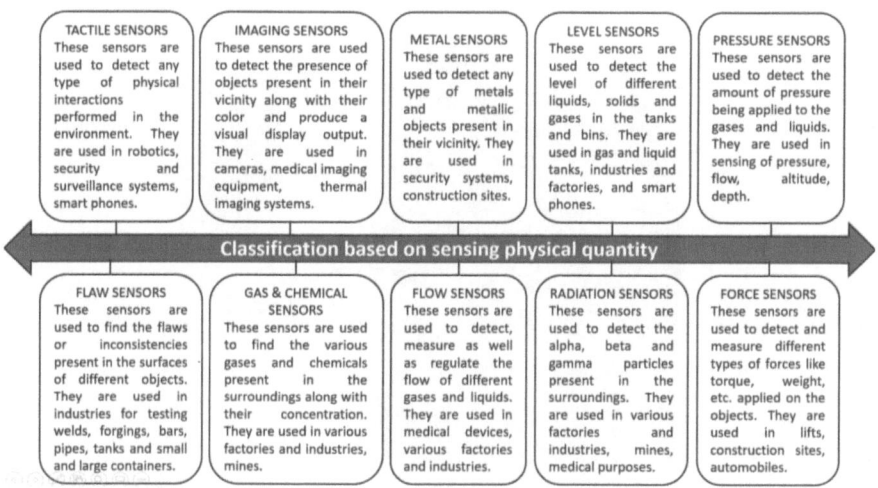

FIGURE 11.1   The different types of sensors based on the physical quantity they can sense.

and humidity sensor for household purposes is the DHT11 sensor. Other available sensors are DHT22, AM2311A, and AHT20.

DHT11 is a low-power, low-cost, highly-stable, and highly-reliable temperature and humidity sensor, which is used to produce the calibrated output. The calibration coefficients are stored in the OTP memory of the sensor. It uses a thermistor, as well as a humidity sensor (capacitive) to determine the temperature and humidity, respectively.

The applications of the temperature and humidity sensor include real-time temperature- and humidity-monitoring systems for different cities, and temperature-monitoring systems in industries and factories for large electrical machines and chambers, automobiles, etc.

Its technical specifications include:

- Operating power supply: 3.3 V–5 V DC
- Measurement range:
- Temperature: 0–50°C
- Humidity: 20–90% RH
- Accuracy:
- Temperature: ± 2°C
- Humidity: ± 5% RH
- Resolution:
- Temperature: 1°C
- Humidity: 1% RH

The schematic diagram in Figure 11.2 represents the interfacing of a DHT11 sensor with Arduino UNO. The Vcc pin of the sensor is attached to the +ve terminal of the power supply (5V) and the Gnd pin of sensor is attached to the –ve terminal of the power supply. The output port of the sensor is attached to digital pin number 2 of Arduino.

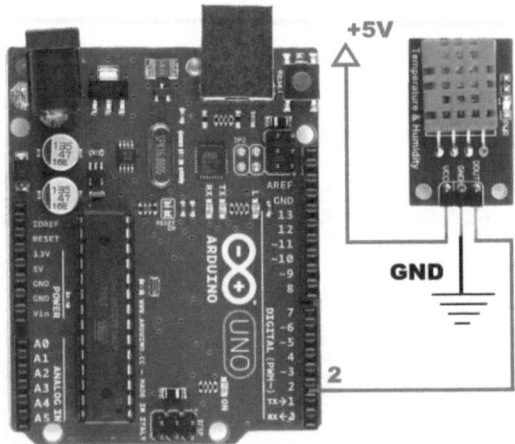

FIGURE 11.2   Schematic diagram of interfacing of a DHT11 sensor with Arduino UNO.

11.2.2 Ultrasonic Sensor

Ultrasonic sensors are sensor modules used to measure the distance of target objects placed in front of them. The ultrasonic sensors emit ultrasonic sound waves (which are not audible to human beings) from the emitter with the help of piezoelectric crystals, and after the sound waves get reflected from the target object, the receiver encounters it.

The equation used to calculate the distance of the target object is

$$D = 0.5 * T * C$$

where:

D = Distance of sensor from obstacle
T = Time of ultrasonic wave to travel to and from target object from and to sensor
C = Speed of sound waves, i.e., 343 m/s

Ultrasonic sensors are capable of detecting objects regardless of their color, transparency, material (except for the very soft materials), and surfaces. They are unsusceptible to interferences from objects like airborne particles, light, gases, as well as smoke present in the atmosphere. However, some common obstructions like dirt, snow, and ice can interfere with the ultrasonic sound waves.

They are most used for determining the level, position, as well as distance of target objects. Applications of ultrasonic sensors are as proximity sensors in automobiles for self-parking systems and collision-avoidance systems, as level sensors in chemical industries to monitor liquid levels, in robotics and drones as obstacle-detection systems, and others.

The technical specifications of the ultrasonic sensor are:

- Operating power supply: 5 V DC
- Operating current: 15 mA
- Operating frequency: 40 Hz
- Angle covered: 15°
- Measurement range: 2 cm ~ 400 cm
- Accuracy: ± 3 mm

The schematic diagram in Figure 11.3 represents the interfacing of the Arduino UNO and HC-SR04 ultrasonic sensor. The Vcc pin of the sensor is attached to the +ve terminal of the power supply (5V) and the Gnd pin of sensor is attached to the -ve terminal of the power supply. The Trig pin of the sensor is attached to pin number 10 of the Arduino while the Echo pin of the sensor is attached to pin number 9 of the Arduino UNO.

11.2.3 MQ-2 Gas Sensor

Gas sensors are defined as the sensing electronic devices, which are used to detect as well as identify the different types of gases present in the atmosphere. Besides this, they can be used to detect the composition of the identified gases present in the surroundings. Gas sensors can detect various toxic and flammable gases like methane, smoke, butane, CO, LPG, alcohol, CNG, ethanol, butane, H_2 and many more.

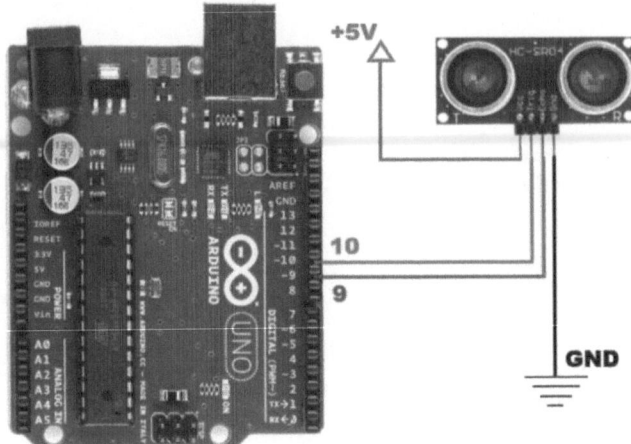

FIGURE 11.3   Schematic diagram of interfacing of ultrasonic sensor (HC-SR04) with Arduino.

The gas sensors are of different types based on:

- Metal oxide
- Optical
- Electrochemical
- Capacitance
- Calorimetric
- Acoustic

Applications of the gas and chemical sensors are detection of toxic and harmful gases in industries and factories, identifying gas leaks and smoke in factories, determining concentration of various gases present in the mines, detecting CO_2 levels in air conditioners, air-quality checking at homes and workplaces, etc. Due to their constant interaction with the different types of gases present in the air, the gas sensors are calibrated more often than any other sensor.

The commonly-used gas sensors are the sensors of the MQ series like MQ-2–MQ-9, MQ-131, MQ-135–MQ-138, MQ-214, MQ-216, MQ-303A, MQ-306A, MQ-307A, and MQ-309A. The MQ sensors have a chemi-resistor, which contacts and reacts with the gases present in the atmosphere. This reaction changes the electrical resistance of the sensors. The changing resistance produces an output voltage, which when measured helps to identify the gases present in the air along with their concentration. The type of gases that can be detected by the sensor depends on the type of sensing material of which the sensor is built.

The technical specifications of the MQ-2 gas sensor are:

- Operating power supply: 5 V DC
- Load resistance: 20 kΩ
- Sensing resistance: 10 kΩ – 40 kΩ

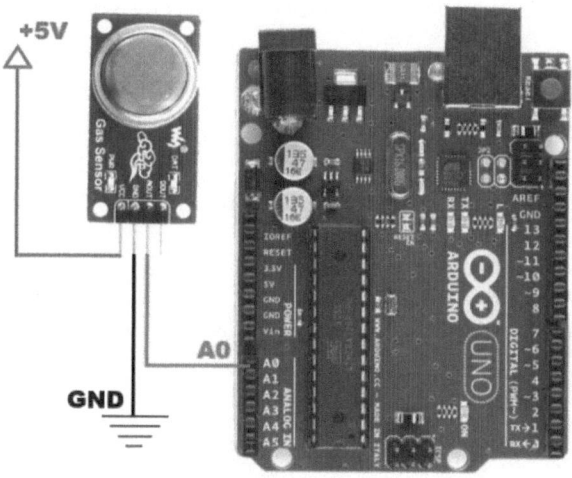

FIGURE 11.4   Schematic diagram of interfacing of MQ-2 gas sensor with the Arduino UNO.

- Concentration range: 200 – 10,000 ppm
- Heater resistance: 33 Ω ± 5%

The schematic diagram in Figure 11.4 represents the connections of the Arduino UNO and MQ-2 gas sensor. The Vcc pin of the sensor is attached to the +ve terminal of the power supply (5 V) and the Gnd pin of sensor is attached to the –ve terminal of the power supply. The output pin of the sensor is attached to the analog pin number A0 of the Arduino.

11.2.4   INFRARED SENSOR

An infrared (IR) sensor is a sensing device, which is used to determine the IR (infrared) radiation in the surrounding. The IR sensors are of two types:

- **Active IR Sensor**: They can both emit the IR radiations with the help of LED and detect the infrared radiations with the help of IR photodiode.
- **Passive IR Sensor**: The passive IR sensors can only detect the infrared radiations but cannot emit the infrared radiations.

The IR sensors are also known as proximity sensors because they are used to detect the obstacles present in the sensor's vicinity. The IR detectors are so made only to detect the IR radiations. The resistance and the output value of IR sensor ∝ the intensity of the received IR light.

The advantages of IR sensors are less power requirements; they require no physical contact with the target object for detection; they are highly immune to noise and corrosion, - as well as they are oxidation-resistant. The disadvantages of IR sensors include line-of-sight requirements, limited range, and they are prone to fog, rain, dirt, and dust. The applications of the IR sensors include obstacle detection systems in robots and drones, remote control devices for electronic appliances, home security systems, etc.

The technical specifications of the IR sensors are:

- Operating power supply: 3 – 5 V DC
- Operating current: 3 mA
- Wavelength detection range: 800 – 1100 nm
- Response wavelength: 940 nm (peak)
- Frequency detection range: 35 – 41 kHz
- Response frequency: 38 kHz (peak)
- Output: LOW on detecting 38 kHz IR radiation, otherwise HIGH
- Range: 10–15 cm

The schematic diagram in Figure 11.5 represents the connections of the Arduino UNO and IR sensor. The Vcc pin of the sensor is attached to the +ve terminal of the power supply (5 V) and the Gnd pin of sensor is attached to the –ve terminal of the power supply. The output of the sensor is attached to pin number 2 of the Arduino.

11.2.5 SOUND SENSOR

The sound sensor is a sensing device used to detect the sound present in its surroundings, as well as measure the intensity of sound. It uses a microphone to feed input to the sensor, which is processed by the LM393 op amp. When the input reaches above the pre-set threshold value, the LED present on the sensor module gets illuminated and output is set to logic LOW.

The main applications of this sensor include security systems in which if the surrounding sensor reaches above the threshold value, the alarm system gets activated.

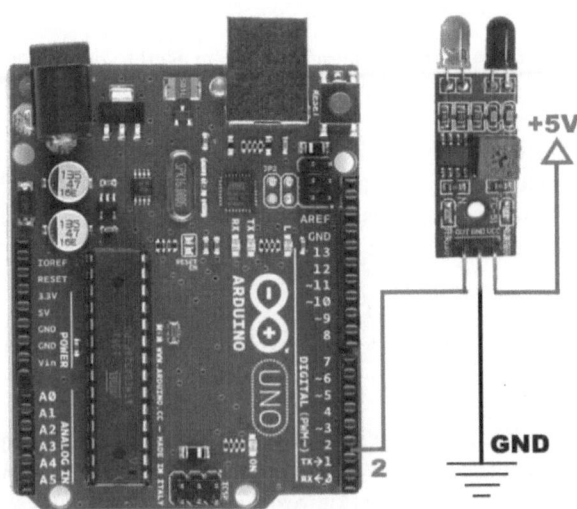

FIGURE 11.5   Schematic diagram of interfacing of IR sensor with the Arduino UNO.

Its other applications include surrounding sound-level monitoring, switching through sound intensity, etc.

The technical specifications of the sound sensor are:

- Operating power supply: 3.3–5 V DC
- Operating current: 4–8 mA
- SNR: 54 dB
- Microphone Sensitivity: 48–52 dB
- Microphone frequency: 16–20 kHz
- Frequency range: 3–6 kHz
- Output: LOW and HIGH
- Impedance: 2.2 kΩ

The schematic diagram in Figure 11.6 represents the connections of the Arduino UNO and sound sensor. The Vcc pin of the sensor is attached to the +ve terminal of the power supply (5V) and the Gnd pin of sensor is attached to the –ve terminal of the power supply. The analog output pin of sensor is attached to the pin number A0 of the Arduino, while the digital output pin of sensor is attached to pin number 2 of the Arduino.

11.2.6 TILT SENSOR

A tilt sensor is defined as the sensor used to measure the degree of tilt done in multiple axes with respect to the reference plane. The tilt sensors do so by measuring the tilting position with respect to the gravity. The tilt sensor uses a rolling ball to form the electrical connectivity between both terminals of the sensor. When the rolling ball settles at the bottom of the sensor, the LED present on the module gets illuminated; otherwise, it stays in logic LOW state.

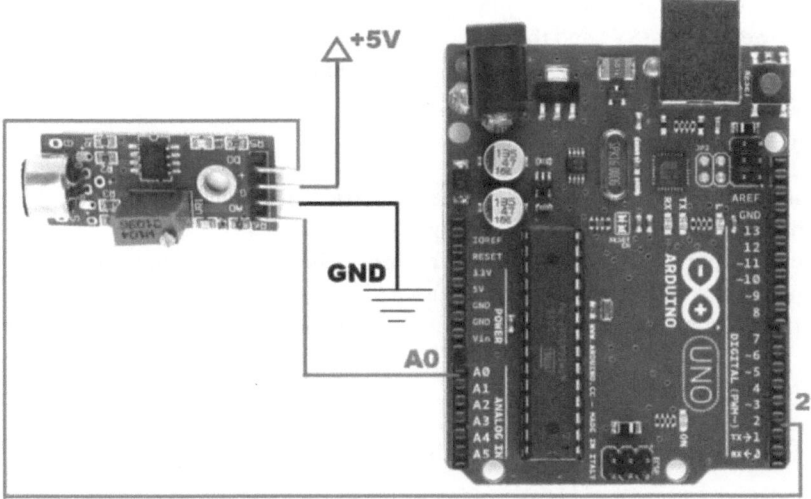

FIGURE 11.6   Schematic diagram of interfacing of sound sensor with the Arduino UNO.

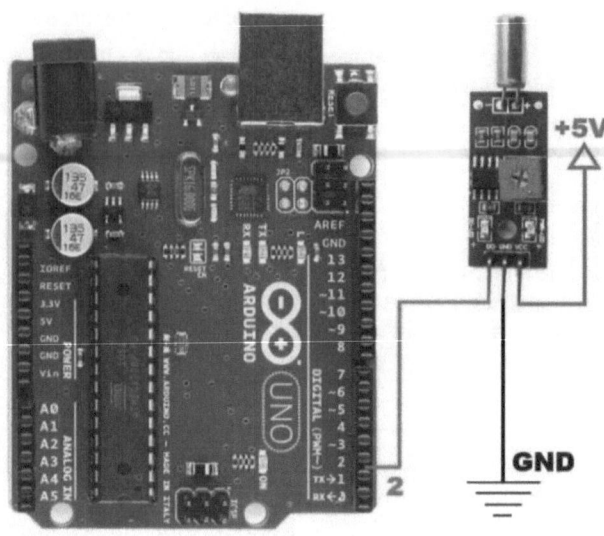

FIGURE 11.7   Schematic diagram of interfacing of tilt sensor with the Arduino UNO.

The advantages of the tilt sensor are high reliability, accuracy and resolution, low power need, ease of use, etc. The applications of the tilt sensor are determining orientation in boats and ships, detecting and measuring the tilt in robotics, tilt switches, detecting orientation to tackle obstacles for aircrafts, indicating roll of boats and ships, detecting position of the game controllers and smart phones, etc.

Its technical specifications are:

• Operating power supply: 3.3–5 V DC

The schematic diagram in Figure 11.7 represents the connections of the Arduino UNO and tilt sensor. The Vcc pin of the sensor is attached to the +ve terminal of the power supply (5V) and the Gnd pin of sensor is attached to the −ve terminal of the power supply. The output of sensor is attached to pin number 2 of the Arduino.

11.2.7   VIBRATION SENSOR

The vibration sensor, also called the piezoelectric sensor, is a sensor used to detect the vibrations present in the surroundings.

The main applications of the vibration sensor include the monitoring as well as measuring the vibration in certain industries. Its other applications are in mining, wind power–generating industry, gas and oil factories, food, beverage, and paper industries, etc.

Its technical specifications are:

• Operating power supply: 3.3–5 V DC

The schematic diagram in Figure 11.8 represents the connections of the Arduino UNO and vibration sensor. The Vcc pin of the sensor is attached to the +ve terminal

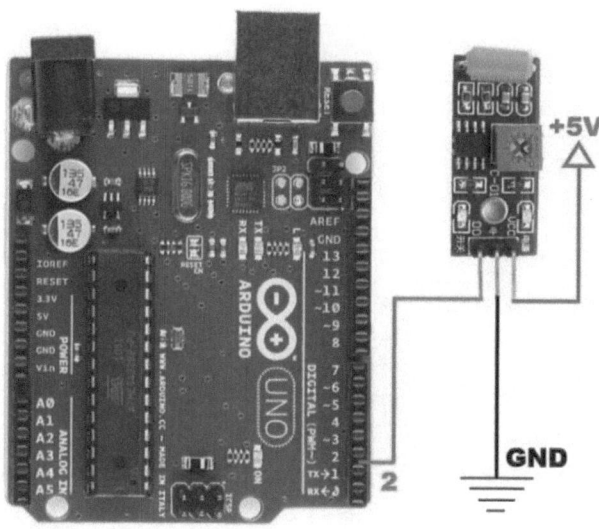

FIGURE 11.8 Schematic diagram of interfacing of vibration sensor with the Arduino UNO.

of the power supply (5 V) and the Gnd pin of sensor is attached to the –ve terminal of the power supply. The output of the sensor is attached to pin number 2 of the Arduino.

11.2.8 MOTION SENSOR

It is a sensing module that can be used to detect the motion nearby to the sensor. The PIR sensors are easy to use, inexpensive, small, and low power-consuming sensing module. The PIR or motion sensor works on the principle that every living thing produces radiations due to thermal energy present in them. The PIR sensors detects the thermal radiations in its vicinity and produces the output.

There are two types of motion sensors:

- **Active Motion Sensors**: These sensors can both transmit and receive the sound waves after they are interfered with by the presence of living bodies.
- **Passive Motion Sensors**: These sensors can only receive the radiations or the waves, which are naturally emitted by the living bodies.

The main application of the sensor is in security systems that deal with the detection of motion in the vicinity of the sensor where it activates the alarm system in case of any unwanted motion. Its other applications include automatic light and fan control, home automation, security, monitoring and surveillance, dispensing systems, and other areas.

The technical specifications of the motion sensor are:

- Operating power supply: 3.3–5 V DC
- Operating current: 1.6 mA
- Sensitivity range: 6 m
- Output: HIGH when motion is detected, otherwise LOW

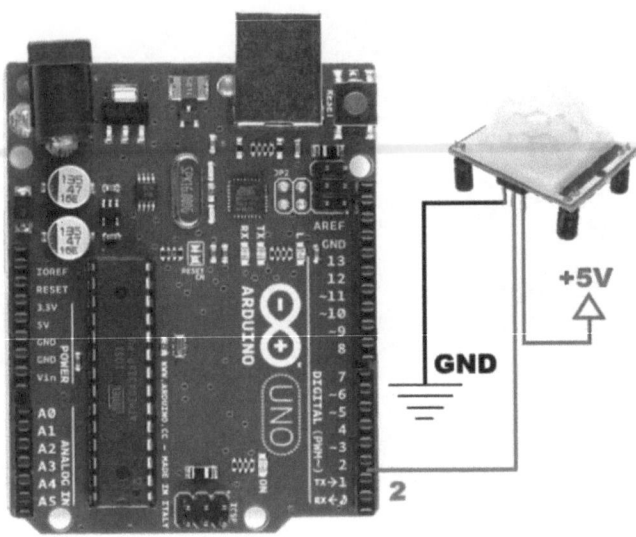

FIGURE 11.9   Schematic diagram of interfacing of motion sensor with the Arduino UNO.

The schematic diagram in Figure 11.9 represents the connections of the Arduino UNO and motion sensor. The Vcc pin of the sensor is attached to the +ve terminal of the power supply (5 V) and the Gnd pin of sensor is attached to the −ve terminal of the power supply. The output of the sensor is attached to pin number 2 of the Arduino.

11.2.9   RAIN SENSOR

The rain sensor is defined as a sensing module that is used to detect and measure the intensity of the rainfall.

The rain sensor consists of two modules:

- The sensing pad made of series of copper traces, which is exposed to the rainfall
- The sensor module, which connects the sensing pad to the Arduino

It works on the principle that resistance is inversely proportional to the amount of water present. The water is good conductor of electricity. As the amount of water on the sensing pad increases, the conductivity of the sensing pad increases and the resistance decreases. The sensor module then measures the voltage across the sensing pad and outputs the resulting voltage to the controller.

The sensor module feeds the output voltage to the LM393 comparator present on it. The LM393 comparator is used to compare the output voltage with the pre-set threshold value and produce the output in digital form. This digital output is too sent to the controller along with the analog one.

The applications of the rain sensor include smart rainwater harvesting system where the rain sensor detects the rainfall and triggers an alarm or activates the

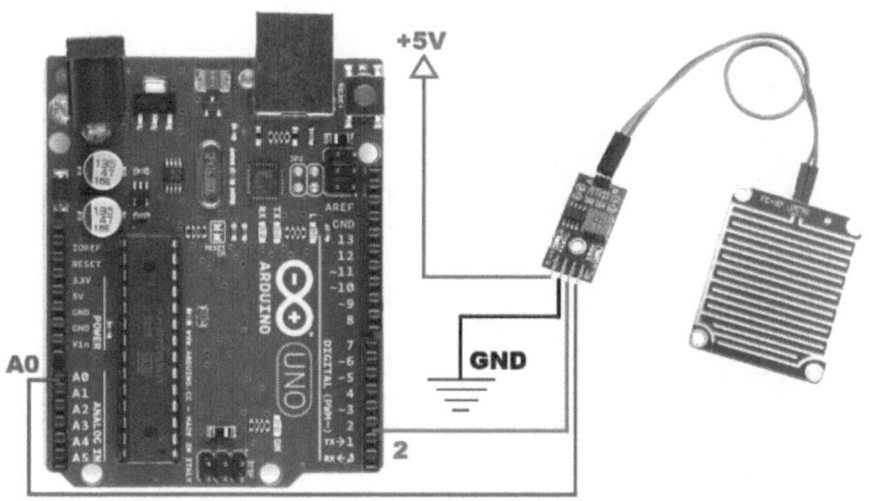

FIGURE 11.10   Schematic diagram of interfacing of rain sensor with the Arduino UNO.

rainwater harvesting system, automatic drain systems to open holes on roads when rain falls, automobiles, irrigation systems, etc.

The technical specifications of the rain sensor are:

- Operating power supply: 5 V DC
- Output: LOW when output voltage > threshold, otherwise HIGH

The schematic diagram in Figure 11.10 represents the connections of the Arduino UNO and rain sensor. The Vcc pin of the sensor is attached to the +ve terminal of the power supply (5 V) and the Gnd pin of sensor is attached to the −ve terminal of the power supply. The digital output of the sensor is attached to pin number 2 of the Arduino. The analog output of the sensor is attached to pin number A0 of the Arduino.

11.2.10   LINE-TRACKING SENSOR

The line-tracking sensor is a highly reliable and accurate sensing device used to iden-tify the no-obstacle path that can be followed by mobile robots. The line-tracking sensor consists of an LED, which emits the infrared light, and a receiver, which detects and receives that infrared light.

The applications of the line-tracking sensor in robotics include line following, anti-collision, anti-edge falling, short-distance obstacle detection, etc.

The technical specifications of the line-tracking sensor are:

- Operating power supply: 3.3–5 V DC
- Operating current: 18–20 mA
- Operating temperature range: 0–50°C

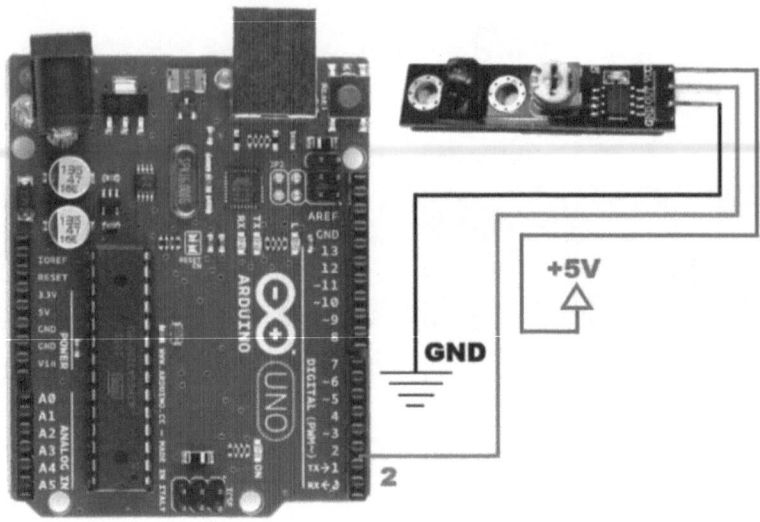

FIGURE 11.11   Schematic diagram of interfacing of line-tracking sensor with the Arduino UNO.

- Detection range: 1–2 cm
- Output: HIGH when no obstacle is detected, otherwise LOW

The schematic diagram in Figure 11.11 represents the connections of the Arduino UNO and line-tracking sensor. The Vcc pin of the sensor is attached to the +ve terminal of the power supply (5V) and the Gnd pin of sensor is attached to the –ve terminal of the power supply. The output of the sensor is attached to pin number 2 of the Arduino.

11.2.11   FLAME SENSOR

The flame sensor is the sensing device used to detect the flames and other light sources whose wavelengths lie in the range 760–1100 nm. It contains an LM393 comparator chip, which compares the detected infrared light wavelength with the pre-set threshold value and produces the digital output.

The applications of the flame sensor include firefighting robots and drones, fire alarms, automatic fire-extinguishing systems, industrial boilers, etc.

The technical specifications of the line-tracking sensor are:

- Operating power supply: 3.3–5 V DC
- Operating current: 15 mA
- Wavelength range: 760–1100 nm
- Detection range: 60°
- Output: HIGH when no flame is detected, otherwise LOW

The schematic diagram in Figure 11.12 represents the connections of the Arduino UNO and flame sensor. The Vcc pin of the sensor is attached to the +ve terminal

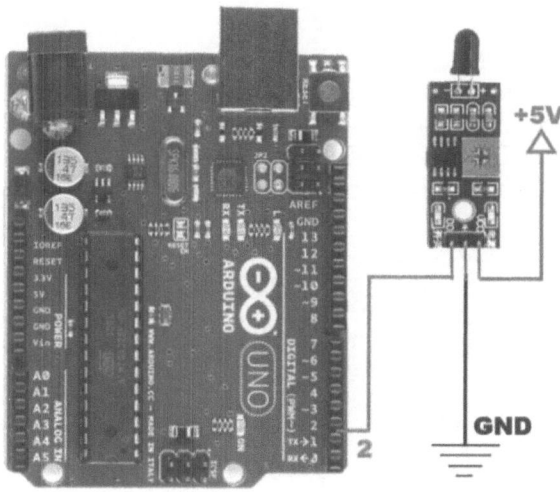

FIGURE 11.12 Schematic diagram of interfacing of flame sensor with the Arduino UNO.

of the power supply (5 V) and the Gnd pin of sensor is attached to the –ve terminal of the power supply. The output of the sensor is attached to pin number 2 of the Arduino.

11.2.12 LIGHT SENSOR

Light sensors are the sensors used to detect and measure the intensity of the visible or infrared light. The light sensors use the light-dependent resistors (LDR), which follow the principle of photoconductivity, i.e., it converts the optical energy into the electrical energy. When the intensity of the light falling on the light sensor increases, the number of photoelectrons increases, which leads to an increase in the conductivity of the light sensor. This leads to a decrease in the resistance across the sensor, and the output voltage increases. The light sensors consist of semiconductor substrates, which are made up of indium antimonide (InSb), lead selenide (PbSe), cadmium sulfide (Cds), and lead sulfide (PbS).

The applications of the light sensor are automatic streetlights, home automation, automatic switching circuits, etc.

The technical specifications of the light sensor are:

- Operating power supply: 3.3–5 V DC
- Operating current: 0.5 A
- Operating temperature: –10°C–40°C

The schematic diagram of Figure 11.13 represents the connections of the Arduino UNO and light sensor. The Vcc pin of the sensor is attached to the +ve terminal of the power supply (5 V) and the Gnd pin of sensor is attached to the –ve terminal of the power supply. The output of the sensor is attached to pin number 2 of the Arduino.

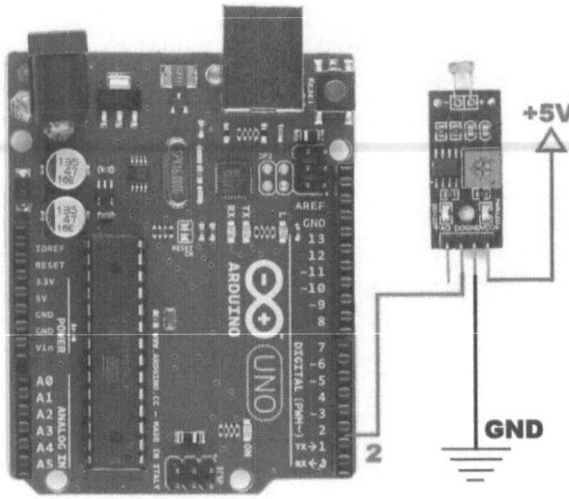

FIGURE 11.13   Schematic diagram of interfacing of light sensor with the Arduino UNO.

11.2.13   FLEX SENSOR

The flex sensor is a form of variable resistor that has the capability to variate its resistance based on the degree of bent. The flex sensor is used to measure the degree of bent or deflection. The flex resistance is directly proportional to the amount of bent.

The flex sensors consist of carbon strip arranged on a plastic strip surface. It comes in two different sizes: 2.2-in long as well as 4.5-in long. Based on the type of material, the flex sensors can be differentiated as:

- Carbon resistive
- Capacitive
- Fiber-optic
- Conductive ink

The applications of the flex are door sensors, gloves, robotics, gaming, security systems, musical instruments, etc.

The technical specifications of the flex sensor include:

- Operating power supply: 0–5 V DC
- Operating temperature: −45°C to +80°C
- Power rating: 0.5 W (continuous), 1 W (peak)
- Flat resistance: 25 kΩ
- Bend resistance range: 45 kΩ–125 kΩ
- Resistance tolerance: ± 30%

The schematic diagram in Figure 11.14 represents the connections of the Arduino UNO and flex sensor. The anode of the sensor is attached with a 10 kΩ resistor,

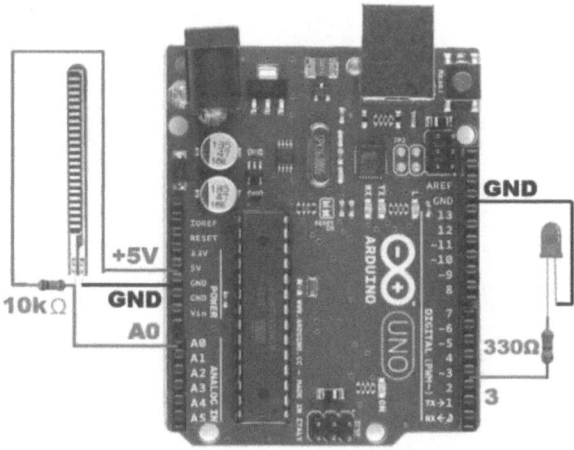

FIGURE 11.14   Schematic diagram of interfacing of flex sensor with the Arduino UNO.

whose other end is attached to the 5 V pin of Arduino, whereas the cathode of sensor is attached to GND pin of Arduino. The anode of sensor is also attached to pin number A0 of the Arduino for output. The anode of LED is attached to a 330 Ω resistor, whose other end is attached to pin number 3 of the Arduino (because pin 3 of the Arduino supports the PWM) and the cathode of LED is attached to GND of the Arduino.

11.2.14   HALL SENSOR

Hall sensor is defined as a sensor used to detect the magnetic field in the surroundings. The magnetic sensor converts the magnetic energy into the electrical energy. The hall sensors get activated whenever they detect the magnetic fields in their vicinity. When the strength of the magnetic field crosses the pre-set value, it produces an output.

The advantages of using the hall sensors are that they can be used for the switching actions and are more reliable than the mechanical switches. They are resistant to dust, dirt, and vibrations. It is the contactless switch, i.e., they need not be required to be touched by the magnetic sources to perform action. The main applications of the hall sensors include the measuring the position and distance, as well as speed. Its other applications are automatic door locks, various purposes-sensing systems, automobiles, home automation, electronic appliances, consumer goods, speedometers, medical instruments, etc.

The technical specifications of the hall sensor are:

- Operating power supply: 4.5–28 V DC
- Operating current: 25 mA
- Operating temperature: −45°C–85°C
- Switching time: 2 µs
- Can be used to detect both the poles

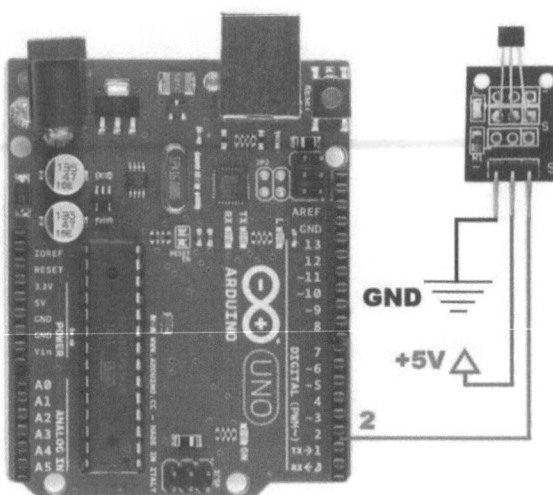

FIGURE 11.15   Schematic diagram of interfacing of hall sensor with the Arduino UNO.

The schematic diagram in Figure 11.15 represents the connections of the Arduino UNO and hall sensor. The Vcc pin of the sensor is attached to the +ve terminal of the power supply (5 V) and the Gnd pin of sensor is attached to the −ve terminal of the power supply. The output of the sensor is attached to pin number 2 of the Arduino.

11.3   ACTUATORS AND OUTPUT MODULES

Besides providing input to the microcontrollers, it is important also to display and show the different information processed by the microcontrollers to the users and outside world by converting the electrical output into the physical quantities like motion, movements, sounds, voice, etc., which can be sensed and observed by the users for future decision-making processes. The process of producing these sensible outputs helps the user to interact with the IOT system more easily and effectively. These types of outputs are produced with the help of different modules or devices known as actuators or output devices.

Actuators are defined as the machines or subsystems that help automate the tasks that are related to the linear and angular movement or motion. There are different types of actuating mechanisms, which work to produce different types of linear and angular motions and movements. The different actuating devices differ in various parameters like operating mechanisms, such as pneumatical or electric or hydraulics, input power supply type such as AC or DC, load capacity, degree of freedom, and other parameters.

Besides actuators, there are a variety of different output devices available which differs from each other on the basis of the type of physical quantity they can produce as well as other technical and non-technical specifications. The different categories of output devices are shown in Figure 11.16.

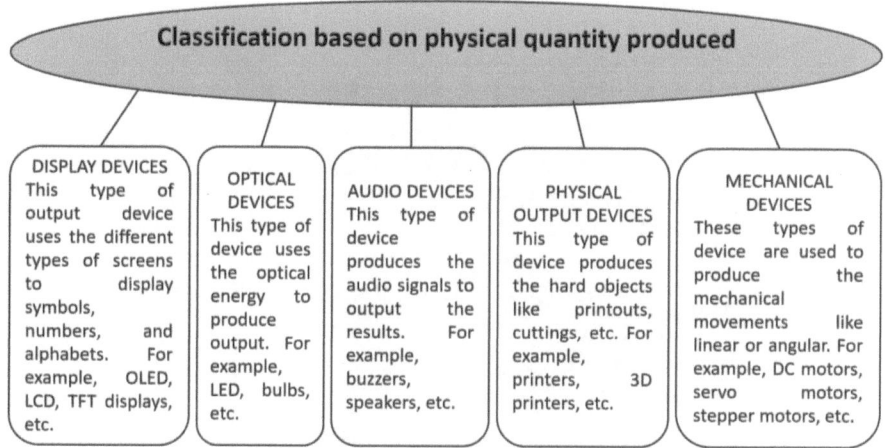

FIGURE 11.16 Different types of output devices based on the physical quantity they can produce.

Let us discuss some of the output devices and how they are interfaced with the Arduino UNO.

11.3.1 LED

Light-emitting diodes (LEDs) are one of the most basic and most used output devices. They are semiconductor devices that have replaced conventional fluorescent and incandescent bulbs. This is due to their properties such as:

- Low power hunger
- Long life
- Fast switching time
- Availability in various colors
- Low heat emission

The technical specifications of the LED are:

- Operating power supply: 1.8–2.4 V DC
- Operating current: 30 mA
- Operating temperature: –30°C–85°C
- Luminous intensity: 20 mcd
- Reverse voltage: 5 V

The schematic diagram in Figure 11.17 represents the interfacing of an LED with the microcontroller Arduino UNO. The cathode of the LED is attached to a 330 Ω resistor whose other end is attached to GND of the Arduino and the anode is attached to pin number 13 of the Arduino. Whenever the logic HIGH is applied at pin 13, the LED glows up; otherwise it stays in the OFF state.

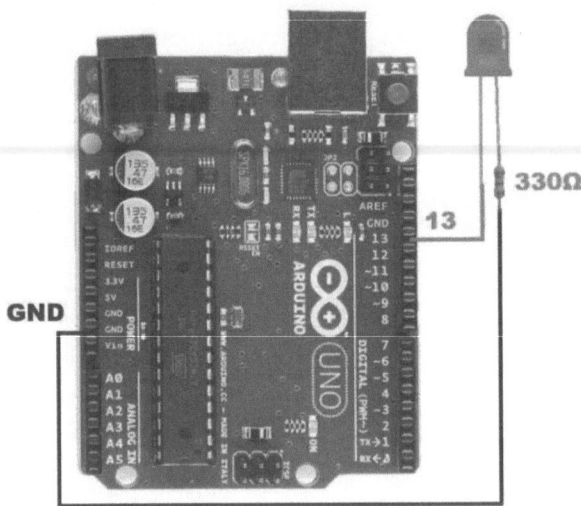

FIGURE 11.17   Schematic diagram of interfacing of LED with the Arduino UNO.

11.3.2   DC Motor

The DC motor is defined as an electrical machine used to produce the rotary motion. The DC motor converts electrical energy (DC voltage) into mechanical energy (rotary motion). The DC motors either work on the electromechanical mechanisms or the electronic mechanism. The DC motor works on the principle that when a current-carrying conductor is placed in the magnetic field, a torque is formed which results in motion. When the electric current flows through the DC motor, the magnetic field gets developed around the shaft, which accelerates the motion of the shaft.

DC motors have various parameters which need to be considered before using them like torque, load capacity, power requirement, etc. The speed of shaft can be changed by various methods like variating the input voltage, changing current in the field windings, changing the resistance of the circuit, etc. The applications of the DC motor are industries, mills and factories, automobiles, tools and machinery, toys, electrical and electronic appliances, etc.

The technical specifications of the DC motor are:

- Operating power supply: 4.5–9 V DC
- Operating current: 70 mA (no load), 250 mA (full load)
- No load speed: 9000 rpm

The technical specifications of the L293D motor driver are:

- Operating power supply for IC: 4.5–7 V DC
- Operating power supply for motors: 4.5–36 V DC
- Operating current for motors: 600 mA (continuous); 1.2 A (peak)

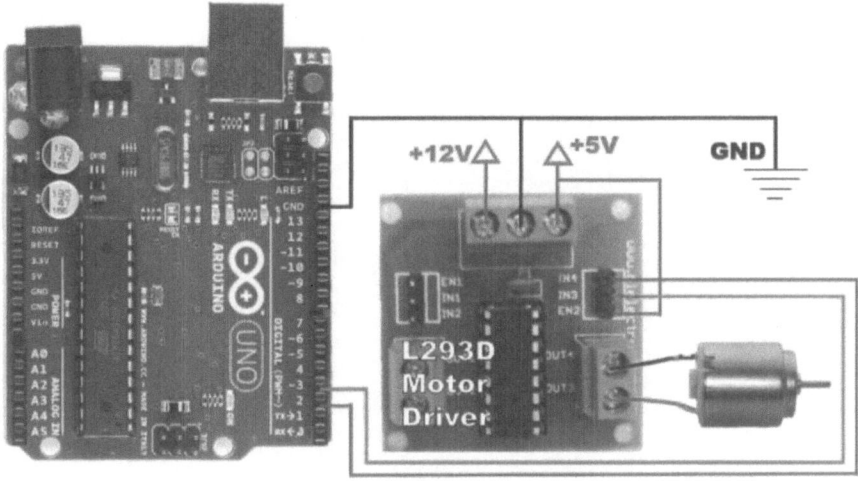

FIGURE 11.18   Schematic diagram of interfacing of DC motor with the Arduino UNO.

- Transition time: 300 ns
- Can be used to control the speed and direction of two motors with a single IC

The schematic diagram in Figure 11.18 represents the interfacing of the Arduino UNO and DC motor. The accessing of the DC motor through the Arduino requires a motor driver so as to enable safe and smooth operation. Here, the motor driver used is L293D. The two input pins IN 3 and IN 4 of the motor driver are connected to the Arduino digital pins 2 and 3, which controls the switching and direction of motor. The enable and IC input supply pins are provided with a +5V voltage, whereas the 12 V input pin is connected to the 12 V power supply. The ground of the IC, the power supply, as well as the Arduino are connected together. The two terminals of the DC motor are connected to the output pins "OUT 3 and OUT 4 of the motor driver.

11.3.3   BUZZER

The buzzer is defined as the output device, which is used to produce electrical energy into the sound energy. It is used to produce the beep sounds.

The applications of the buzzers include alarm devices, timers, mini projects, electronic appliances, recreation, automobile equipment, communication equipment, etc.

The technical specifications of the buzzer are:

- Operating power supply: 4–8 V DC
- Operating current: < 30 mA
- Resonant frequency: 2300 Hz

The schematic diagram in Figure 11.19 represents the interfacing of the Arduino UNO and buzzer. Here LED is used to show the result of the working circuit.

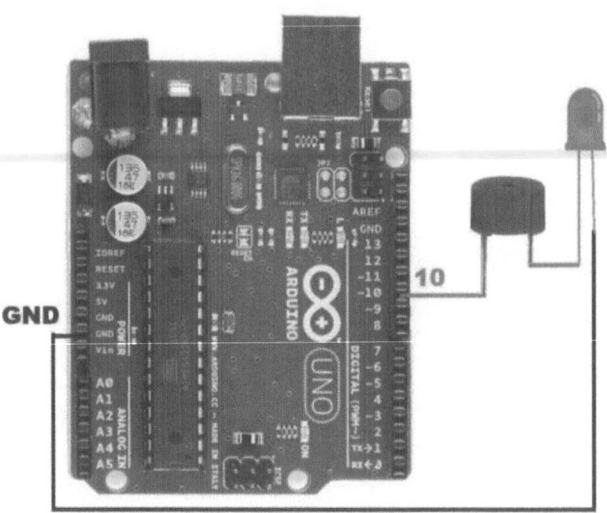

FIGURE 11.19   Schematic diagram of interfacing of buzzer with the Arduino UNO.

The anode of the buzzer is attached to pin number 10 of the Arduino while cathode is attached to the anode of LED. The cathode of LED is connected to GND of the Arduino UNO. Whenever the logic HIGH is supplied to the pin 10, the buzzer sounds up and LED lightens; otherwise both stay in the switch OFF state.

11.3.4 SERVO MOTOR

The servo motor is an actuating device that is used to produce the angular motion. The unique feature of the servo motor that makes it differentiable from the DC motor is that the servo motor is highly precise and accurate. It has the capability to rotate at different angle values with great efficiency. The servo motor uses a sensor coupled in the motor to provide the position feedback.

The two types of servo motors are DC servo motor and AC servo motor, which are defined by the type of input supply that is required for them to function. Both the motors differ in operating characteristics, as well as gear arrangements.

The main advantages of the servo motor are that it can rotate with high efficiency and great precision irrespective of light or heavy load conditions. The applications of the servo motor are robotics and drones, tools and machinery, manufacturing and production processes, conveyor belts, etc. The servo motor can be used at the situations where the motor needs to be rotated at a particular angle or position.

The technical specifications of the servo motor are:

- Operating power supply: 4.8–6 V DC
- Speed: 0.12–0.14 s/60° (for large motors); 0.10–0.12 s/60° (for small motors)
- Torque: 2.0–2.5 Nm (for large motors); 1.5–2.0 Nm (for small motors)

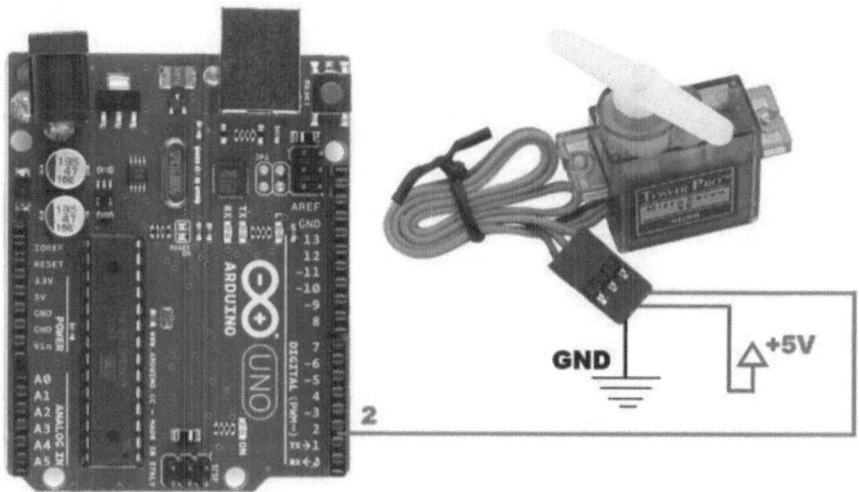

FIGURE 11.20   Schematic diagram of interfacing of servo motor with the Arduino UNO.

The schematic diagram in Figure 11.20 represents the connections of Arduino UNO and servo. The Vcc of the servo motor (i.e., RED wire) is connected to +ve terminal of the +5 V of voltage supply and the Gnd pin of the servo (i.e., BLACK or BROWN wire) is attached to –ve terminal of the voltage supply. The input pin of the servo (i.e., ORANGE or YELLOW wire) is attached to pin number 2 of the Arduino.

11.3.5   16 × 2 LCD Display

The liquid crystal display (LCD) is an electronic-displaying output device used to display a variety of characters, symbols, alphabets, and numbers.

The advantages of LCDs include ease of use, inexpensiveness, easy program-mability, etc. The applications of the LCD include mobile phones, electronic devices and circuits, electronic appliances, etc.

The LCD comes in various sizes with different numbers of rows and columns like 8 × 1, 8 × 2, 16 × 1, 16 × 2, etc.

The pin configuration of the 16 × 2 LCD is as follows:

- **PIN 1 → VSS:** It is GND pin, which needs to be connected to the GND of the power supply.
- **PIN 2 → VCC:** It is the VCC pin, which needs to be attached to the +5 V power supply.
- **PIN 3 → VEE:** This pin is connected to a potentiometer, which is used to control the contrast of the LCD.
- **PIN 4 → RS:** It is the register select pin. This pin is used to select between data mode (logic '0' or LOW) and command mode (logic '1' or HIGH).
- **PIN 5 → RW:** It is the read/write pin. This pin is used to switch between write mode (logic '0' or LOW) and read mode (logic '1' or HIGH).

- **PIN 6 → E:** It is the Enable pin. The LCD will work only when this pin is set to logic '1' or HIGH.
- **PINS 7–14 → D0–D7:** These are the data pins that are connected to the microcontroller to enable data communication. The pins can be connected to the microcontroller in either 4-wire mode or 8-wire mode.
- **PIN 15 → LED+:** This pin is connected to the 5 V power supply to light up the backlight of the LCD.
- **PIN 16 → LED−:** This pin is connected to the GND of the power supply.

The technical specifications of the LCD are:

- Operating power supply: 4.7–5.3 V DC
- Operating current: ~ 1 mA
- Wiring modes: 4-wire mode and 8-wire mode

The schematic diagram in Figure 11.21 represents the connections of the Arduino UNO and 16×2 LCD Display. The pin connections for above schematic are explained as follows:

- Pin 1 → GND
- Pin 2 → +5 V
- Pin 3 → Potentiometer or GND
- Pin 4 → Digital pin 7
- Pin 5 → GND
- Pin 6 → Digital pin 8
- Pin 7 → Unconnected

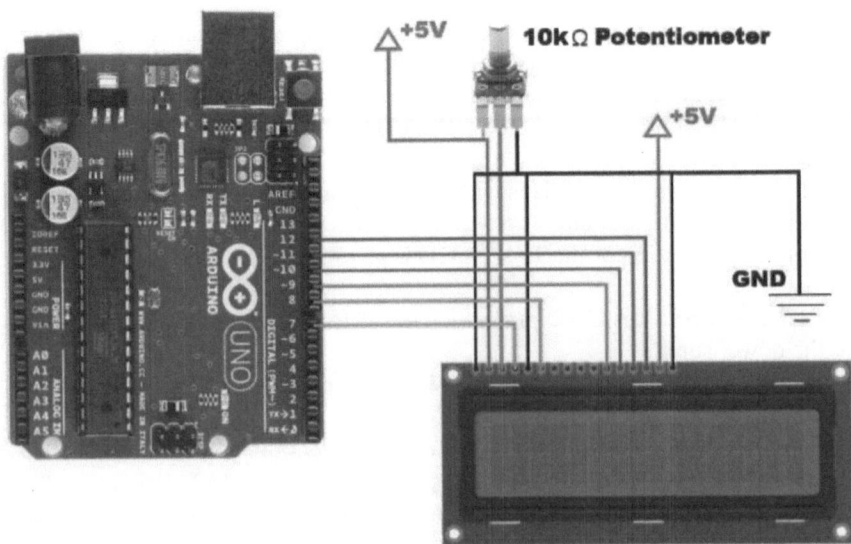

FIGURE 11.21 Schematic diagram of interfacing of a 16×2 LCD display with the Arduino UNO.

- Pin 8 → Unconnected
- Pin 9 → Unconnected
- Pin 10 → Unconnected
- Pin 11 → Pin 9
- Pin 12 → Pin 10
- Pin 13 → Pin 11
- Pin 14 → Pin 12
- Pin 15 → +5 V
- Pin 16 → GND

11.3.6 KY-008 LASER EMITTER

The laser emitter is a device used to produce the laser beam. The laser is very much useful due to its good directivity as well as high-energy concentration features. The applications of the laser are in medical treatment, military, and many other key areas. Its other applications include robotics and drones, laser-pointing robots, etc.

The technical specifications of the laser-emitter module are:

- Operating power supply: 5 V DC
- Operating current: < 40 mA
- Operating temperature: –10°C–40°C
- Output wavelength: 650 nm
- Output power: 5 mW

The schematic diagram in Figure 11.22 represents the connections of the Arduino UNO and KY-008 laser emitter. The Vcc pin of the laser is attached to the +ve terminal

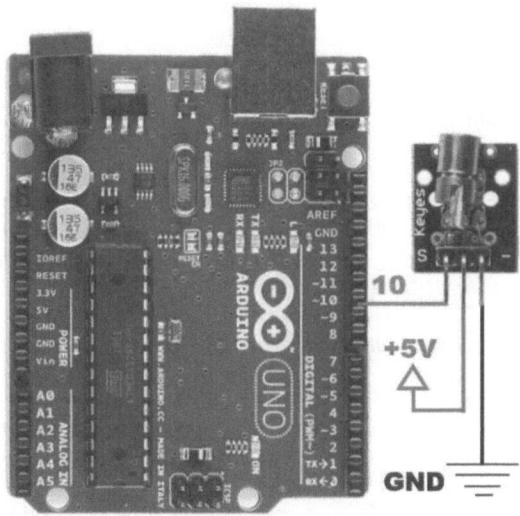

FIGURE 11.22 Schematic diagram of interfacing of KY-008 laser emitter with the Arduino UNO.

of the power supply (5 V) and the Gnd pin of laser is attached to the –ve terminal of the power supply. The input of laser is attached to pin number 2 of the Arduino.

11.4   CONCLUSION

The chapter covers some of the widely used sensors and output devices that are incorporated in a variety of different applications in the field of IOT and robotics. We have seen a few of them interfacing with the widely supported development board Arduino UNO. The sensors and output devices come in a variety of ranges depending upon their applications, whether personal or industrial, technical specifications, and other factors such as cost. In the field of robotics and drones, the actuators and different output devices have made a huge impact on the society, allowing them to make more compact and reliable robots and drones with greater functionality and degrees of freedom. The development in the field of sensors and output devices is still in the research and development phase and many tech-giants, scientists, and researchers are trying to make more technically powerful electronic sensors and output devices, which can offer not only low complexity and cost but also higher precision and reliability, which can lead to an increase in the power and functionality of robots and drones in the future.

REFERENCES

1. Wadhwani, S., Singh, U., Singh, P., and Dwivedi, S. (2018). Smart home automation and security system using Arduino and IOT. *International Research Journal of Engineering and Technology (IRJET)*, 5(2), 1357–1359.
2. Kumar, N.S., Vuayalakshmi, B., Prarthana, R.J., and Shankar, A. (2016, November). IOT based smart garbage alert system using Arduino UNO. In *2016 IEEE Region 10 Conference (TENCON)* (pp. 1028–1034). IEEE.
3. Patnaik Patnaikuni, D.R. (2017). A comparative study of Arduino, raspberry pi and ESP8266 as IoT development board. *International Journal of Advanced Research in Computer Science*, 8(5), 1–2.
4. Vimal, P.V., and Shivaprakasha, K.S. (2017, July). IOT based greenhouse environment monitoring and controlling system using Arduino platform. In *2017 International Conference on Intelligent Computing, Instrumentation and Control Technologies (ICICICT)* (pp. 1514–1519). IEEE.
5. Nayyar, A., and Puri, V. (2016, March). A review of Arduino boards, Lilypad's & Arduino shields. In *2016 3rd International Conference on Computing for Sustainable Global Development (INDIACom)* (pp. 1485–1492). IEEE.
6. Mahalakshmi, G., and Vigneshwaran, M. (2017). IOT based home automation using Arduino. *International Journal of Engineering and Advanced Research and Technology*, 3(8), 1–6.
7. Rana, A.K., Krishna, R., Dhwan, S., Sharma, S., and Gupta, R. (2019, October). Review on artificial intelligence with internet of things-problems, challenges and opportunities. In *2019 2nd International Conference on Power Energy, Environment and Intelligent Control (PEEIC)* (pp. 383–387). IEEE.
8. Rana, A.K., and Sharma, S. (2021). Contiki Cooja Security Solution (CCSS) with IPv6 routing protocol for low-power and lossy networks (RPL) in Internet of Things applications. In *Mobile Radio Communications and 5G Networks* (pp. 251–259). Springer, Singapore.

9. Rana, A.K., and Sharma, S. (2019). Enhanced Energy-Efficient Heterogeneous Routing Protocols in WSNs for IoT Application.

10. Ahmed, E., Islam, A., Ashraf, M., Chowdhury, A.I., and Rahman, M.M. (2020). Internet of Things (IoT): Vulnerabilities, Security Concerns and Things to Consider.

11. Veeramanickam, M.R.M., and Mohanapriya, M. (2016). IOT enabled futurus smart campus with effective e-learning: i-campus. *GSTF journal of Engineering Technology (JET)*, *3*(4), 8–87.

12. Cho, S.P., and Kim, J.G. (2016). E-learning based on Internet of Things. *Advanced Science Letters*, *22*(11), 3294–3298.

13. Kumar, A., and Sharma, S. Demur and routing Protocols with application in underwater wireless sensor networks for smart city. In *Energy-Efficient Underwater Wireless Communications and Networking* (pp. 262–278). IGI Global.

14. Abbasy, M.B., and Quesada, E.V. (2017). Predictable influence of IoT (Internet of Things) in the higher education. *International Journal of Information and Education Technology*, *7*(12), 914–920.

15. Kumar, A., Salau, A.O., Gupta, S., and Paliwal, K. (2019). Recent trends in IoT and its requisition with IoT built engineering: A review. In *Advances in Signal Processing and Communication* (pp. 15–25). Springer, Singapore.

16. Charmonman, S., Mongkhonvanit, P., Dieu, V., and Linden, N. (2015). Applications of internet of things in e-learning. *International Journal of the Computer, the Internet and Management*, *23*(3), 1–4.

17. Vharkute, M., and Wagh, S. (2015, April). An architectural approach of internet of things in E-Learning. In *2015 International Conference on Communications and Signal Processing (ICCSP)* (pp. 1773–1776). IEEE.

18. Li, H., Ota, K., and Dong, M. (2018). Learning IoT in edge: Deep learning for the Internet of Things with edge computing. *IEEE Network*, *32*(1), 96–101.

19. Dalal, P., Aggarwal, G., and Tejasvee, S. (2020). Internet of Things (IoT) in Healthcare System: IA3 (Idea, Architecture, Advantages and Applications). *Available at SSRN 3566282.*

20. Rana, A.K., and Sharma, S. Industry 4.0 manufacturing based on IoT, cloud computing, and big data: Manufacturing purpose scenario. In *Advances in Communication and Computational Technology* (pp. 1109–1119). Springer, Singapore.

21. Wang, Q., Zhu, X., Ni, Y., Gu, L., and Zhu, H. (2020). Blockchain for the IoT and industrial IoT: A review. *Internet of Things*, *10*, 100081.

22. Rana, A.K., Salau, A., Gupta, S., and Arora, S. (2018). A Survey of Machine Learning Methods for IoT and their Future Applications.

23. Kumar, K., Gupta, E.S., and Rana, E.A.K. Wireless Sensor Networks: A Review on "Challenges and Opportunities for the Future world-LTE".

24. Sachdev, R., (2020, April). Towards security and privacy for edge AI in IoT/IoE based digital marketing environments. In *2020 Fifth International Conference on Fog and Mobile Edge Computing (FMEC)* (pp. 341–346). IEEE.

25. Simic, K., Despotovic-Zrakic, M., Đuric, I., Milic, A., and Bogdanovic, N. (2015). A model of smart environment for e-learning based on crowdsourcing. *RUO. Revija za Univerzalno Odlicnost*, *4*(1), A1.

26. Kim, T., Cho, J.Y., and Lee, B.G. (2012, July). Evolution to smart learning in public education: A case study of Korean public education. In *IFIP WG 3.4 International Conference on Open and Social Technologies for Networked Learning* (pp. 170–178). Springer.

12 Fault Detection in Robotic Arms

Tarun Jaiswal, Manju Pandey, and Priyanka Tripathi
Department of Computer Applications,
NIT Raipur, Raipur, India

CONTENTS

12.1 INTRODUCTION

Robotic arms or artificial limbs can be substituted for human body parts which have been lost because of the illness, trauma, or other reasons. The robotic arms or artificial limbs that a person could use basically depends on various factors, such as the cause of limb loss/amputation and the positioning of lost limbs [1]. These non-natural parts have been used for a long time; these include parts such as wooden legs, metal arms, hooks for hands, etc. They provide some resemblance of human limbs, but they were often uncomfortable, difficult to use, and they also had poor functionalities and were very unattractive, as well as they do not inherit the fault identification

DOI: 10.1201/9781003181613-12

capabilities. Today, we have developed lighter, smaller, better-controlled, more efficient, and attractive limbs. The robotic arms we discuss use the current technology and combine diverse fields, such as electronic computing, that are interdisciplinary sciences, and this combination provides precision and functionality, such as an artificial human arm. However, these artificial limbs provide the functionalities, features, and motion similar to natural human limbs, but these artificial limbs or robotic arms are more complicated than it sounds. For example, a robotic arm might need to move its forearm so that it is in the proper place to pick up objects; the person may need to switch their forearm to the desired position or angle so that fingers can grasp that object without damaging it [2]. For instance, we can imagine constructing a robotic arm or artificial limb that can do such functions quickly and without interruption or fault; this is a challenging task. Unfortunately, if a fault exists in this system, then identifying such faults is a challenging task since the system compromises various components. If we know which component is malfunctioning, then it will be easy for technicians or a person to rectify it in a timely fashion so that a person's work is not interrupted [3]. Similarly, when a person feels pain, then the nervous system and brain identify the exact place of pain analogously and the proposed system also tries to identify the malfunctioning location. In that case, it will take a moment to consider how you will actually do this and also consider the worst-case where system will experience some malfunctioning. In case of the developing countries such as those situated in Asia like India, these prosthetic limbs are too expensive and if some person has access to this system, then he needs to exercise extreme care while using this so that it's functioning is proper, and in case of some mishappening in it, it will cost a huge amount of money to identify the fault in it.

A robotic arm is nothing but an electro-mechanical device (a robot manipulator), which can be programmed to perform the function or work similar to the human arm. In other words, robotic arms mimic the human hand's functionality, and they perform similar functions that human hands can do. A comparison of a human arm with a robo-arm is seen in Figure 12.1.

A robotic arm is constructed with a microcontroller to govern the motion of the arm, and it also contains servomotor that provides the task for the arm, and through the gripper, arm can pick or hold any object.

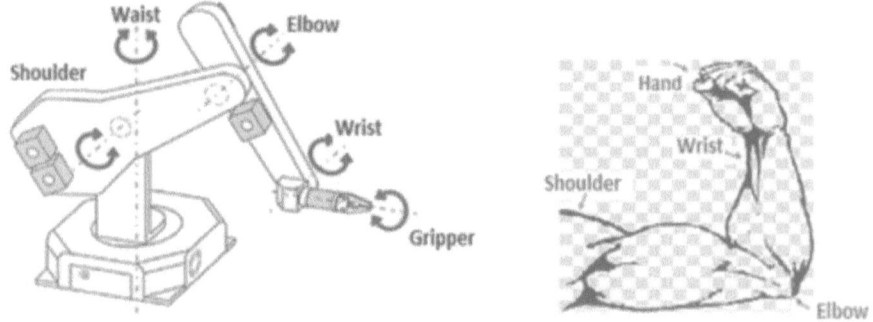

FIGURE 12.1 Comparison of robot arm and human arm.

Since the arm is constructed very precisely and it also contains the dedicated part such as tiny diameter wire and other electrical components. So, if there is any fault in this system, then it is difficult to identify the failure in the system, i.e., arm; integral wiring and components are more critical for working of the whole system.

For example, if this system is deployed for amputees, then its proper working is necessary since it is crucial for that person's daily life, and any fault in the system will severely impact their day. Thus, finding the fault in the system is very important to work in such system.

For constructing a robotic arm that is governed according to human arm movement, and for that information acquired by the variable resistor, fault detection is proposed. For building this system, we used the variable resistor (potentiometer) and the ATmega328p microcontroller for data processing [4]. Ultimately, This robotic arm will be capable of tackling a wide range of real-world tasks, including holding or picking up hazardous objects such as explosives, which are extremely dangerous to the user. The system should allow for more natural human-computer interconnection and a concise manipulation of the robotic arm. The combination of human and robot is very helpful in the context of assisted living, where some use-case plan was introduced [5].

Ever since robotics became a subject of science, so for the last twenty years, research and development has been focused only on static robotic manipulators, due to their industrial applications. Treaded and legged locomotions have been studied, in which the numbers of robots that are equipped with wheels for their movements are more [6].

The robotic arm is attached to such robots so that they are capable of performing a variety of tasks like grasping and placing. These robotic arms have some degree of freedom that involves kinematics and dynamic analysis [7].

A robotic arm is a confluence of diverse fields that includes many areas such as electrical, mechanical, computer science, math and science, electronics, etc. Robots are extremely important and necessary in various manufacturing industries. Apart from this, if they are programmed to do any work, robots can perform tasks continuously without stopping and with high precision that eliminates the need of humans in many cases [8]. We need actuators in the robotic arm in order for it to act like a human arm. Actuators are devices that convert electrical energy into physical motion [9].

Robotic arms pave the way for people who have severe problems related to motion, such as people with various disabilities [10].

The robotic arm uses a servomotor for its operation, since this servomotor is lightweight and provides superior operation. To mimic the human hand, we connect this to the clamp, and with the help of this, we construct the proposed robotic arm. We use grippers to pick up items and place them somewhere else; this gripper is called the end-effector [11].

The proposed system provides the functionality of a robot arm as well as identifies the fault in the system. The fault is also shown or displayed via cloud server to the mobile application, which is, in the proposed system, an IoT MQTT Panel App.

For fault identification purposes, we use the ESP8266 Wi-Fi module, and this module uses the feedback path of the system. If there is any fault in the system, such

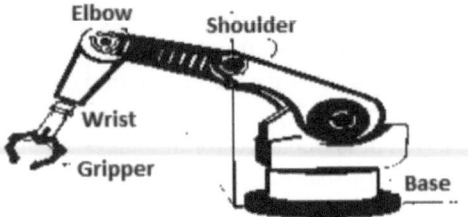

FIGURE 12.2   Simple robot arm.

as wire breakage, then it is displayed on the mobile application. A robotic arm consists of several parts as shown in Figures 12.2 and 12.3 for the accurate and proper functioning of the system.

The parts used in the proposed robotic arms are:

- Controllers
- Arms
- End-effectors
- Drivers
- Sensors

12.1.1   CONTROLLERS

The controller is the central component of the arm, and this controller behaves as the brain for the whole system. This controller can be programmed to perform automatically, or it can get direction from the technician to work in manual mode. The controller can be chosen according to the processing power requirement. Some controllers are very intricate and large, and they are deployed basically in massive

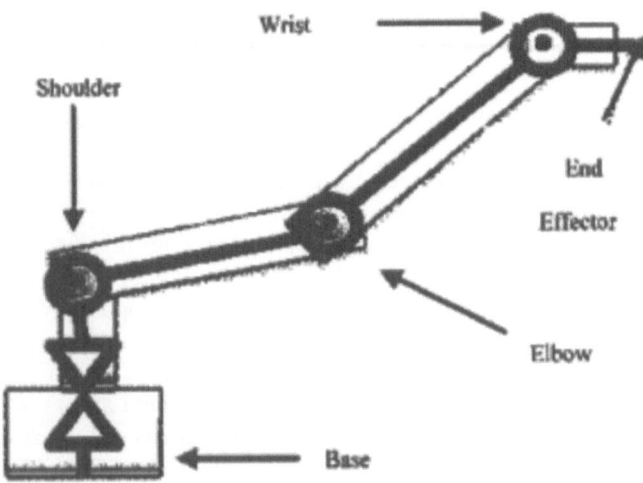

FIGURE 12.3   Free body diagram of the robot arm.

manufacturing works. In contrast, others are the straightforward types that are used in science projects, or in game consoles, etc.

12.1.2 ARMS

For the robotic arm system, the arm is the most important section or part, and it comprises three main parts: shoulder, wrist, and elbow. These are all joints. Where the shoulder rests on the base of the arm, it is attached to the controller, and it can move back and forth and also spin. The elbow is situated in the middle thus allowing the arm's upper section to move frontward or backward freely of the lower part. Finally, the wrist is situated at the end of the upper arm and attaches to the end effector.

12.1.3 END-EFFECTORS

End-effectors are also known as the hand of the robotic arm. This end-effector consists of two or more claws, and this operates through the command. This end-effector provides the rotational movement, so that movement of objects and material is quite easy.

12.1.4 DRIVERS

Drivers are basically motors located between joints. They provide control movement and maneuvers.

12.1.5 SENSORS

Sensors are internal parts of the most advanced robots. Some of the robot sensors allow the system to sense their environment and respond accordingly. For example, a sensor can detect the temperature of a car cabin and adjust the temperature of the car to suit the driver's choice.

The links of systems are combined by the joints, and these joints provide the maneuvering capabilities to the arm, and the links these manipulators make a kinematic chain [12]. The business end of the kinematic chain of the manipulator is called the end effector and it is very similar to the human hand, as shown in Figure 12.4.

End-effectors can be constructed in such a way that they can be used for various purposes, such as gripping objects, performing welding tasks, and other tasks, etc. The robotic arm can be controlled with the help of a human, or it may be self-governing so that it can perform the various tasks with exceptional accuracy. The robotic arm system may be mobile, or it may be static, and these can be well-suited for either domestic or industrial use. This chapter deals with a system that uses a servomotor and potentiometer to mimic human hand movement. This method of control provides greater flexibility in controlling the robotic arm versus using a controller where each actuator is controlled individually. The processing unit takes care of each actuator's control signal according to the inputs from the potentiometer, in order to replicate the movements of the human arm [4].

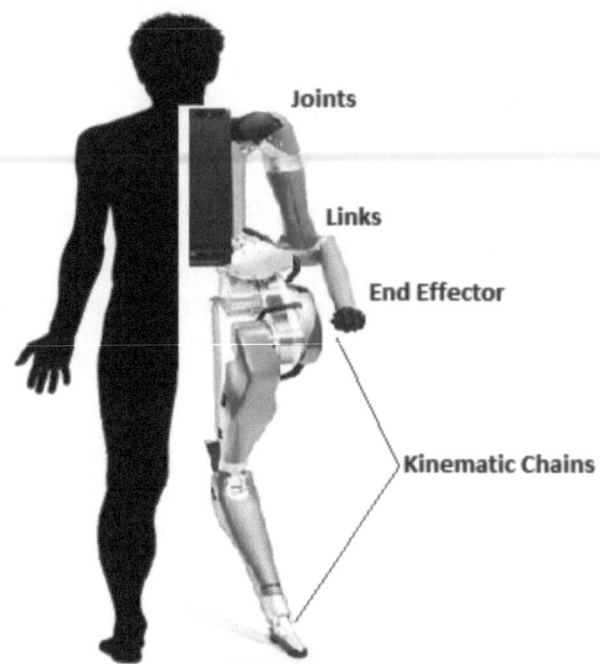

FIGURE 12.4 Illustration of joints, links, end-effectors, and kinematic chains.

12.2 RELATED WORKS

Fault detection in any system, particularly in a robotic arm, is a major concern because if this malfunctioning is not identified or detected, it can have serious consequences for the system and person [13]. Deploying the ESP-8266 module for this system fulfills the requirement of fault detection in the system. Various methods have been implemented in this regard for this purpose in the IoT scenario.

Kinematics as well as dynamics are major parameters for study of manipulators that are controlled automatically. These manipulators or robotic arms are very useful for various kinds of tasks, such as object handling or in a lab. The authors used the robotic arm with five degrees of freedom and **Denavit–Hartenberg (DH)** parameters for representation of joints and links [14].

The authors [15] presented the manipulator with two degrees of freedom for object handling in small industries. The authors also performed the synthetization over dynamic dimensional and inverse kinematics, as well as rigid body dynamics that offer an index of dynamic performance. This index is associated with lowering the high-driving force of single links.

In [16], the author presented two methodologies for fault identification in autonomous robots. The first approach is a method for synthesizing software components that let a robot identify the fault in itself. In the second method, the author studies a method that permits multi-robot systems to detect fault in each other.

In [17], the robotic arm PUMA 560 is used as the case study for combined design. This combined design gives the generic setting to study essential transactions existing in a concurrent control and diagnostic system. The authors also offered favorable outcomes for sudden payload variation by seeing them as a problem of failure diagnostic.

In 2019, Cho et al. [18] proposed a fault detection and isolation algorithm in their work. They used an artificial neural network (ANN) for fault detection. and this method provides an efficient way to identify and diagnose the fault. They find that their fault detection and identification method increases the reliability of robotic systems. The damage to robots as well as injuries to the human is also reduced if fault, malfunctioning, or collision is detected early. The author proposed a neural network–based fault detection and identification algorithm. This algorithm is very efficient and does not require supplementary sensors and extra understanding of the system.

Myint et al. [19] developed a robotic arm for object grasping and placing. In their work, they used microcontroller for generating pulse with modulation for controlling servos. The authors investigated the inverse kinematics of the three-degrees-of-freedom robot. They presented the mathematical model, which is used for solving the joint angles.

The authors in [20] surveyed the various methods of fault detection and fault tolerance in robotic systems, and they also highlighted robotics controls and kinematics. They also identified that fault tolerance in robots is a rapidly growing area of research, especially due to movement of the robots in isolated or hazardous areas, so the aim is to develop a robot that is fully sustainable and can survive harsh environments without failure.

In [21], the authors described robotic arm implementation; this robotic arm can be used for placing the object as well as for picking also. The proposed arm can work for several hours, and it needs less power. This designed arm is best suited for operations in dangerous environments.

12.3   METHODOLOGY

The robotic arm can be controlled with the help of various available technology such as WiFi, Bluetooth, etc., but when the robotic arm is involved in any operation, then angles made by the robotic arm must be controlled accurately and effectively. In this proposed system, we used the variable resistor also called the potentiometer.

This potentiometer is used to acquire the angle made by the arm system. In addition to this, we also employed the ESP8266 module; this module is used to sense the status of the system at all times, and it keeps an overall eye on the system. If any faults occur in the system (in this case, the robo arm), then it notifies to the technician, as shown in Figure 12.5. This allows preemptive action to be taken at the correct time so that fault can be corrected instantly.

The proposed system detects and identifies the fault as well as notifies it to the concerned person. In this proposed system, we have servomotor, which controls the arm of the system, and this motor is controlled through the positioning of the variable resistor. The variable resistor reads this angle and transfers the information to the microcontroller, which in turn governs the movement of the arm. In a similar way, feedback from the system is supplied to the ESP module to keep track of the status of the system. And it continuously senses the whole system for work-worthy condition and proper functioning. If, for example, any wire is broken in the system, then it immediately shows the message to the concerned person through the mobile application, which in the proposed system is the IoT MQTT Panel App.

The flowchart of the proposed model is illustrated in Figure 12.6.

FIGURE 12.5   Fault detection in robotic arm.

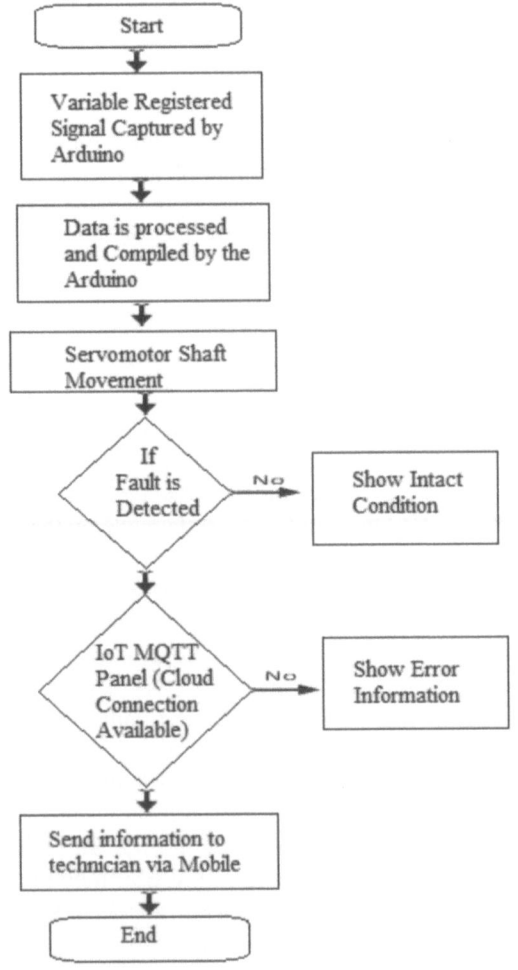

FIGURE 12.6   Flowchart for the system.

The procedure includes: (a) the Arduino captures the generated signals; (b) data is processed and compiled by the Arduino (microcontroller); (c) according to the potentiometer movement, the servomotor shaft is positioned; (d) the fault is identified in the system, and if this condition is true then the system proceeds further, and if not, then it displays the intact condition; and finally (e) the cloud connection is available and if this condition is true then information is sent to the concerned person. If not, then the system displays the error.

Further details of the system are described next.

12.4   SYSTEM ARCHITECTURE

12.4.1   VARIABLE RESISTOR SIGNAL PROCESSING

We have an important parameter, i.e., joint angle, which we identify and analyze. For this purpose, we can use the variable resistor at that particular joint position. A variable resistor is essentially a voltage-divider, and it is used for computing the voltage. The voltage variation can be obtained with the help of a variable resistor as a simple formula, i.e., $r * v$, which is applied on the variable resistor. When we use the variable resistor, which is rotatory, the internal resistance is obtained by turning the variable resistor handle from the initial value. This internal resistor is used to obtain the approximate angle established by the variable resistor. This variable resistor is in the form of knob, and it is connected to the system with the joint of the proposed system.

12.4.2   DATA PROCESSING BY THE ARDUINO

The Arduino (microcontroller ATmega328p) is the board that is equipped with the microcontroller, and it also has the digital input/output pin, as well as an analogue pin, so that various other components and expansion boards can be attached to it.

This system also provides the serial communication functionality, and the program can be uploaded into the Arduino board via USB. The program, which is written in the Arduino IDE environment, is known as Sketch. And its writing is very similar to C++ [22].

As shown in Figure 12.7, the signal that is obtained from the variable resistor is fed to the embedded system that is the Arduino for further processing. The variable resistor value is changed according to the movement of the hand, i.e., as the hand moves, the variable resistor values change accordingly [23]. When this internal gain is found out, this value is rendered by the embedded system (in the proposed case, it is Arduino). In Arduino UNO, these values are broken down into 1024 levels of

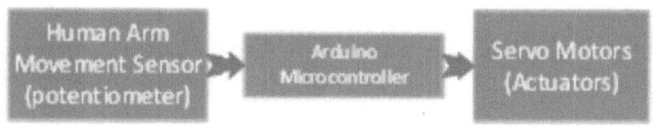

FIGURE 12.7   Information-processing by Arduino.

voltage (with 0–1023), and we call this process mapping. The Arduino measures the values from 0–5 V and 0–1023. The values now can be mapped from 0 to 180. These values can be considered as the angle turned by the potentiometer from the zero levels.

12.4.3 Servomotor Control

The Tower Pro SG90 motor obtains its digital input through the Arduino microcontroller. The position of the servomotor changes according to the position of the variable resistor. This servomotor has its own feedback mechanism to govern its position and movement. All of this is obtained through the encoder. In the absence of this encoder, we have to calibrate the motor whenever we perform any operation on it [24].

The encoder gets its input from the feedback of the motor, which causes an error if the current position of the motor is not the same as the desired position. This helps in deciding the final position of motor in which we want to place it.

12.4.4 ESP8266 Module

The ESP8266 module is the system's on-chip module. This module is used for IoT applications, but while working with this module, caution must be exercised because it runs on 3.3 V and any access voltage above it would kill it. So to reduce or lower this power supply, we can use the IC that converts 5 V to 3.3 V, or we can opt out the AMS1117 regulator. Table 12.1 shows the pin configuration of the module.

TABLE 12.1
ESP8266 Pin Configuration

Pin Number	Pin Name	Alternate Name	Normally Used for	Alternate Purpose
1	Ground	–	Connected to the ground of the circuit	–
2	TX	GPIO – 1	Connected to Rx pin of programmer/u C to upload program	Can act as a general-purpose input/output pin when not used as TX
3	GPIO-2	–	General-purpose input/output pin	–
4	CH_EN	–	Chip enabled–Active high	–
5	GPIO – 0	Flash	General purpose Input/output pin	Takes module into serial programming when held low during startup
6	Reset	–	Resets the module	–
7	RX	GPIO - 3	General purpose Input/output pin	Can act as a general-purpose Input/output pin when not used as RX
8	Vcc	–	Connect to +3.3 V only	–

It is a low cost and cheap device used for the accurate transmitting and receiving of the messages from both sender and receiver. In this proposed system, we take connection (via wire) from the actuators and connect it to the input and out-pin of the Wi-Fi module so that we can continuously sense the status of the system. And if there is any blockage in the wires, then it immediately identifies them and sends the message to the concerned person.

12.4.5 ESP8266 PIN CONFIGURATION

The ESP8266 pin configuration is shown in Table.12.1.

12.5 DETERMINING THE POSITION OF THE ARM USING KINEMATICS

In robotic arms, the location of the arm is most vital since it requires precision and accuracy to obtain the required position. For that purpose, the desired joint position and its related angles are more important to govern the position of the arm, and it's governed by the kinematics [25]. Forward kinematics in the robotic arm gives all the angles or translation degrees of motion of the arm, such as all the motors and actuators locations, and the combination of those positions leads to the gripper position. So, forward kinematics will go from your joint space to the physical position in the system's arm, and inverse kinematics will go from your joint space to the physical position in the system's arm. If you want your robot to be in a specific position and orientation in the world, you'll need to figure out how to position all of your motors [26].

Inverse kinematics is a more difficult problem than forward kinematics [27]. Hence, it involves some sort of iterative algorithm and functionalities. MATLAB makes it easy, and we are here first to define the variables for link length, such as $L-1 = 20$, $L-2 = 50$, $L-3 = 40$, which are defined for the link length. It determines the length of the link and then we proceed for the DH parameter [26].

We set this parameter for each link. For that, we use the Link command and enter the DH parameter in the form of a vector. Here, the parameter ordering is more important for the link; then we proceed for the type of join; by default, the join is the rotation join. And here, we add the type of join after adding the required parameter. Now, we define L as a vector by typing $L(1)$, which means the first element in vector L; then $L(2)$, $L(3)$ for the second and third link, and so on. We start with the first link, $L(1)$, and then enter the DH parameter first, which is theta; theta gives its initial value as zero, and MATLAB will make its join variable. Then we have the second parameter, which in the proposed scenario is $L-1$, and the third parameter for the first link is zero, and finally, the third parameter is pi/2 (since MATLAB takes angles in radians). Similarly, we do the same thing for the rest of the parts. We define the value of $L-1$, $L-2$, $L-3$ in the first line of code, then we connect all links in vector L using the command called Serial Link. The MATLAB command creates an object variable of this system and passes the vector L to the command Serial Link. Now we have the object for the robot (system) in MATLAB (as shown in Figure 12.8) and we have the DH table as output (as shown in Table 12.2). The first

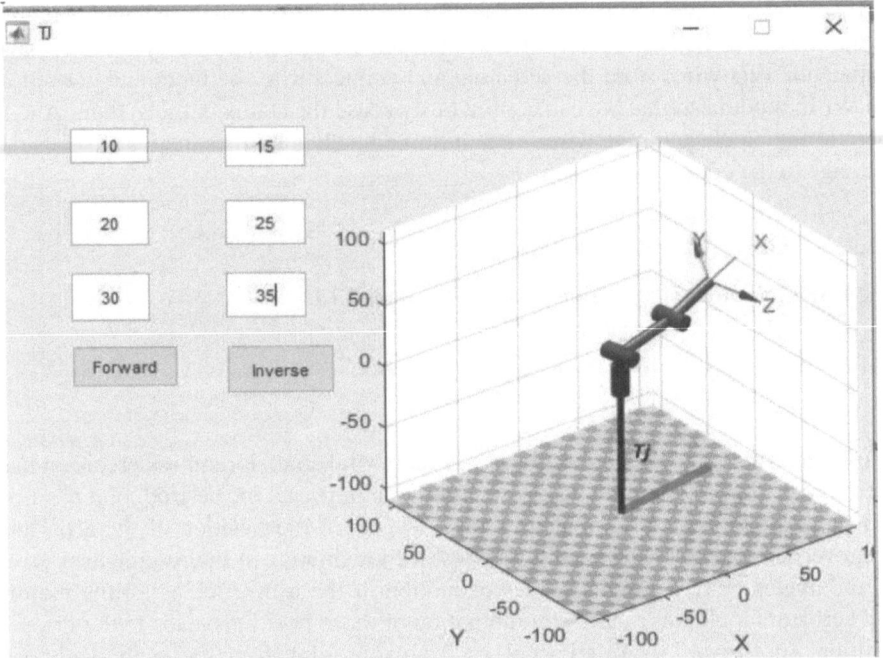

FIGURE 12.8   Simulation of robotic arm.

column is theta, and for each link, we have q_1, q_2; this means that we have obtained the required join variable.

$$
\begin{pmatrix}
\cos\theta & -\sin\theta\cos\alpha & \sin\theta\sin\alpha \\
\sin\theta & \cos\theta\cos\alpha & -\cos\theta\sin\alpha \\
0 & \sin\alpha & \cos\alpha
\end{pmatrix}
\begin{pmatrix}
r\cos\theta \\
r\sin\theta \\
d
\end{pmatrix}
\\
\overline{\qquad\qquad 0\ 0\ 0 \qquad\qquad\qquad 1}
\tag{12.1}
$$

After obtaining the required parameter, we can use the following equation to calculate the forward kinematics for proposed system.

TABLE 12.2

Showing the Relationship between Link and *DH* Parameter

Link	Theta	D	a	Alpha
1	0	L − 1	0	Pi/2
2	0	0	L − 2	0
3	0	0	L − 3	0

TABLE 12.3
Showing the Forward Kinematics

0.9545	−0.1875	0.0989	1.9187
0.0978	−0.0198	−0.995	0.1977
0.1987	0.9801	0	0.1987
0	0	0	1

Here, we multiply all of the matrices together, starting with the first joint till the end effector, and we acquire the final matrix that will contain the orientation (position) of the end-effector. This will govern the robotic arm position to hold any object. Table 12.3 shows the values for the forward kinematics.

For forward kinematics, we obtain the following forward kinematic equation:

$$\text{Robot.fkine}\left(\left[\text{Th-1 Th-2 Th-3}\right]\right);$$

For a given join angle, for example, radian $0.2(L-1.A(0.2))$, we have the transform matrix shown in Table 12.4.

12.6   FORCE NEEDED TO GRIP AN OBJECT

There are many types of forces that act on the body; when we raise an object with a robotic arm, one of the main forces is frictional force. The gripping surface of the robotic arm is made of a soft material, and its frictional coefficient is quite high, so as not to cause object damage. The robotic arm should bear the forces in the following subsections.

12.6.1   OBJECT'S WEIGHT

Acceleration and motion (produced by the movement of the object) are two forces. The given formula is used to detect the force that may be used to catch or grip an object.

$$F = \frac{ma}{\mu n} \tag{12.2}$$

TABLE 12.4
Showing the Transform Matrix for the System

0.9801	0	0.1987	0.9801
0.1987	0	−0.9801	0.1987
0	0	1	0
0	0	0	1

where:
- F = force needed to grip the object
- M = mass of object
- a = object acceleration
- μ = friction coefficient
- n = total number of fingers in the gripper

A complete equation would account for the direction of movement. For instance, whenever the body moves upward, against gravity force, the force required will be more than that towards the gravitational force. Hence, one more term is introduced, and the formula becomes:

$$F = \frac{m(a+g)}{\mu n} \tag{12.3}$$

In this formula, g indicates acceleration due to gravity, which is caused by movement. For several tangible cooperative manipulation tasks, like writing, holding a wrench, etc., a task-related grasp measure may be applied to pick grasps that are suitable to meet the precise task requirements.

12.6.1.1   The General-dynamic Algorithm for the Robotic Arm

The following expression [8] gives the general algorithm, which describes the manipulator equations of motion for the torque or force F_i acting at joint i :

$$\sum_{i=0}^{n}\left\{\sum_{i=0}^{n}\left[\text{Trace}\left(U_{jk}J_jU_{ji}^{T}\right)\ddot{q}_k\right]\sum_{k=1}^{j}\sum_{p=1}^{j}\left[\text{Trace}\left(U_{jkp}J_jU_{ji}^{T}\right)\dot{q}_k q_p\right]-m_jGU_{ji}P_j\right\} \tag{12.4}$$

$$= F_{i=1,2,2,\ldots,n}$$

Where T (superscript) describes the transpose matrix Uji.

- F_i = torque force performing on the joint i
- q_i = joint variable $i = 1\ldots, j\ldots, k\ldots$, and n, where n represents DOF
- $qi'. qi$ = acceleration and velocity of joint variable i

The constructing blocks $m\,i, pj, G, Uji, Ujkp,$ and Jj of Eq. (12.1) are defined as follows:

- m_j = mass of body j in the chain of n bodies (links)
- Pj = mass center vector of the body (link)j in the coordinate system; fixed in the same body, given as a 4×1 vector with components
- G = acceleration of gravity
- U_{ji} = first fractional derivative of the T_0^j transformation matrix with respect to qi

12.7   IMPLEMENTATION AND RESULTS

In this chapter, we have a potentiometer and servomotor. The movement of the variable resistor controls the servomotor. The potentiometer is tuned from an initial 0° position up to 180° angle position. This helps the variable resistor give out a range of values, which are eventually mapped from a 0° to 180° angle.

The data is afterwards obtained and sent to the Arduino UNO microcontroller for further processing of the uncooked data. The angle of the robotic arm servomotor is according to the movement of the potentiometer. The desired position of the servo is to send in the form of a pulse-width modulation (PWM) signal by the microcontroller.

A PWM signal is an electrical signal of which the voltage periodically generates pulses. The servo position width is determined by the pulse width. The Command () function is used to change the width of pulses, thereby we can change or obtain the desired position for the servo. For steering, the servomotor, PWM signal is used, and it has period 20 ms, and the pulse width varies between 0.7 ms and 2.3 ms. By moving the hand, we rotate the variable resistor. In this way, we give variable voltage to the ADC channels of Arduino UNO. Therefore, the digital value of Arduino UNO is under the control of the user.

These digital values are mapped so that the position of the servomotor can be adjusted, thus the servo position is under control of the user. By the movement of the hand, the user or person can change the place of the joint of the robotic arm and thereby can grasp or pick up any object. We also give the connection of the servo signal to the Wi-Fi module that continuously monitors the signal value. In other words, we provide feedback of the signal to ESP8266 so that if there is any wear in the wire or the wire is broken, then it immediately sends a signal to the technician.

12.8   CONCLUSION

This chapter proposed fault identification in robotic arms using the ESP-8266 WiFi module. The solution is to identify the fault in the robotic arm using the ESP-8266 WiFi module. The robot arm is constructed with servomotors that provide links between arms and achieve the arm movements. A microcontroller is responsible for driving the servomotor. Programming for the microcontroller is done using the Arduino IDE. A potentiometer is used to identify the angle of rotation, and its signals are sent to the UNO microcontroller. Thus, we can control the movement of the robotic arm. In this system, we have used the robotic arm, and the objective of this robotic arm, which can be controlled via the potentiometer, has been achieved. We also identify the fault in the arm system with the help of ESP8266 module.

The ESP8266 module is used to monitor the arm system in case any unwanted condition occurs, such as a broken wire or system broke condition. If so, then it sends a message to the concerned person. The proposed system not only mimics human arm movement, but also it identifies the fault in the system.

REFERENCES

1. Munjed, M., Ridgewell, E.: Bionic Limbs – Curious. pp. 1–10 (2016), https://www.science.org.au/curious/people-medicine/bionic-limbs
2. Fletcher, M. J.: Problems in design of artificial hands. Orthopedic & Prosthetic Appliance Journal, 9(2), 59–68 (1955).
3. McDermid, W., Black, T., Gamblin, R.: Repair of a damaged 300 MVA machine. 2010 IEEE International Symposium on Electrical Insulation, San Diego, CA, pp. 1–4. (2010).
4. Mathew, E. B., Khanduja, D., Sapra, B., Bhushan, B.: Robotic arm control through human arm movement detection using potentiometers. In 2005 International Conference on Recent Developments in Control, Automation and Power Engineering (RDCAPE), pp. 298–303 (2015).
5. Yusoffa, M. A. K., Saminb, R. E., Ibrahimc, B. S. K.: Wireless mobile robotic arm. Procedia Engineering,. 41, pp. 1072–1078 (2012). https://doi.org/10.1016/j.proeng.2012.07.285.
6. Kadir, W. M. H. W., Samin, R. E., Ibrahim B.S. K.: Internet controller robotic arm. Procedia Engineering. 41, pp.1065–1071 (2012). https://doi.org/10.1016/j.proeng.2012.07.284.
7. Carignan, C. R., Gefke, G. G., Roberts, B. J.: Intro to Space Mission Design: Space Robotics. Seminar of Space Robotics, University of Maryland, Baltimore, 26 March 2002, pp. 16–25 (2002).
8. Olwan, O., Matan, A., Abdullah, M., Abu-Khalaf, J.: The design and analysis of a six-degree of freedom robotic arm. In 2015 10th International Symposium on Mechatronics and its Applications (ISMA), ISMA, pp. 1–6 (2015).
9. Wang, R. J., Zhang, J. W., Xu, J., Liu, H.: The multiple-function intelligent robotic arms. In 2009 IEEE International Conference on Fuzzy Systems, Jeju Island, IEEE, pp. 1995–2000 (2009).
10. Uehara, H., Higa, H., Soken, T.: A mobile robotic arm for people with severe disabilities. In 2010 3rd IEEE RAS & EMBS International Conference on Biomedical Robotics and Biomechatronics, Tokyo, pp. 126–129 (2010).
11. Jegede, O., Awodele, O., Ajayi, A.: Development of a Microcontroller-Based Robotic Arm. In Proceedings of the Computer Science and IT Education Conference, pp. 549–557 (2007).
12. Rakotomanana, L. R.: Geometry and kinematics. In A Geometric Approach to Thermomechanics of Dissipating Continua. Progress in Mathematical Physics, vol 31. Birkhäuser, Springer, Boston, MA,, pp. 5–38 (2004). https://doi.org/10.1007/978-0-8176-8132-6-2.
13. Isogai, M., Arai, F., Fukuda, T.: Study on fault identification for the vibration control system of flexible structures (fault detection of sensors and actuators). In 2000 IEEE International Conference on Industrial Electronics, Control and Instrumentation. 21st Century Technologies, 3, IECON, pp.1509–1514 (2000).
14. Shah, J. A., Rattan, S. S., Nakra, B. C.: End-effector position analysis using forward kinematics for 5 Dof Pravak robot arm. IAES International Journal of Robotics and Automation (IJRA), 2(3), pp.112–116 (2013). https://doi.org/10.11591/ijra.v2i3.2015
15. Lian, B., Song, Y., Dong, G., Sun, T., Qi, Y.: Dimensional synthesis of a planar parallel manipulator for pick-and-place operations based on rigid-body dynamics., ICIRA, pp. 261–270 (2012). https://doi.org/10.1007/978-3-642-33509-9_25
16. Christensen, A. L.: Fault Detection in Autonomous Robots. (2008), PhD, Université Libre de Bruxelles.
17. Valavanis, K. P., Jacobson, C. A., Gold, B. H.: Integration control and failure detection with application to the robot payload variation problem. Journal of Intelligent and Robotic Systems, 4, pp. 145–173 (1991).

18. Cho, C. N., Hong, J. T., Kim, H. J.: Neural network based adaptive actuator fault detection algorithm for robot manipulators. Journal of Intelligent & Robotic Systems, 95(1), pp. 1–11 (2019). https://doi.org/10.1007/s10846-018-0781-0

19. Myint, K. M., Min, Z., Htun, M., Tun, H. M.: Position control method for pick and place robot arm for object sorting system. Position Control Method Pick Place Robot Arm Object Sorting System, 5(6), 57–61 (2016).

20. Visinsky, M. L., Cavallaro, J. R., Walker, I. D.: Robotic fault detection and fault tolerance: A survey. Reliability Engineering & System Safety, 46(2), 139–158 (1994). http://dx.doi.org/10.1016/0951-8320(94)90132-5.

21. Mar Myint, W.: Kinematic control of pick and place robot arm. International Journal of Engineering and Technology, 1, (4), 63–70 (2015).

22. Padmanaban, K., Sanjana Sharon, G., Sudharini, N., Vishnuvarthini, K.: Detection of underground cable fault using Arduino. International Research Journal of Engineering and Technology, 4(3), 2451–2455 (2017).

23. Dodds, G., Wilson, G., Zatari, A.: Cooperative arm tasks and sensor fusion. (1995). http://dx.doi.org/10.1049/ic:19950441.

24. Ghosal, A.: Manipulator kinematics. Handbook of Manufacturing Engineering and Technology, pp. 1777–1808 (2015).https://doi.org/10.1007/978-1-4471-4670-4_90

25. Zhiyong, Z., Dongjian, H., Lei, T. J., Lingshuai, M.: Picking robot arm trajectory planning method. Sensors and Transducers, 162, 11–20 (2014).

26. Kucuk, S. Bingul, Z.: Robot Kinematics: Forward and inverse kinematics. In Industrial Robotics: Theory, Modelling and Control, Pro Literatur, Verlag, Germany/ARS, Austria (2006).https://doi.org/10.5772/5015

27. Li, X., Liu, M., Wang, W., Li, Y., Li, M.: Fault-tolerant control method of robotic arm based on machine vision. In 2018 Chinese Control and Decision Conference (CCDC), pp. 484–489 (2018).

13 Artificial Intelligence Advanced Technology IoT, 5G Robots, and Drones to Combat COVID-19

K. R. Padmaand*, and K. R. Don#
*Department of Biotechnology, Sri Padmavati Mahila
VisvaVidyalayam (Women's) University,
Tirupati, AP, India
#Department of Oral Pathology, Saveetha
Dental College, Saveetha Institute of Medical
and Technical Sciences, Saveetha University,
Velappanchavadi, Chennai, Tamil Nadu, India

CONTENTS

DOI: 10.1201/9781003181613-13

13.1 INTRODUCTION

The origin of coronavirus first reported in Wuhan City, China, in December 2019, has distressed more than 213 countries plus territories globally with 28,692,285 million coronavirus cases and greater than 920,46 deaths as of September 12, 2020 [1]. With this mounting cataclysm, enterprises and researchers globally had to consider means to mitigate the transmission of this virus, as well as to develop a vaccine for this disease. However, the science sector progression at peaks showed some ways for researchers and enterprises to fight this baffling battle. China was the first to focus on artificial intelligence (AI) to detect infected persons by relying on deep learning, smartphones, and drones with built-in cameras. The drone technology was very helpful in tracing infected patients, image processing, and CT scanning, and drones were employed to disinfect public places [2].

However, several governments and the public, along with health organizations, struggled to stop the spread of coronavirus. While the emerging AI was very helpful, they are far from replicating human intelligence. The applications of AI are (a) to monitor and track the outbreak, (b) include smartphones with thermal sensors helping to diagnose patients, and (c) include drones built by AI technology helping in disinfecting areas as well as monitoring large populations [3–4]. Although, approximately millions of cases progressed due to this pandemic and depicted transmission from one to another [5]. However, at present, the clinical data demonstrates COVID-19 infected patients exhibiting minor symptoms for the first four days, exemplifying the covert spreading potential of this infectious disease. Nevertheless, researchers as well as scientists have demonstrated that Covid-19 is far more contagious and lethal compared to the common flu [6].

According to the World Meter Report as on September 14, 2020, so far, 29,182,605 million total cases have been reported worldwide, with 7,227,163 active cases and 928,281 deaths [7]. Although, the total recovered is 21,027,161, death is highest among health complication cases as well as older people in comparison to young ones. Figure 13.1 depicts the novel coronavirus statistics for the top ten highest cases and deaths in countries and regions globally.

Our current chapter portrays the different applications of AI technology such as Internet of Things (IoT), unmanned aerial vehicles (UAVs), smartphones and 5G robots especially employed to support patients for home care, and dissemination of drugs, food, etc. Therefore, this chapter helps the policy-makers and decision-makers with the latest insights on AI technology to tackle the current pandemic. This chapter is organized and divided into five sections.

13.2 DESIGN OF STUDY

As authors, both of us were determined to contribute our perception and conception to bring awareness to society about AI technology utilized to tackle the COVID-19 pandemic. We have referred to several research articles, systematic review articles, pilot study, IEEE papers, and Elsevier, Wiley, and Springer articles for procuring guidance as well as for obtaining comprehension about vital topics to be emphasized in our current study. Moreover, insights about current AI technology and its

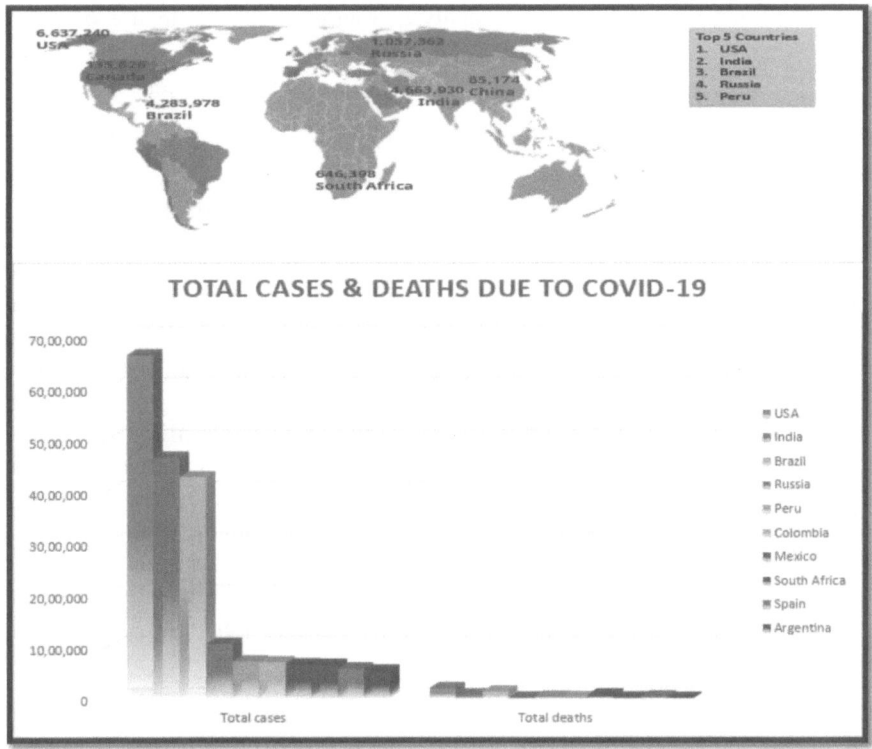

FIGURE 13.1 Reported top ten countries of COVID-19 cases and deaths (September 12–14, 2020).

implementation during the pandemic caused relief to health care workers, medical professionals, and several government agencies and it is emphasized in our manuscript as a novel theme.

We have divided our study design into the following sections to give a distinct vision to readers to explore what is happening around us. The information regarding AI was collected online using verified sources along with literature reports as reference for supporting our study. The literature survey reports are collected from online search engines such as Google Scholar, Wiley, Springer, Elsevier, PubMed, EMBASE, and IEEE. The beginning of our study is a concise abstract, which gives a broad outline of our review articles followed by background study. Then, we divide sections into different stages. Section 1 gives a clear understanding about the 5G robots. Section 2 provides insight on drone technology in detection of diseases. Further, Section 3 includes the application of AI for the benefit of society during the pandemic of COVID-19. Section 4 highlights AI utilization in E-learning and social distancing.

In order to gather sufficient and relevant information about AI, the prime keyword searches used were, artificial intelligence in healthcare, 5G robots, applications of drones, E-learning technology in Elsevier, IEEE, and PubMed platforms for which the initial screening is performed to analyze whether it is relevant for our study or not.

13.3 5G ROBOTS IN HEALTH CARE

The employment of 5G robots by several government agencies and medical organizations is primarily used to help the world struggling to lessen the robust transmission of COVID-19. Table 13.1 shows the adoption of AI robot deployment throughout the world, particularly to aid in the handling of patients and mitigating the anxiety levels faced by medical professionals. In addition, these robots are trained to disinfect contaminated areas with noncontact ultraviolet (UV) surface disinfection methods. This practice of usage of robots in decontamination purposes supports health care workers from the risk of contracting the virus [8].

According to the World Health Organization (WHO), several million people have died during the COVID crisis due to enhancement of pre-existing diseases and loss of immunity [9]. Consequently, as the pandemic situation rises, it is important to practice hand sanitizing, wear N-95 masks, and to follow social distancing to prevent transmission of the virus. Medical and healthcare workers are at risk, hence they can use AI devices comprising medical robots and telemedicine systems that can manage the dissemination of infection and mitigate to a large population [10].

13.4 APPLICATIONS OF 5G ROBOTS IN DIFFERENT SECTORS

The first application of IoT technology and 5G robots in health care and medical systems occurred as lockdowns led to decrements in economy and subsequent shortages of personal protection equipment (PPE), N95 masks, and ventilators, which caused fear among health care and medical professionals. AI offered health care workers relief during the introduction of 5G robots, as well reduced their stress levels [11–14]. As COVID-19 remains, its ambush throughout the world continues, which instigates employment of a multitude of technological approaches like IoT, AI, blockchain,

---

TABLE 13.1

Adoption of AI Robots Deployment throughout the World

Report Metric	Details
Market size available for years	**2016–2030**
Base year considered	**2017**
Forecast period	**2018–2030**
Segments covered	**Product, Application and Region**
Geographies Covered	**North America (US, Canada), Europe (Germany, UK, France, Italy, Spain), Japan, China, Australia, Asia, Latin America, and the Middle East Africa**
Companies covered	**Intuitive surgical (US), Stryker cooperation (US). In India, a Kerala-based startup named Asimov Robotics, Mazor Robotics (Israel)**

---

along with next-generation telecommunication networks like 5G to the frontline [15, 16].

AI robots are chiefly categorized based on their functions in related areas into different types, especially in healthcare and other related fields. A receptionist robot's main function is to guide patients and visitors to the right physician, as well as to entertain children to reduce their stress [17].

Nurse robots are appointed in Japan for taking care of elderly people, especially due to overload of patients. It's difficult for health care workers to take care of patients as they feel stress in hospitals [18].

Ambulance robots are employed for emergency purposes. As per European Union statistics, 800,000 people suffered from cardiac arrest [19]. In order to provide prompt treatment to victims, life-saving strategies such as emergency medication and cardiopulmonary resuscitation through ambulance robots are made possible as they can be delivered via a flying drone and can reach the emergency site promptly. Thus flying drones acting as ambulance robots help in speeding up emergency response and fast recovery [20, 21].

The concept of a telemedicine robot is based on Internet of Medical Things (IoMT) technology, which facilitates remote monitoring of patients as well as allows clinicians to assess and heal patients through online interactions without any corporeal interaction. Subsequently, since the outburst of COVID-19, telemedicine is the best way and rapid way to treat trauma victims [22, 23]. However, a few Medicare rules have been imposed by the Office of Civil Rights (OCR) and the Centers for Medicare and Medicaid Services (CMS) in the United States to exert their medical expertise via telehealth platforms [24]. A few robots were used in hospitals for cleaning hospital rooms [25], spraying/disinfestation [26], and for surgery [27, 28]. These are broad applications of IoT, which are a boon for present pandemic-hit areas.

13.5 SECTION 2: DRONES IN DETECTION OF DISEASES AND THEIR DIVERSE APPLICATIONS

Currently, AI technology-made drones are widely used to detect persons infected with COVID-19. The in-built thermal sensor camera in drones instinctively detects each person with high-precision infrared. However, the drones are fed with information services for detecting any changes in body temperature, or breathing rate from a distance [29]. The UAVs were first used by China during the pandemic for distribution of medicines and disinfecting the streets. AI is considered machine learning (ML), natural language processing, and as a computer vision application. Through this AI technology, several lives can be saved, as well as curbing the expansion of diseases. Drones possess diverse AI applications, and today even medical radiologists use it for quicker diagnoses. The other purposes of UAVs include X-rays and CT scans [30]. Table 13.2 gives a distinct UAV AI application during COVID-19 pandemic. The chest x-rays, with the help of UAV technology utilization, can detect changes in lungs especially in COVID-19 patients [31]. Hence, AI devices were recognized using inception migration neural networks and proved to achieve 90% accuracy [32, 33].

TABLE 13.2

Modern Artificial Intelligence Technology Functions During COVID-19 Pandemic

S. No.	Functions of AI	AI Models	References
1.	Telemedicine platform UAV fed with different geographical locations		[34]
2.	UAV in detecting Corona hit victims by scanning with the help of thermal sensors		[35–36]
3.	AI technology has considerable potential to improve image-based medical diagnosis, such as CT scan and X-rays.		[37]
4.	Integrated AI-based drug discovery pipeline to generate novel drug compounds	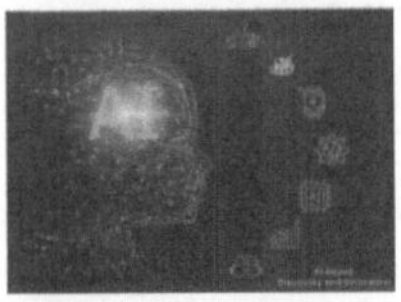	[38–39]
5.	Drones built with AI technology can detect abnormal respiratory patterns and contribute to large-scale screening of people infected with COVID-19, and also help in disease tracking	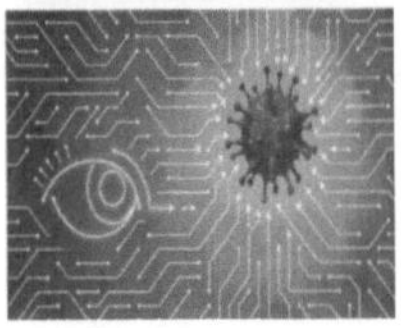	[40]

13.6 SECTION 3: BROAD APPLICATIONS OF AI AND ITS BENEFITS IN PANDEMIC TIME

The launch of AI is a landmark for technological advancement since it was revealed to be an extremely effectual tool during the combat of the coronavirus pandemic [41]. The wide applications of AI and its benefits for battling the novel COVID-19 pandemic are described concisely:

- Disease surveillance
- Risk prediction
- Spraying/disinfecting
- Medical diagnosis
- Vaccine/drug development
- Host recognition/identification

- Bursting bogus news
- Implementing lockdown measures

13.6.1 DISEASE SURVEILLANCE

Well-timed surveillance as well as forecast of diseases is crucial to curb the transmission of disease. The Blue Dot, a Toronto-based health surveillance enterprise, was the first worthy AI technology in reporting forthcoming outbursts of coronavirus nine days before the WHO [42]. BlueDot's AI model helps track SARS-CoV2 outbursts, as it been processed with natural language learning and ML to forecast outbreaks even before epidemiologists.

13.6.2 RISK PREDICTION

The AI application in risk prediction of COVID-19 can be illustrated by [43]: (a) predicting the persons infected from amass; (b) how to treat an infected person is predicted by AI technology; (c) ML stratagems to predict the susceptibility index for individuals vulnerable to corona virus [44]; (d) ML algorithms that help to foresee the result of drugs on a group of populations by correlation method; (e) the AI model of risk prediction shown to have 70%–80% accuracy (Figure 5.2).

13.6.3 SPRAYING/DISINFECTING

In contaminated regions, drones play a critical role in helping, especially, the frontline workers to curb the transmission of virus. After the onset of the virus and its subsequent rise, India as well as China depended on UAVs to spray disinfectant in public places [45]. The drone capacity load was made for 16L, which can disinfect about one-tenth of a kilometer per hour [46] (Figure 13.3).

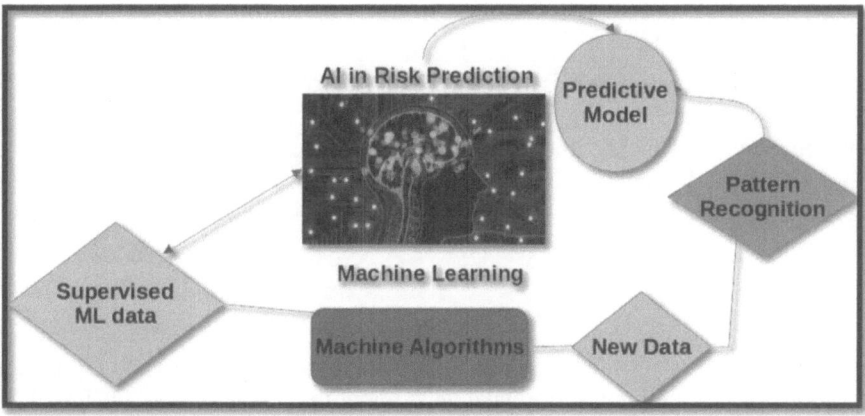

FIGURE 13.2 The utility of AI in risk prediction.

FIGURE 13.3 AI drones employed for disinfecting contaminated zones.

13.6.4 Medical Diagnosis

A neural network algorithm with 3D deep learning was constructed for scanning lungs, especially in patients with symptoms of community-acquired pneumonia (CAP), as well as other lung diseases [47]. For coronavirus diagnosis, this 3D deep learning was regarded as the best in medical diagnosis, offering 80%–90% precision [48]. Any specific abnormality based on symptoms of this virus in the lungs of the patient are observed using a deep learning method [49].

13.6.5 Vaccine/Drug Development

Vaccine/drug development with AI-based ML has shown its benefits in the course of drug development compared to earlier health emergencies. For instance, in the earlier epidemic of Ebola, Bayesian ML models were employed to instigate ascertaining molecular inhibitors against the virus [50]. The adoption of ML-assisted virtual screening and energy-based scoring helped to trigger the discovery of viral inhibitors against the avian H7N9 virus, which was responsible for recurring influenza epidemics in China [51]. Thus, the AI subset of ML is a potential applicant for medication due to its novelty and efficacy.

13.6.6 Host Identification

The ML models have supervised and trained sets of data which can be practiced effectually to relate the known viral genomes and also to identify connections between them [52]. In [53], the influenza-A virus is identified by use of the AI algorithm known as the random forest algorithm. Another article examined such an AI algorithm for detecting SARS-CoV-2 [54]. Thus, ML models help in identifying host species.

13.6.7 Bursting Fake News

Social media companies, such as Facebook, Instagram, YouTube, and Google, are developing AI technology to stop fake news about the outbreaks of COVID-19. The uncertain fake news is integrated with various myths and conspiracy theories as shown in Figure 13.4 [55].

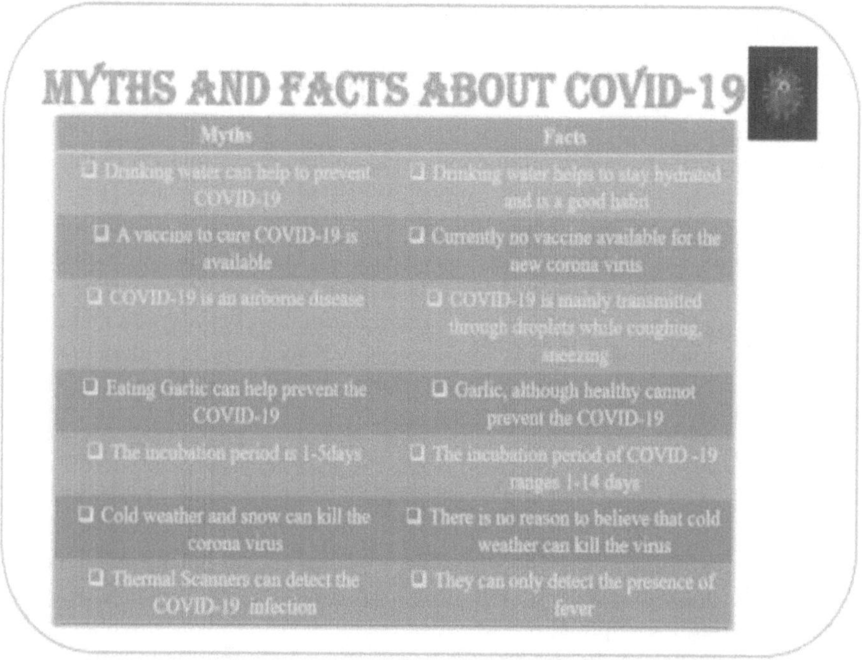
MYTHS AND FACTS ABOUT COVID-19

Myths	Facts
❑ Drinking water can help to prevent COVID-19	❑ Drinking water helps to stay hydrated and is a good habit
❑ A vaccine to cure COVID-19 is available	❑ Currently no vaccine available for the new corona virus
❑ COVID-19 is an airborne disease	❑ COVID-19 is mainly transmitted through droplets while coughing, sneezing
❑ Eating Garlic can help prevent the COVID-19	❑ Garlic, although healthy cannot prevent the COVID-19
❑ The incubation period is 1-5days	❑ The incubation period of COVID -19 ranges 1-14 days
❑ Cold weather and snow can kill the corona virus	❑ There is no reason to believe that cold weather can kill the virus
❑ Thermal Scanners can detect the COVID-19 infection	❑ They can only detect the presence of fever

FIGURE 13.4 Myths and facts about COVID-19.

13.6.8 Enforcing Lockdown Measures

Due to the outbreak of SARS-COV-2, several countries around the globe, including China, India, the United States, and the United Kingdom, have implemented usage of AI in enforcing restriction measures. In China, Baidu is considered the most effective AI-based Internet company around the globe, which has developed computer vision (CV)– powered infrared cameras in order to scan public places. The purpose of this CV camera system is to examine if the crowds are following the social distancing measures [56].

13.7 SECTION 4: AI UTILIZATION IN E-LEARNING AND SOCIAL DISTANCING

The whole world faced a drastic change throughout 2020 due to the pandemic effect of the disease. The most prevailing issue is that humanity encountered several million cases of COVID-19 around the globe [57]. This pandemic has changed our lives in terms of economy, culture, and even brought extreme changes in our day-to-day lives. To prevent spread of disease almost all countries have implemented lockdowns, social distancing, and quarantines to curb the spread of the novel disease [58]. Since it is transmitted in air from one person to another, this necessitated social distancing by introduction of lockdown measures [59]. This pandemic shook the entire world and led to the shutting of schools, college, and universities. To train young minds, the shift from a face-to face-educational system to digital online classes occurred [60, 61]. AI-based online classes have been relevant in these times and they are the most convenient and efficient method. Therefore, AI-based technology is of utmost relevance amidst the current pandemic and will be essential in implementing the new education system [62, 63].

Indian universities have started implementing online education to curb the spread of the virus. Several educational institutes started to rely upon apps like Zoom, Google Meet, and Microsoft Teams for all schools, both undergraduate and post-graduate courses, for online means. Nevertheless, Internet-learning satisfies all requirements of today's learners and has proven fruitful [64]. Through online classes, the lecturers, teachers, and professors can share their teaching-learning materials in different layouts such as slideshows, whiteboards, MS Word, audio, video, PDFs, and so on. Thus AI-based E-learning provides a platform for the learners and helps them to cover almost all topics and solve their doubts [65].

13.8   CONCLUSION

In this chapter, we have offered information about emerging machineries like IoT, UAVs, and 5G robots as effective in battling the latest in SARS-CoV. We began this chapter explaining the applications of AI in exploring its clinical features, curbing transmission mechanisms, as well as diagnosis procedures. Following this, we discussed the myths and facts regarding COVID-19. During this pandemic, AI-based E-learning was very supportive for young minds to build up their knowledge with the help of educational apps while maintaining social distancing, which prevents the spread of the disease. Thus, society has employed AI drones and 5G in reducing the influence of the COVID-19 pandemic until the period for remedy against this infection is available.

REFERENCES

1. Ruiz Estrada, MA. The uses of drones in case of massive epidemics, contagious diseases, relief, humanitarian aid: Wuhan-COVID-19 crisis. SSRN Electron J. 2020. https://doi.org/10.2139/ssrn.3546547 (February).
2. Nguyen, TT; Waurn, G; Campus, P. Artificial intelligence in the battle against coronavirus (COVID-19): A survey and future research directions. 2020. https://doi.org/10.13140/RG.2.2.36491.23846.
3. Chaturvedi A. Top 10 popular smartphone apps to track Covid-19; 2020. Available from: https://www.geospatialworld.net/blogs/pop ular-apps-covid-19/. [Last accessed on 2020 May 01].
4. Government of India, Aarogya Setu app. Available from: https://www. mygov.in/aarogya-setu-app/. [Last accessed on 2020 May 01].
5. A; Rocklöv, J. The reproductive number of COVID-19 is higher compared to SARS coronavirus. J Travel Med 2020, 27, 1–4.
6. Bai, Y; Yao, L; Wei, T; Tian, F; Jin, D-Y; Chen, L; Wang, M. Presumed asymptomatic carrier transmission of COVID-19. JAMA 2020, 323, 1406.
7. Worldometer, The coronavirus COVID-19 is *affecting 213 countries and territories* around the world and 2 international conveyances. https://www.worldometers.info/coronavirus/ september12th -14th 2020.

8. Yang, G-Z; Nelson, BJ; Murphy, RR; Choset, H; Christensen, H; Collins, SH; Dario, P; Goldberg, K; Ikuta, K; Jacobstein, N. Combating COVID-19—The role of robotics in managing public health and infectious diseases. Sci Robot Mar. 2020, 5, 40, Art. no. eabb5589, doi: 10.1126/scirobotics.abb5589.

9. Roser, M; Ritchie, H; Ortiz-Ospina, E; Hasell, J. Coronavirus (COVID-19) Deaths. Available online: https://ourworldindata.org/covid-deaths (accessed on 27 May 2020).

10. Yang, GZ; Nelson, BJ; Murphy, RR; Choset, H; Christensen, H; Collins, SH; Dario, P; Goldberg, K; Ikuta, K; Jacobstein, N, et al. Combating COVID-19—The role of robotics in managing public health and infectious diseases. Sci Robot vol. 33, pp. 234-244, 2020.

11. Beusekom MV. Doctors: COVID-19 pushing Italian ICUs toward collapse, Univ. Minnesota (UMN), Mar. 2020. [Online]. Available: https://www.cidrap.umn.edu/news-perspective/2020/03/doctors-covid19-pu%shing-italian-icus-toward-collapse.

12. Hockaday J. Spain's healthcare system on verge of collapse as another 655 die of coronavirus, Mar. 2020. [Online]. Available: https://metro.co.uk/2020/03/26/spains-healthcare-system-vergecollapse-%another-655-die-coronavirus-12459204/.

13. Yamaguchi, M; Kageyama Y. Coronavirus: Japan's medical system on verge of collapse, doctors say, Global News, Apr. 2020. [Online]. Available: https://globalnews.ca/news/6836522/coronavirusjapan-medical-system/.

14. Feuer W. WHO officials warn health systems are "collapsing" under coronavirus: "This isn't just a bad flu season," CNBC, Mar. 2020. [Online]. Available: https://www.cnbc.com/2020/03/20/coronaviruswho-says-health-systems-col%lapsing-this-isnt-just-a-bad-fluseason.html.

15. Ting, DSW; Carin, L; Dzau, V; Wong, TY. Digital technology and COVID-19. Nature Med Apr. 2020, 26, 4, 459–461.

16. World Health Organization. Digital technology for Covid-19 response, Apr. 2020. [Online]. Available: https://www.who.int/newsroom/detail/03-04-2020-digital-technology-for-%covid-19-response.

17. Karabegovi´c, I; Dole˘cek, V. The role of service robots and robotic systems in the treatment of patients in medical institutions. Micro Electron. Telecommun Eng 2016, 3, 9–25.

18. Kumar, B; Sharma, L; Wu, S-L. Job allocation schemes for mobile service robots in hospitals. In Proceedings of the 2018 IEEE International Conference on Bioinformatics and Biomedicine (BIBM); IEEE: Madrid, Spain, 3–6 December 2018; pp. 1323–1326.

19. Samani, H; Zhu, R. Robotic automated external defibrillator ambulance for emergency medical service in smart cities. IEEE Access 2016, 4, 268–283.

20. Momont A. Ambulance drone. Available online: https://www.tudelft.nl/en/ide/research/research-labs/ applied-labs/ambulance-drone/ (accessed on 20 March 2020).

21. Scudellari M. Drone beats ambulance in race to deliver first aid to patients. IEEE Spectrum. Available online: https://spectrum.ieee.org/the-human-os/biomedical/devices/drone-vs-ambulance-drone-wins (accessed on 20 March 2020).

22. AMD Telemedicine. Telemedicine defined. Accessed: Apr. 20, 2020. [Online]. Available: https://www.amdtelemedicine.com/telemedicineresources/telemedicine-def%ined.html.

23. Hornyak T. What America can learn from China's use of robots and telemedicine to combat the coronavirus. CNBC. Mar. 2020. [Online]. Available: https://www.cnbc.com/2020/03/18/how-china-isusing-robots-and-telemedic%ine-to-combat-the-coronavirus.html.

24. Hinkley, G; Briskin, A. U.S. waives Medicare and HIPAA rules to promote telehealth. Pillsbury Law, Mar. 2020. [Online]. Available: https://www.pillsburylaw.com/en/news-and-insights/uswaives-medicare-an%d-hipaa-rules-to-promote-telehealth.htm.

25. Prassler, E; Ritter, A; Schaeffer, C; Fiorini, P. A short history of cleaning robots. Auton. Robot. 2000, 9, 211–226.

26. Meisenzahl, M. These robots are fighting the coronavirus in China by disinfecting hospitals, taking temperatures, and preparing meals. Business Insider. Available online: https://www.businessinsider.com/see-chinese-robots-fighting-the-coronavirus-in-photos-2020-3# hangzhou-china-is-yet-another-city-using-robots-to-disinfect-large-areas-6 (accessed on 16 March 2020).

27. Grespan, L; Fiorini, P; Colucci, G. The route to patient safety in robotic surgery. In Springer Proceedings in Advanced Robotics; Springer International Publishing: Basel, Switzerland, 2019; pp. 25–35.

28. Da Vinci Surgical Robots. Available online: https://www.intuitive.com/en-us/products-and-services/davinci (accessed on 18 March 2020).

29. Chandler, PR; Pachter, M; Swaroop D, et al. Complexity in UAV cooperative control. Proceedings of the 2002 American control conference; 2002 May 8–10; Anchorage, AK, USA: IEEE Press; 2002. pp. 1831–1836.

30. Jin, C; Chen, W; Cao, Y; Xu, Z; Zhang, X; Deng, L. Development and evaluation of an AI system for COVID-19 diagnosis. 2020. pp. 1–23.

31. Wang, L; Wong, A. COVID-net: A tailored deep convolutional neural network design for detection of COVID-19 cases from chest radiography images. 2020. http://arxiv.org/abs/2003.09871.

32. Wang, S; Kang, B; Ma J, et al. A deep learning algorithm using CT images to screen for Corona Virus Disease (COVID-19). 2020. pp. 1–28.

33. Li, L; Qin, L; Zeguo X, et al. Artificial intelligence distinguishes COVID-19 from community acquired pneumonia on chest CT; 2019.

34. Harnett, BM; Doarn, CR; Rosen J, et al. Evaluation of unmanned airborne vehicles and mobile robotic telesurgery in an extreme environment. Telemed e-Health 2008; 14(6):539–544.

35. Marr B. Robots And Drones Are Now Used To Fight COVID-19. Forbes, Mar. 2020. [Online]. Available: https://www.forbes.com/ sites/bernardmarr/2020/03/18/how-robots-and-dron%es-are-helping-tofight-coronavirus/#2a8bfbca2a12.

36. Cyient. Cyient provides drone-based surveillance technology to support Telangana state police in implementing COVID19 lockdown, Apr. 2020. [Online]. Available: https://www.cyient.com/prlisting/corporate/cyient-provides-drone-based %surveillance-technology-tosupport-telangana-state-police-in-implementing-covi%d-19-lockdown.

37. Naude W. Artificial intelligence against covid-19: An early review, Medium, Apr. 2020. [Online]. Available: https://towardsdatascience.com/artificial-intelligence-against-covid19%-an-early-review-92a8360edaba.

38. Zhavoronkov, A; Aladinskiy, V; Zhebrak A, et al. Potential COVID-2019 3C-like protease inhibitors designed using generative deep learning approaches, vol. 2; 2020. https://doi.org/10.26434/chemrxiv.11829102.v2.

39. Makhzani, A; Shlens, J; Jaitly, N; Goodfellow, I; Frey, B. Adversarial autoencoders. 2015 (November). http://arxiv.org/abs/1511.05644.

40. Wang, Y; Hu, M; Li, Q; Zhang, X-P; Zhai, G; Yao, N. Abnormal respiratory patterns classifier may contribute to large-scale screening of people infected with COVID-19 in an accurate and unobtrusive manner. 2020. http://arxiv.org/abs/2002.05534.

41. Wittbold, KA; Carroll, C; Iansiti, M; Zhang, HM; Landman, AB. How hospitals are using AI to battle Covid-19, Harvard Business Review, Apr. 2020. [Online]. Available: https://hbr.org/2020/04/howhospitals-are-using-ai-to-battle-covid-19.

42. Hollister M. COVID-19: AI can help—But the right human input is key, World EconomicForum, Mar. 2020. [Online]. Available:https://www.weforum.org/agenda/2020/03/covid-19-crisis-artificialintel%ligence-creativity/.

43. Schmitt M. How to fight COVID-19 with machine learning—towards data ccience. Medium. Apr. 2020. [Online]. Available: https://towardsdatascience.com/fight-covid-19-with-machine-learning1d1106192d84.

44. DeCaprio, D; Gartner, J; Burgess, T; Kothari, S; Sayed, S; McCall CJ. Building a COVID-19 vulnerability index, 2020, arXiv:2003.07347. [Online]. Available: http://arxiv.org/abs/2003.07347.

45. Pan C. Spain's military uses DJI agricultural drones to spray disinfectant in fight against Covid-19, South China Morning Post Apr. 2020. [Online]. Available: https://www.scmp.com/tech/ gear/article/3077945/spains-military-uses-dji%-agricultural-dronesspray-disinfectant-fight.

46. Sharma, M. How drones are being used to combat COVID19, Geospatial World, Apr. 2020. [Online]. Available: https://www.geospatialworld.net/blogs/how-drones-are-being-usedto-comb%at-covid-19/.

47. Xu, X; Jiang, X; Ma C, et al. Deep learning system to screen coronavirus disease 2019 pneumonia. 2020. p. 1–29. http://arxiv.org/abs/2002.09334.

48. Huang, L; Han, R; Ai T, et al. Serial quantitative chest CT assessment of COVID19: deep-learning approach. Radiol Cardiothorac Imaging 2020; 2(2) :e200075. https://doi.org/10.1148/ryct.2020200075.

49. Ai, T; Yang, Z; Xia, L. Correlation of chest CT and RT-PCR testing in coronavirus disease. Radiology 2020; 2019: 1–8. https://doi.org/10.14358/PERS.80.2.000.

50. Ekins, S; Freundlich, JS; Clark, AM; Anantpadma, M; Davey, RA; Madrid, P. Machine learning models identify molecules active against the ebola virus in vitro, FResearch Jan. 2016, 4, 1091.

51. Zhang, L; Ai, H-X; Li, S-M; Qi, M-Y; Zhao, J; Zhao, Q; Liu, H-S. Virtual screening approach to identifying influenza virus neuraminidase inhibitors using molecular docking combined with machine learning-based scoring function, Oncotarget Oct. 2017, 8, 47, 83142.

52. Fludb.org. Influenza Research Database—Influenza Genome Database With Visualization and Analysis Tools. Accessed: Apr. 8, 2020. [Online]. Available: https://www.fludb.org/brc/home.spg?decorator=influenza.

53. Eng, CL; Tong, J; Tan, T. Predicting host tropism of influenza a virus proteins using random forest, BMC Med. Genomics 2014, 7, 3, S1.

54. Babayan, SA; Orton, RJ; Streicker, DG. Predicting reservoir hosts and arthropod vectors from evolutionary signatures in RNA virus genomes, Science Nov. 2018, 362, 6414, 577–580.

55. Obeidat S. How artificial intelligence is helping fight the COVID19 pandemic, Entrepreneur, Mar. 2020. [Online]. Available: https://www.entrepreneur.com/article/348368.

56. Naude W. Artificial intelligence against covid-19: An early review, Medium, Apr. 2020. [Online]. Available: https://towardsdatascience.com/artificial-intelligence-against-covid19%-an-early-review-92a8360edaba.

57. Cohut M. COVID-19 global impact: How the coronavirus is affecting the world, Medical News Today, April 24, 2020, Retrieved from https://www.medicalnewstoday.com/articles/covid-19-global-impacthow-the-coronavirus-is-affecting-the-world.

58. Social Distancing, Center for Disease Control and Prevention, July 15, 2020, Retrieved from https://www.cdc.gov/coronavirus/2019- ncov/prevent-getting-sick/social-distancing.html.

59. Bass,G;Lawrence-Riddell,M.Aneducationalshift:encouragingmission-drivenonlinelearning, Faculty Focus, May 20, 2020, Retrieved from https://www.facultyfocus.com/articles/online-education/aneducational-shift-encouraging-mission-driven-online-learning/.

60. Agoncillo, J; Aurelio, JM. Class opening reset amid virus, preparation woes, Inquirer, August 15, 2020, Retrieved form https://newsinfo.inquirer.net/1321901/class-opening-reset-amid-viruspreparation-woes.

61. Maragakis LL. The new normal and coronavirus, Johns Hopkins Medicine, 2020, Retrieved from https://www.hopkinsmedicine.org/health/conditions-anddiseases/coronavirus/coronavirus-new-normal.

62. Simari, G; Rahwan, I. Argumentation in artificial intelligence, Springer, Boston, MA, 2018, Retrieved from https://link.springer.com/book/10.10072F978-0-387-98197-0.

63. Karal, H; Nabiyev, V; Erumit, AK; Arslan, S; Cebi, A. Students' opinions on artificial intelligence based distance education system (Artimat), Procedia—Social and Behavioral Sciences, July 2014, https://www.researchgate.net/publication/273428989.

64. Colchester, K; Hagras, H; Alghazzawi, DM; Aldabbagh, G. A survey of artificial intelligence techniques employed for adaptive educational systems within E-learning platforms. J Artif Intel Soft Comp Res 2017, 7, 47–64.

65. Bajaj, RV; Sharma, V. Smart education with artificial intelligence based determination of learning styles. Procedia Comp Sci 2018, 132, 834–842.

14 E-learning with Internet of Things

Architecture and Its Applications

Arun Kumar*, Sharad Sharma#, Sachin Dhawan*, Nitin Goyal†, Suman Lata Triphati‡, and Sandeep Kajal‡‡
*Panipat Institute of Engineering and Technology, Samalkha, Panipat, India
#Maharishi Markandeshwar (Deemed to be University), Mullana, India
†Chitkara University Institute of Engineering and Technology, Chitkara University, Punjab, India
‡Lovely Professional University, Punjab, India
‡‡North Carolina State University, Raleigh, North Carolina, US

CONTENTS

DOI: 10.1201/9781003181613-14

14.1   INTRODUCTION

The Internet of Things (IoT) is a common infrastructure that offers physical and virtual objects for connectivity and collaboration. It increases in size and dimension as it progresses, affecting multiple facets of our lives, such as education [1]. Kevin Ashton brought IoT to the world in 1999 when he worked in the Auto-ID laboratory at MIT, explaining IoT-driving innovations such as radio frequency identification devices (RFID) and wireless sensor networks (WSNs) [2]. Based on the primary concept of IoT, all artifacts that acquire an IP address will be able to communicate physically and virtually with each other [3]. The primary structure of the Internet of Things is based on data sensed by sensors, tags, or actuators and sent to a cloud system through a gateway [4]. E-learning IoT is designed to help gather data from sensors and share it with other sensors that are used for a successful smart campus e-learning program [5]. This chapter, therefore, offers a detailed study on how to design those IoT systems [6]. IoT is a recent paradigm of communication that envisages a future that will provide everyday objects the use of microcontrollers, digital transceivers, and protocol stacks, which allow them to communicate and become a component of the Internet [7]. By operating with various gadgets and devices, such as a camera, an audio player, intelligent watches, Google Glass, digital large screens, sensors, etc., the IoT can enhance the learning experience using the enormous subject data created by these artifacts to provide dynamic services to teachers, learners, and even to others, such as new campus content creators. Smart campuses help us to use IoT methodologies to make them available in the network area for classroom notes. E-learning with IoT provides knowledge of Industry IoT, Internet of Underwater Things, and Internet of Health Care of Things, etc. [24–28]. We aim to facilitate the sharing of notes using software web-based apps that enable us to share IoT [8].

The main purpose of this chapter is to explore how IoT technologies can be used for smart campuses with modern IoT classrooms, in which the collecting of data can be carried out using e-learning devices. We describe the following: the importance of the intelligent classroom for data recovery in real-time with IoT-enhanced computers. We survey the literature and the approach to the resources offered by e-learning education for students by demonstrating the contributions and the workability of the program to smart campus environments.

14.2   OVERVIEW OF IoT-CONNECTED E-LEARNING

Over the Internet, information can be used to collect and transmit analyzed data, and the particular methods with interconnection with various items can serve the capability. Examples of Web 2.0 apps include YouTube, Flickr, Facebook, and many others [9]. Web 3.0 is currently defined to Internet users as a semantic web for specialized browsing. Knowledge availability can be viewed intellectually in the network 3.0 present on computers. Web 3.0 will disseminate the information provided by its data sources, as well as build it. The Internet can be protected, as the outlets offered that are useful for linking and helping people communicate with each other helps to exchange, disseminate, and obtain fruitful knowledge, so it can be used for transmitting the information, and its availability can even be served from advanced-version video conferences. In addition to person-to-person connectivity, Web 3.0 promotion will be useful for human-to-machine conversation. Network 3.0 also satisfies machine-to-machine connections. That is considered as an example that establishes a machine-to-machine relationship, although it could mean that the interacting things signify an air conditioner and a temperature sensor.

14.3 SMART ENVIRONMENTS

As part of a wider concept, the IoT, the primary purpose of smart environments is to make daily life simpler. For example, people want to be able to get the latest updates on road conditions and traffic congestion, to change the radio channels they listen to, etc., as they drive cars. With the aid of modern sensors, actuators, and smart devices that combine all of this, with their voices alone, people can check the weather on the Internet, see nearby traffic incidents, and know which roads are better for the least traffic. Smart environments include smart homes, intelligent schools, intelligent workplaces, and other smart areas. Due to recent developments in the use of smart energy delivery in the form of smart grids, the number of smart devices and ecosystems is expected to rise exponentially. In the smart world, three primary objectives are to understand, reason, and predict. In other words, when an event or signal occurs, smart environments need to understand how their environment operates and thinks and to know exactly how to respond. It is possible to characterize a smart environment (SE) as one that can acquire and apply information about the environment and its inhabitants to enhance human experiences in that environment. Speaking of algorithms and protocols used in smart environments and with IoT, there are several protocols aimed at balancing energy consumption, making systems function faster, and making them more effective. ZigBee, DECT, IEEE, and others are some of the protocols that are popular in smart environments. The ZigBee protocol has become one of the most commonly-used home wireless protocols because it is a perfect solution for low data rate devices and devices that need a long battery life, which is very common in home-network applications. One of ZigBee's big issues is that it introduces problems of coexistence, often dropping to meet the answer period needed by applications for the home network. When the ZigBee transmission is underway, this issue can be resolved by monitoring the WLAN.

14.4 IoT E-LEARNING ARCHITECTURE

Figure 14.1 displays a simple IoT architecture scheme. As shown in Figure 14.1, the core architecture of IoT is divided into three distinct layers: application, network, and perception layers. The relation between nodes and gateway is given by the network layer. To receive sensed data on the sensor nodes of the perception layer and to transfer information to a cloud system, the gates are used as an intermediary between the device and perceptible levels. In comparison, the layer of interpretation contains physical objects or sensors capable of detecting events or objects. Sensing (RFID, WSN) and storage of sensed data is a limited detection device within this layer [10].

Figure 14.2 shows the basic structure of IoT education in which IoT devices detect in the perception-layer activities, object detection, or any details. In a small cloud

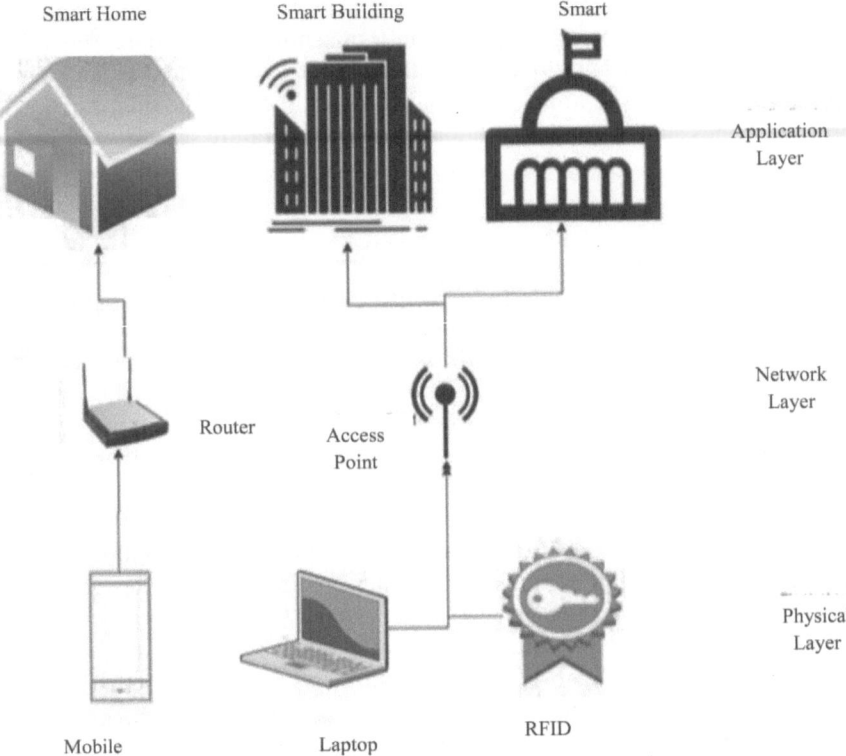

FIGURE 14.1   IoT architecture.

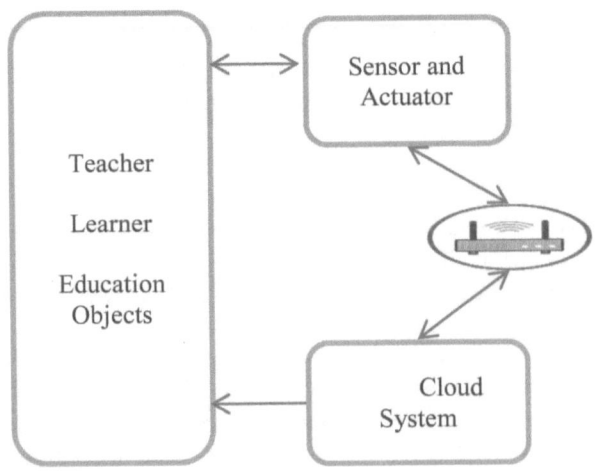

FIGURE 14.2   E-learning IoT architecture.

system, the sensory data are forwarded to the gateway. Data may be used for further judgments after processing.

Although each IoT framework is different, both the technological foundation and the ultimate data process flow of each IoT is roughly the same. First, it consists of things that are Internet-linked devices that can detect the world surrounding them and capture information that is then passed on to IoT portals through their built-in sensors and actuators. The next step is IoT data collection systems and gates, which gather vast volumes of unprocessed data, turn it into digital streams, filter it, and pre-process it to be analytical.

14.5   APPLICATION OF IoT-BASED E-LEARNING EDUCATION

The IoT, then, is all that is related to the Internet, and there are just a couple of electronics that are not part of the IoT in the real world we live in. In the coming years, this divide will also be bridged, as a report estimates that the number of devices that will be part of the IoT (linked to the IoT) by 2025 will hit about 75 billion. How does IoT impact e-learning, though? Technology is used by e-learning, and IoT is used by that application. The IoT is, therefore, used for e-learning. Let us now learn how e-learning has been impacted by the IoT, and how it will change e-learning in the future. Some IoT-based learning education uses are shown in Figure 14.3.

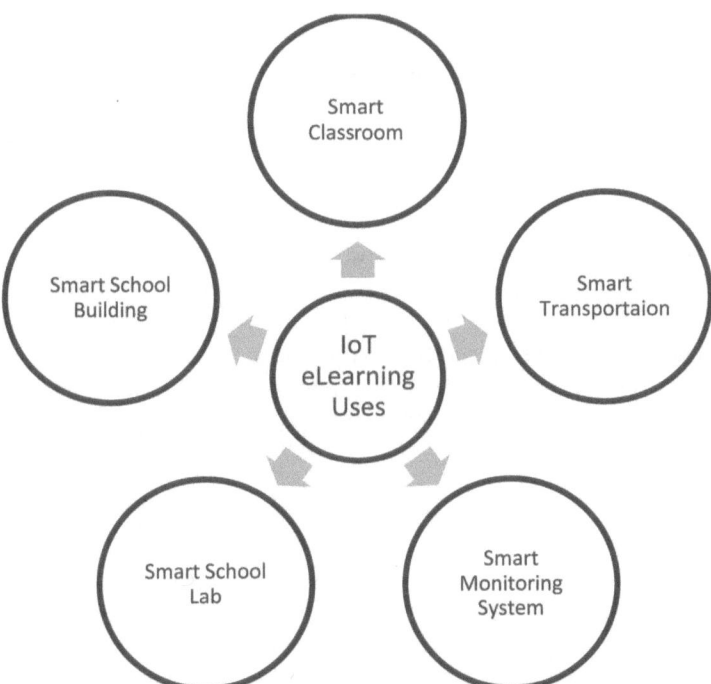

FIGURE 14.3   IoT-based learning education uses.

14.5.1 IoT E-learning to the Learner

What is a mobile? It is a device that makes use of the Internet. It is thus a portion of the IoT. The mobile has made it possible for real-time education to be available, wherever it is needed by learners, whenever they need it. It also made it possible for teachers from the safety of their homes to educate students sitting at home. Mobile learning provides learners with the full freedom of time and place to navigate e-learning lessons, and all of this is only possible thanks to the IoT.

14.5.2 Transformed Content

Material on the Internet is incredibly visual. Videos and pictures, often by users, are favored over text, and videos are no longer just a convenient recording of visual and audio elements, but immersive pieces of content that reach the viewer 40% more than basic videos. In the future, this is simply going to escalate. This has also modified the content of e-learning, making it extremely visual, as new learners have become used to seeing such content on the Internet every day and do not interact in anything other than that method.

14.5.3 Transform Tests and Examinations

The IoT will soon make it so that with a range of wearables, including sneakers, glasses/shades, pens, etc. will have 24 × 7 access to the Internet, and who knows what else? Teachers would have to revert to modifying the learning format from "remembering and writing things" to "reading things on the Internet and building on them" with such 24 × 7 connectivity open to any pupil. E-learning assessments will still have to shift, and e-learning practitioners will most likely be the first educators in an environment in which IoT-connected computers are prevalent to discover new forms of learning.

14.5.4 Life Easier and Convenient

With artificial intelligence (AI) only a few years away from being ubiquitous, IoT would offer increased home security, better use of resources, and quicker streaming of video and data along with AI. It would also modify the lives, aspirations, and habits of tomorrow's people. In addition to lifestyle changes, there may be changes in schooling or, rather, e-learning. On a greater range of devices, e-learning will be available, will become cheaper to create, be more engaging, and help draw more learners.

14.5.5 Create Jobs

Just like machine learning (ML) and AI now have advanced courses and degrees for those who wish to train and develop a career, IoT will soon be devoting curricula and main courses to learners who are interested in them and guess what the distribution process will be in the future? E-learning! The future will need people with IoT technology expertise, and e-learning will provide them with the education they need in an inexpensive, accessible, easy, and successful manner. For teachers and students, new technology, information, and communication systems have offered great improvements. Compared to fifty years ago, teaching and learning in class-rooms today are entirely distinct processes [20]. New technology allows learners to learn quicker and develop better knowledge and helps teachers to teach students easily. The best thing about smart classrooms is that they enable teachers to see how learners want to learn, and smart classrooms enable teachers to give learners the skills they want [21]. It supports teachers and students alike. Another important thing about smart classrooms is that they encourage students to see a specific goal of using technology and make it easier for all to learn when they see a real goal using technol-ogy. Students can get data in a few seconds, thanks to the Internet, laptops, and smart devices [22]. This is not enough, however. The next step is to get the students' infor-mation before they click the button and type in how they feel. This can be carried out in smart classrooms with the help of modern sensors and actuators. With smart classrooms, intelligent things such as cameras, microphones, and many other sensors can evaluate how happy students are with the things they learn about the world.

14.6   SMART CLASSROOMS IN E-LEARNING

Today, classroom management is found to be very easy and also it is now easier to comprehend with smart devices than what we do without them. The word "classroom management" is a manner in which a teacher achieves order in his or her classroom. Teachers can know when to take a break with the help of smart devices, when to talk louder, and when the concentration of students drops. They can use temperature sensors, sensors for walking in front of the table that check whether the teacher is present in the classroom, and if so, some actuators switch on the projector. There could be noise sensors that display how students talk, and so on. These are some examples of sensors and actuators that can be used in smart classrooms. Strong and exciting smart classrooms could be designed with some inexpensive sensors and a lot of creativity, which can help students learn better.

14.7   A NEW INSTRUCTIONAL COMPETENCY MODEL

A new modern technological environment in schools is needed because of the intro-duction of the e-learning framework. Furthermore, this technology brings many chal-lenges to the instructor/teacher and forces him/her to keep pace with new changes, whether developments in schooling, technology, or actions. The new educational environment, based on the introduction of the role of the teacher, is strengthened by the investment in technology in the educational process. Also, the nature of the role of the instructor will be impacted by significant new aspects of the technical climate. New instructional competency model is shown in Figure 14.4. We summarize these aspects in the following subsections.

14.7.1   AWARENESS AND CULTURAL SKILLS

Information and cultural factors influence the personality of the instructor/teacher in the e-learning world. These dimensions are not limited to teaching but are surpassed by a large inflow of information across computer networks, the Internet, other forms

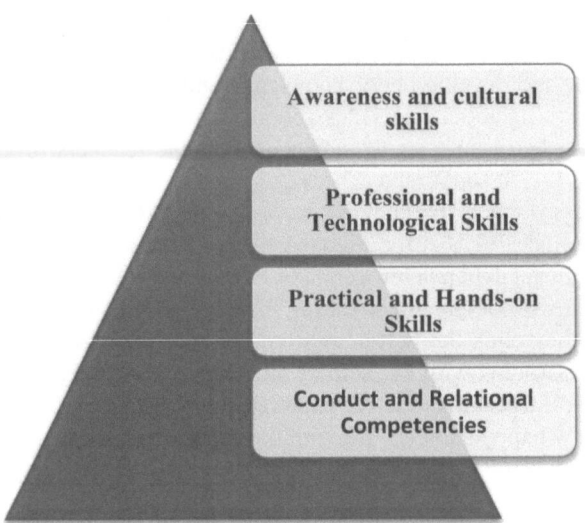

FIGURE 14.4   New instructional competency model.

of learning, and e-libraries. The role of the teacher in the environment for e-learning based on technology is like both historic and modern communications: to promote, direct, and supervise the process of education. Also, the instructor/teacher is active in the instructional design process in such a setting that he or she plans the educational and science materials and the necessary teaching methods.

14.7.2   PROFESSIONAL AND TECHNOLOGICAL SKILLS

In addition to various display devices and interactive whiteboards, the integrated e-learning environment allows teachers to communicate with computers, computer networks, the Internet, and learning management systems. It is, therefore, critical for teachers to be skilled in how to use these techniques and resources in the classroom setting, and how to adapt those tools to serve the curriculum and help students use those techniques.

14.7.3   PRACTICAL AND HANDS-ON SKILLS

This expertise focuses on the current reality imposed by e-learning and the educational paradigm of the twenty-first century, which offers many modern and technical instruments that enrich the learning process. These new technological advances are used in e-learning environments to create an engaging atmosphere for both teachers and students, where they can use these tools to connect with the teacher and/or other students, access the e-curriculum, and enhance the learning experience. In the educational sector, practical and hands-on elements have different forms and require many skills and abilities. The teacher must have the skills and abilities necessary to use and gain from these resources and use them to connect with the students and the students' classroom program.

14.7.4 Conduct and Relational Competencies

As instructors/teachers encourage and guide the learning process and communicate with their students, they can play various roles in the e-learning environment. Furthermore, by understanding the desires, actions, and skills of the learners, instructors can achieve their objectives and goals. Instructors should also be able to communicate with the learners and instill healthy attitudes by motivating and directing them to follow ethics and take our faith into account when engaging with the e-learning world.

14.8 INTERNET OF EVERYTHING (IoE) IN E-LEARNING

Energy independence, context awareness, and enhanced power processors can be applied as certain kinds of items that can be processed as IoT and then become the Internet of Everything (IoE). In almost every sector, IoT can be suggested as a ground-breaking notion. Within the conceptual base of education, transport, agriculture, healthcare, industry, management, and many more, its applications can be fulfilled [14–18]. IoT programs can be organized in the following way:

1. Patient surveillance, physician surveillance, staff tracking, monitoring of real-time patient health status, information on predictive skills to assist physicians and practitioners.
2. Smart Inventory Management, supply chain monitoring, software for intelligent shopping, object tracking, and fleet tracking.
3. On-demand traffic data can be segmented via a dynamic real-time assumption that can be assumed via intelligent transport. This includes optimization via the traveled path through the shortest time period covered.
4. Intrusion detection systems, water use, applications of remote control, energy use.
5. Waterways, monitoring of industry, monitoring of noise, and air pollution.
6. Control of soil moisture, manure, irrigation management.

Because of the education system, some of them are claimed to be linked to the IoT around the world [19]. IoE is used around the world to make a direct connection with the education system through its ability to be served. In the academic area, IoT serves the need by providing the student with a basic knowledge of participating in the learning process through certain motivational values. To fight against abuse through recognizing the potential threats to this system, both educators and policymakers need to be well-prepared. Global System for Mobile Communications Association (GSMA) claimed that the aid of IoT will allow life-enhanced services. Some argue that the significance of IoT can be described as mobile-enabled solutions in the education system. The teachers' learning process that satisfies the individual needs of each student to be met is necessary and beneficial. When it is related to physical or virtual classrooms, all above-competency levels can be increased. In physical or virtual classroom learning, IoT allows productivity that offers more comfort and accessibility.

14.9 CONCLUSION

This chapter examined a succinct analysis of the benefits of the Internet. IoT transforms the real structure of education. It offers an incredibly powerful physical and interactive connection between objects. It also makes possible a previously unfeasible connection between the real world and the Internet. IoT facilitates the worldwide relation of physical objects between multiple points, centers, institutes, hospitals, libraries, individuals, associations, businesses, and agencies. To examine the principal benefits of IoT in the e-learning environment and populations, a theoretical review has been established. IoT advances in e-learning grow very rapidly, so all interested parties can check for up-to-date information on Google periodically.

REFERENCES

1. Khan, M.A. and Salah, K., 2018. IoT security: Review, blockchain solutions, and open challenges. *Future Generation Computer Systems*, *82*, pp. 395–411.
2. Rana, A.K., Krishna, R., Dhwan, S., Sharma, S. and Gupta, R., 2019, October. Review on artificial intelligence with Internet of Things—Problems, challenges and opportunities. In *2019 2nd International Conference on Power Energy, Environment and Intelligent Control (PEEIC)* (pp. 383–387). IEEE.
3. Rana, A.K. and Sharma, S., 2021. Contiki Cooja Security Solution (CCSS) with IPv6 routing protocol for low-power and lossy networks (RPL) in Internet of Things applications. In *Mobile Radio Communications and 5G Networks* (pp. 251–259). Springer, Singapore.
4. Rana, A.K. and Sharma, S., 2019. Enhanced Energy-Efficient Heterogeneous Routing Protocols in WSNs for IoT Application.
5. Ahmed, E., Islam, A., Ashraf, M., Chowdhury, A.I. and Rahman, M.M., 2020 Internet of Things (IoT): Vulnerabilities, Security Concerns and Things to Consider.
6. Veeramanickam, M.R.M. and Mohanapriya, M., 2016. IoT enabled futurus smart campus with effective e-learning: i-campus. *GSTF Journal of Engineering Technology (JET)*, *3*(4), pp. 8–87.
7. Cho, S.P. and Kim, J.G., 2016. E-learning based on Internet of Things. *Advanced Science Letters*, *22*(11), pp. 3294–3298.
8. Kumar, A. and Sharma, S., 2021. Demur and routing protocols with application in underwater wireless sensor networks for smart city. In *Energy-Efficient Underwater Wireless Communications and Networking* (pp. 262–278). IGI Global.
9. Abbasy, M.B. and Quesada, E.V., 2017. Predictable influence of IoT (Internet of Things) in the higher education. *International Journal of Information and Education Technology*, *7*(12), pp. 914–920.
10. Kumar, A., Salau, A.O., Gupta, S. and Paliwal, K., 2019. Recent trends in IoT and its requisition with IoT built engineering: A review. In *Advances in Signal Processing and Communication* (pp. 15–25). Springer, Singapore.
11. Charmonman, S., Mongkhonvanit, P., Dieu, V. and Linden, N., 2015. Applications of internet of things in e-learning. *International Journal of the Computer, the Internet and Management*, *23*(3), pp. 1–4.
12. Vharkute, M. and Wagh, S., 2015, April. An architectural approach of Internet of Things in E-Learning. In *2015 International Conference on Communications and Signal Processing (ICCSP)* (pp. 1773–1776). IEEE.
13. Li, H., Ota, K. and Dong, M., 2018. Learning IoT in edge: Deep learning for the Internet of Things with edge computing. *IEEE network*, *32*(1), pp. 96–101.

14. Dalal, P., Aggarwal, G. and Tejasvee, S., 2020. Internet of Things (IoT) in Healthcare System: IA3 (Idea, Architecture, Advantages and Applications). *Available at SSRN 3566282*.

15. Rana, A.K. and Sharma, S., 2021. Industry 4.0 manufacturing based on IoT, cloud computing, and big data: manufacturing purpose scenario. In *Advances in Communication and Computational Technology* (pp. 1109–1119). Springer, Singapore.

16. Wang, Q., Zhu, X., Ni, Y., Gu, L. and Zhu, H., 2020. Blockchain for the IoT and industrial IoT: A review. *Internet of Things*, *10*, p. 100081.

17. Rana, A.K., Salau, A., Gupta, S. and Arora, S., 2018. A Survey of Machine Learning Methods for IoT and their Future Applications.

18. Kumar, K., Gupta, E.S. and Rana, E.A.K., 2018. Wireless Sensor Networks: A review on "Challenges and Opportunities for the Future world-LTE".

19. Sachdev, R., 2020, April. Towards security and privacy for edge AI in IoT/IoE-based digital marketing environments. In *2020 Fifth International Conference on Fog and Mobile Edge Computing (FMEC)* (pp. 341–346). IEEE.

20. Simic, K., Despotovic-Zrakic, M., Đuric, I., Milic, A. and Bogdanovic, N., 2015. A model of smart environment for e-learning based on crowdsourcing. *RUO. Revija za Univerzalno Odlicnost*, *4*(1), p. A1.

21. Kim, T., Cho, J. Y. and Lee, B. G., 2012, July. Evolution to smart learning in public education: a case study of Korean public education. In *IFIP WG 3.4 International Conference on Open and Social Technologies for Networked Learning* (pp. 170–178). Springer, Berlin, Heidelberg.

22. An, S., Lee, E. and Lee, Y., 2013. A Comparative Study of E-Learning System for Smart Education. International Association for Development of the Information Society.

23. Kumar, Arun, Sharma, Sharad, Goyal, Nitin, Singh, Aman, Cheng, Xiaochun, and Singh, Parminder, 2021. Secure and energy-efficient smart building architecture with emerging technology IoT. *Computer Communications*, *176*, pp. 207–217.

24. Gupta, O., Goyal, N., Anand, D., Kadry, S., Nam, Y. and Singh, A., 2020. Underwater networked wireless sensor data collection for computational intelligence techniques: issues, challenges, and approaches. *IEEE Access*, *8*, pp. 122959–122974.

25. Gupta, O., Kumar, M., Mushtaq, A. and Goyal, N., 2020. Localization schemes and its challenges in underwater wireless sensor networks. *Journal of Computational and Theoretical Nanoscience*, *17*(6), pp. 2750–2754.

26. Goyal, N., 2020. Architectural analysis of wireless sensor network and underwater wireless sensor network with issues and challenges. *Journal of Computational and Theoretical Nanoscience*, *17*(6), pp. 2706–2712.

27. Kumar, A. and Sharma, S., 2021. Internet of Things (IoT) with energy sector-challenges and development. In *Electrical and Electronic Devices, Circuits and Materials* (pp. 183–196). CRC Press.

28. Goyal, N., Sandhu, J. K. and Verma, L. 2020. CDMA-based security against wormhole attack in underwater wireless sensor networks. In *Advances in Communication and Computational Technology* (pp. 829–835). Springer, Singapore.

15 A Review on the Internet of Robotic Things Concepts, Added Values, Applications, and Issues

Rajiv Dey and Pankaj Sahu
BML Munjal University, Haryana, India

CONTENTS

15.1   INTRODUCTION

Robotic technology has brought incredible changes in various social economic aspects of human society compared to earlier decades. Today, industrial robotic manipulators are widely used and deployed in almost all types of industries to perform complicated, repetitive tasks efficiently that are difficult for a human to perform. These robotic systems are factory programmed and are highly efficient in successfully implementing industrial tasks, e.g., product packaging, welding, pick and place tasks, etc. This is made possible due to the high accuracy and precision of these robotic system. So far, the objectives of the Internet of Things (IoT) and robotics are complementary, e.g., IoT deals with the remote data acquisition of sensors, monitoring, and communication, whereas robotics deals with producing action and autonomous behavior [1]. Due to this, the combination of these two technologies creates a sturdy additional value [2]. Earlier the signs of IoT and robotics in conjunction can be seen in multiple applications, such as distributed, heterogeneous robotic systems like networked robotics. According to the IEEE Robotic and Automation Society, the networked robotic system can be defined as a group of individual robots connected via wire or wireless medium [3]. The application of networked robotic systems can be further categorized as remote-operation robots, such as a robot controlled and monitored remotely using human-driven commands via some communication network or collective robotic system working in a distributive network for performing

DOI: 10.1201/9781003181613-15

a given task by exchange of data between each other. The limitation of these robotic system are communication latency, limited memory capacity, deficiency of intelligence, and less execution speed [2]. These boundaries of networked robotics systems have inspired the research community to explore new robotic technology such as cloud robotics, which will be efficient and can overcome the discussed limitations. James Kuffner of Google coined the term cloud robotics in 2010. It is a domain that utilizes cloud technology, such as cloud computing, processing power, analysis, and storage for its support. Therefore, it can be concluded that a single standalone device does not require huge processing power and storage to be incorporated into the system. One common example of cloud robotics is Google's self-driving car. According to the latest ABI Research report, the concept of the Internet of Robotics Things (IoRT) belongs to a class of intelligent devices that monitors events and acquires data using sensor fusion from a variety of sources that utilize local and distributed intelligence to take necessary action [4].

The IoRT is created by the blend of two trending technologies—cloud computing and IoT; the idea is shown in Figure 15.1. Cloud computing provides on-demand

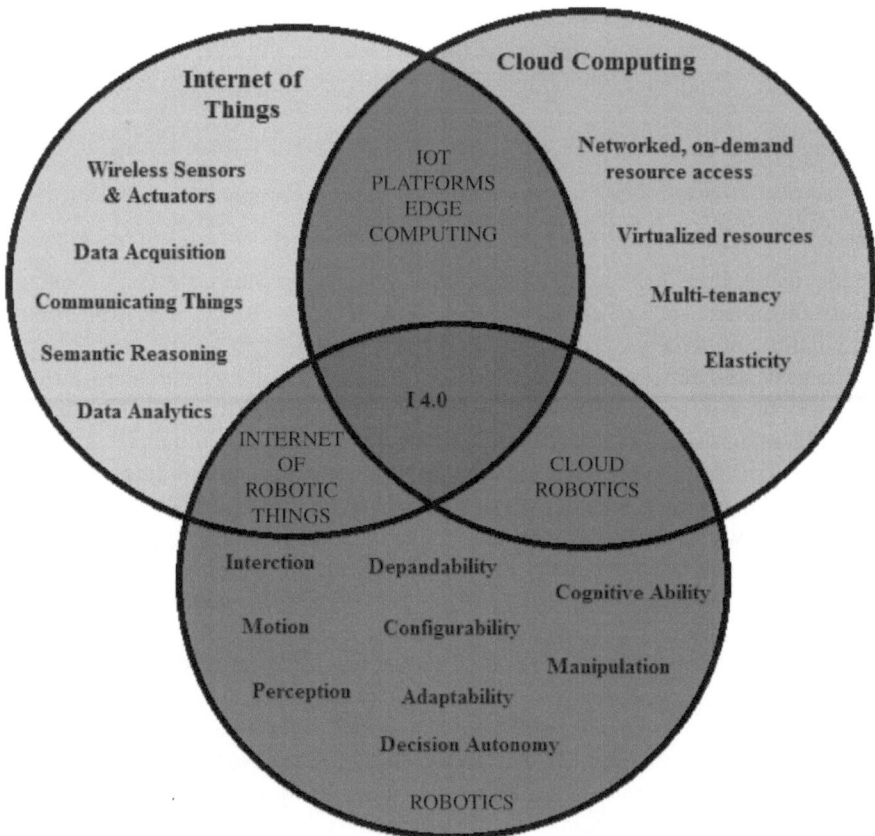

FIGURE 15.1   A representation of overlapping technologies [1].

access to the network resources to a group of virtual hardware for data storage and processing. IoT technology facilitates the integration of remotely located sensors using machine-to-machine communication. The cloud-based infrastructure has been extensively used by most of the IoT community to deploy scalable IoT applications that provide access to the sensor data [5]. The huge data generated from billions of remotely-situated devices come with the concern of latency, bandwidth, processing power, data filtering, storage, and its security, etc. Fog computing brings flexible computational capability at the edge of the network near the data source [6]. Big data analytics provides ways to systematically analyze the data mode available from huge data sets to extract their meaning. From this discussion, it is clear that the IoRT intersects with cloud robotics where cloud robotics is focused toward providing network capability, processing power, and data storage. On the other hand, IoRT is oriented toward machine-to-machine communication. This chapter presents a review on the IoRT, its tentative architecture, concepts, additional benefits of blending IoT with IoRT, and its technological challenges.

15.1.1 INTERNET OF ROBOTIC THINGS ARCHITECTURE

In this section, we present definitions of IoT and IoRT and a five-layer architecture of IoRT.

1. **Definition of IoT**: The concept of IoT was first coined by Mark Weiser in his article on ubiquitous computing called "The Computer for 21st Century." Later on, in 1999, Kevin Aston, the executive director of the Auto-ID Center at that time, coined the term Internet of Things. According to some researchers, IoT associates people, process with sensors and actuators. The overall association of people in terms of communication, collaboration, and data analytics enable real-time decisions. According to the International Telecommunication Union (ITU), IoT can be defined as
"A global infrastructure for the information society, enabling advanced services by interconnecting (physical and virtual) things based on existing and evolving interoperable information and communication technologies."
The reason behind multiple perceptions and definitions of IoT is due to the fact that IoT is not a novel technology but rather a new way of depicting the developing business model that integrates a group of technologies together to run the business in a combined manner.

2. **Definition of IoRT** [2]:
"A global infrastructure for the information society enabling robotic services by intersecting robotic things based on, existing and evolving, interoperable information and communication technologies where cloud computing, storage, and other prevailing internet technologies are focussed around the benefits of the converged cloud infrastructure and shared services that allows robots to take advantage from the influential computational, storage, and communications resources of modern data centres attached with the clouds, while eliminating overheads for maintenance and updates, and enhancing independence on the custom cloud based middleware platforms, entailing additional power requirements which

may reduce the operating duration and constrain robot mobility by covering cloud data transfer rates to offload tasks without hard real time requirements."

Within the IoT structure, the IoRT is mainly focused on the integration of smart space ability and autonomous agent (robots) where smart spaces are the applications, such as smart home, smart city, smart factory, etc. [7]. The main functionality of smart space is to monitor and control the states using sensors and actuators and process it within a defined zone. Its other functionality is to maintain certain environmental conditions such as temperature, humidity, etc., using special types of air conditioning systems, fulfilling certain desired conditions such as turning on or off a light bulb/device when some condition is met. As such, smart spaces can perform power management tasks, i.e., turning on the loads of a smart home when people are present and switching off the loads in their absence. In spite the availability of monitoring and control facilities in the smart space using sensors and actuators, no moving agents such as robots (assistive robots, robotic manipulators, robotic vehicles, etc.) are present to perform the task of moving objects or some other specified tasks. The appearance of intelligent agents with the smart space concludes the idea of IoRT. Integration of their functionality enables the smart space to expand its capabilities. A detailed pictorial representation of this concept is shown in Figure 15.2.

According to [8], "The Internet of Robotic Things is a more advanced level of the Internet of Things, allowing to integrate such modern technologies as cloud computing, wireless sensing and actuating, data analysis, distributed monitoring and networking from the Smart Space, as well as decisional autonomy, perception, manipulation, multi-agent control, control and planning and human-robot interaction from the robot side."

3. **Architecture of IoRT** [2]: The architecture of IoRT can be categorized into five layers. The block diagram representation of IoRT architecture is shown in Figures 15.2 and 15.3, and a detailed description of each individual layer is given as follows:

   a. **Hardware Layer**: This is the lowest layer of IoRT architecture, and it's commonly known as the physical layer. This layer consists of mainly the hardware entities, such as robots, sensors, actuators, smartphones, home appliances, etc. In general, physical things come under this layer, which gathers necessary information about its periphery to leverage it to the above layers.

   b. **Network Layer**: This is the connecting layer between the hardware layer and the Internet layer, it is also known as the communication layer as it is responsible for the smooth communication of information between the hardware layer and the network layer. The information transfer is basically done using a variety of communication technologies depending upon the range of applications such as:

      i. **Cellular Connectivity**: Technologies such as 3G, 4G, LTE, etc. are used in this category.

      ii. **Short-range Connectivity**: Technologies such as Wi-Fi, Bluetooth low energy (BLE), broad-band global area network (BGAN), near-field connectivity (NFC), etc. are used in this category

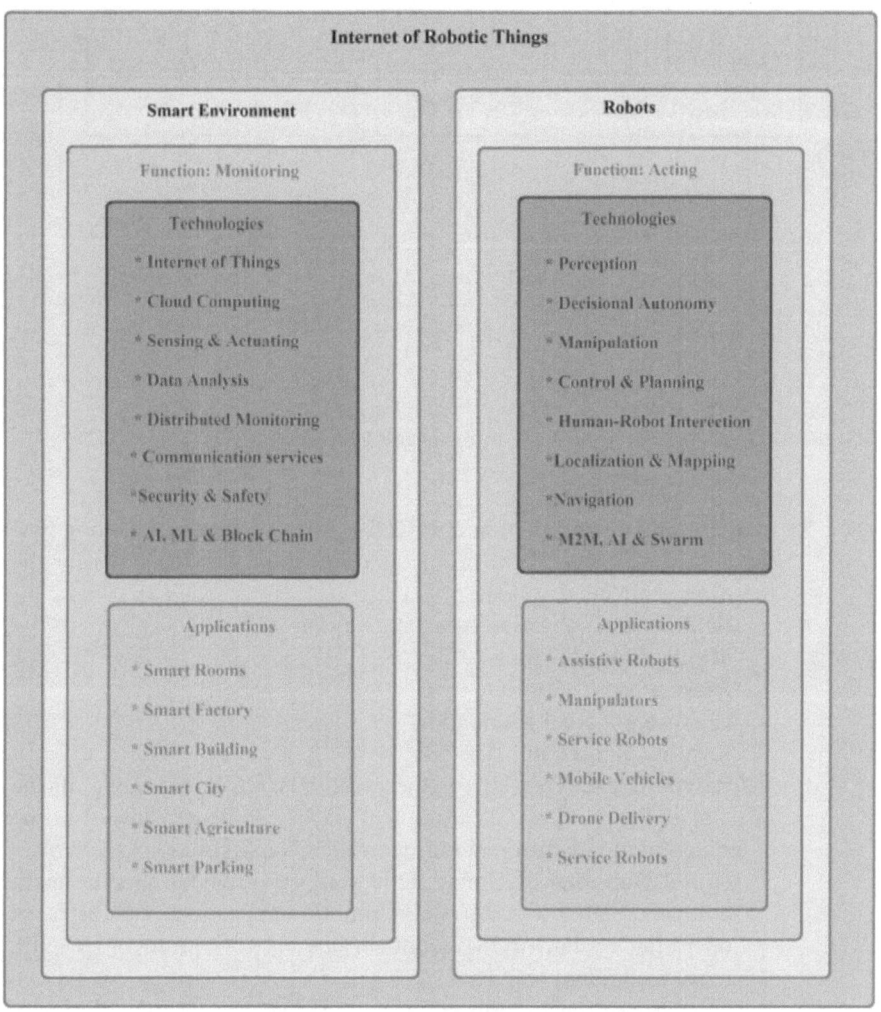

FIGURE 15.2 Internet of Things block scheme [1].

    iii. **Medium Long-range Connectivity**: Technologies such as ZigBee, Z-Wave, low-power wide area network (LoRAWAN), worldwide interoperability for microwave access (WiMAX) are used in this category

  c. **Internet Layer**: In IoRT architecture, Internet connectivity is the fundamental part of the communication system. As it is known through the literature that most of the IoRT-based applications are battery-driven and have limited storage and data-processing capability due to this reason, some IoT-specific protocols are added into this layer, which is more energy efficient and requires less processing power, such as message queuing telemetry transport (MQTT), constrained application protocol

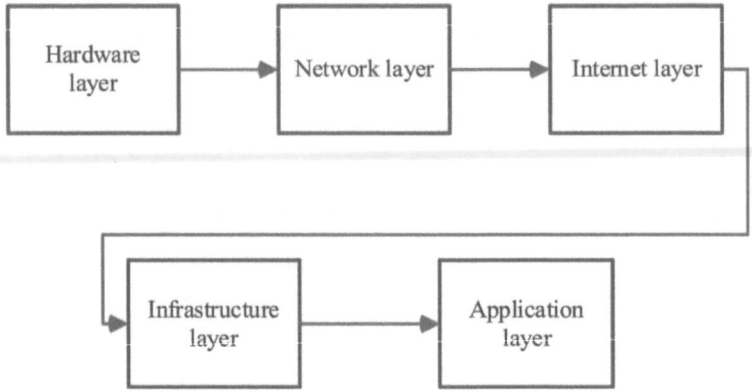

FIGURE 15.3   Representation of five layers of IoRT architecture [2].

(CoAP), user datagram protocol (UDP), advance message queuing protocol (AMPQ), etc. These protocols cover the following tasks for the real-time embedded system:

    i. Publish- and subscribe-based messaging

   ii. Multicast and broadcast

  iii. Packet-switch substitute to TCP

  iv. Lightweight local automation

   v. Security and reliability, etc.

d. **Infrastructure Layer**: This is the most valuable layer among all the other layers of IoRT architecture. This layer can be considered as the combination of five different but interrelated configurations such as:

    i. **Cloud Robotics Platform**: This platform provides specific technologies related to a particular robotic application, such as robot operating system (ROS), robot service network protocol (RSNP), robot technology, etc.

   ii. **Machine-to-machine-to-actuator (M2M2A) Cloud Platform**: M2M2A is a crucial part of an IoRT system, where machine to machine is a combination of many systems that are interconnected to a network for information exchange and optimum control. This technology is meant for leveraging the hands-on solutions, where the sensors and robotic systems combine to integrate the real and virtual world [9].

  iii. **Business Cloud Services of IoT**: This platform basically facilitates the business-related services to the IoRT-based systems. Some of the common business cloud services are Microsoft Azure IoT suite, Google's IoT platform, IBM Watson IoT platform, Amazon AWS IoT platform, etc. Different manufacturers and organizations of IoRT systems benefited from the business IoT platform upon reducing their operational activities using a common encrusted approach.

  iv. **Big Data Services**: This service facilitates the manufacturer of the IoRT systems to deploy Hadoop clusters of all magnitudes with the

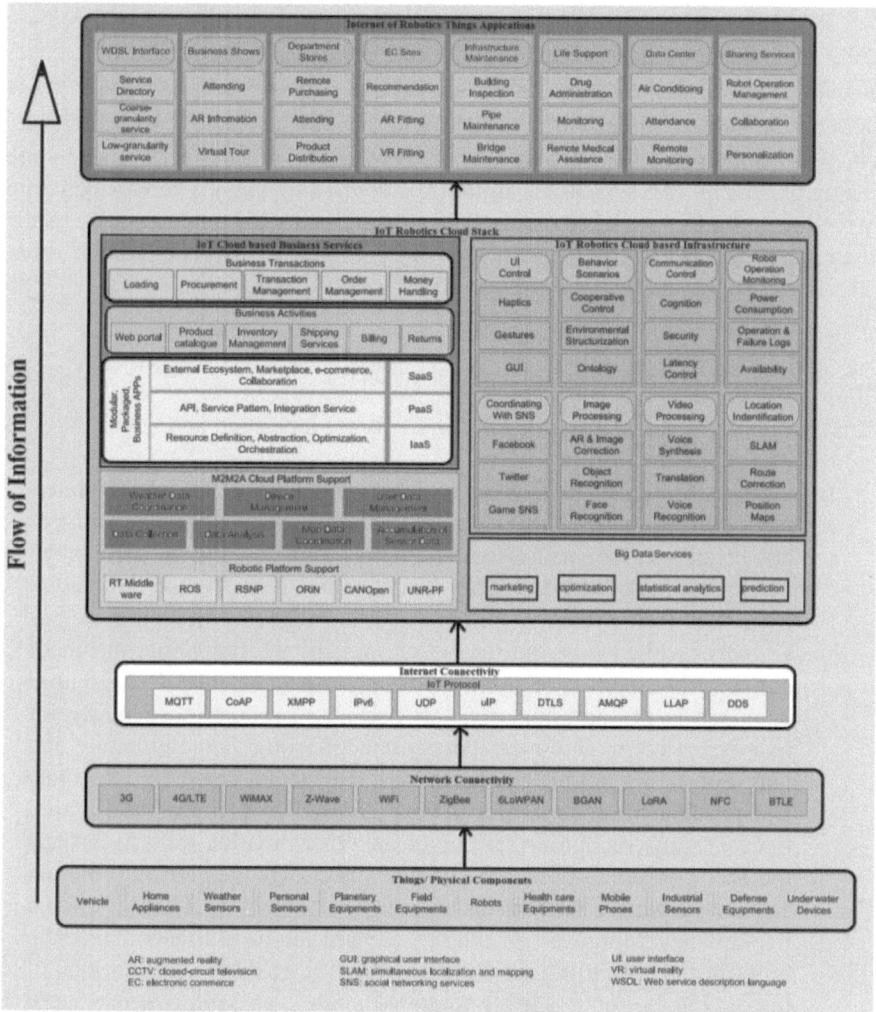

FIGURE 15.4   Conceptual representation of Internet of Robotic Things.

virtual machines ranging from ten CPU to a devoted bare metal environment as shown in figure 15.4 [10].

  v. **Cloud Robotics Services of IoT**: The IoT-driven cloud enables the IoRT systems to be empowered with services, such as communication control, image-processing control, user-interface control, video processing, location detection, etc.

 e. **Application Layer**: This is the top layer of the IoRT architecture. At this layer, the execution and implementation of the user-defined programs and applications are done to observe, control, and process the IoRT system parameters, sensors, and actuators.

15.2   BASIC ABILITIES OF THE INTERNET OF ROBOTIC THINGS

One of the agendas of this chapter is to inspire researchers to explore the potential of blending IoT and IoRT technology, hence, in this section, the focus of our discussion is on the system capabilities readily found in robotic systems, regardless of a particular robotic personification or application domain. In this section, we mainly focus on exploring nine system capabilities of defined by the EU-robotics community roadmap [5, 11]. As a matter of fact, these abilities are also meticulously related to the robotics issues and challenges acknowledged in the EU robotics and US robotics roadmap [11, 12], as shown in Figure 15.5.

15.2.1   Elementary Abilities of IoRT Systems

In this section, the basic abilities of IoRT systems are presented and are as follows:

1. **Interaction Ability of IoRT System**: This is the ability of a robot to interact physically with the user, operator, or other system intellectually or socially [11]. Here, the focus is mainly on exploring how the IoT technology can benefit human–robot interaction, functionally, socially, and remotely.

   a. **Functional**: The presence of persistent IoT sensors, which are used as a feedback mechanism in the system, makes the functional means of human-robot interaction more robust as shown in Figure 15.6. Natural language is the optimum way of instructing robots, particularly for non-expert users. However, these instructions are vague in nature and may consist of implicit assumptions. IoT technology can facilitate and provide necessary information required to process the instructions. Gesture recognition is one of the other possible ways to instruct the robot by pointing to a particular object. The gesture information acquired from the sensors on board the robot can function in a fixed field of view. Moreover, an external camera can be used to enhance the robot performance in order to recognize the gesture. Another way of instructing the robot is using wearable sensors, e.g., sleeve sensors that can be used to detect the forearm muscle movement of a human user to instruct the robot [13].

   b. **Social**: Body signals such as gestures and face expressions can be used to detect the emotional status of a user and make the robot respond. This estimation can be further improved by measuring physiological signals upon integration of body-worn IoT sensors [13, 14]. Some researchers have also used heart rate and electro-dermal signals to estimate the engagement and alertness during the human–robot interaction [15], and some used these signals for autism therapy [16].

   c. **Remote Interaction**: Today, robots have been widely used for remote interaction with the human user, e.g., Al-Taee et al. [17] used humanoid robots to monitor a diabetes patient remotely by reading a set of medical sensors, e.g., glucose, blood pressure sensors, connected wirelessly as shown in Figure 15.7.

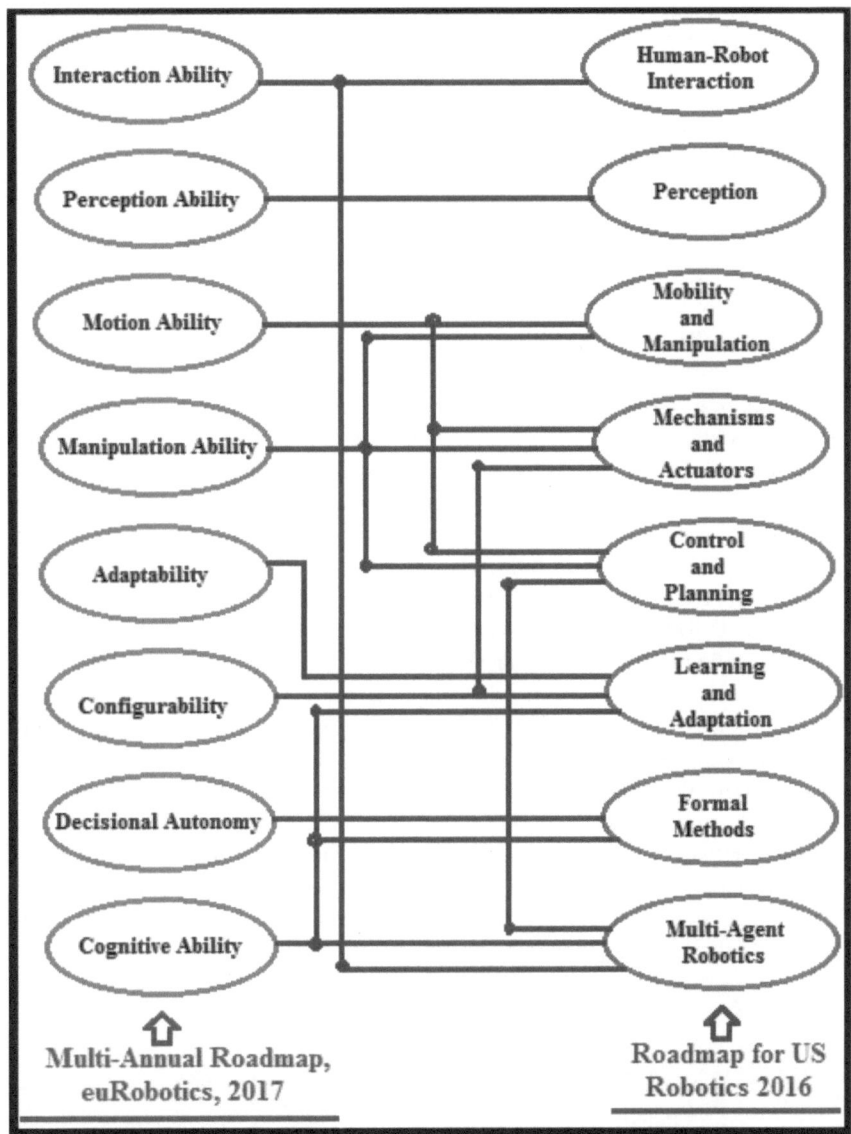

FIGURE 15.5 Mapping between the abilities defined by the EU robotics [11] and US robotics [12].

2. **Perception Ability of IoRT System**: With the integration of IoT-based sensors and data analytics technologies, the scope of robots has been extensively enhanced compared to the local onboard sensors. An important component of perception ability of a robot is to acquire the knowledge of its own position, i.e., building and updating models of the environment, and act accordingly [18]. Different approaches are used for self-localization purpose, e.g.,

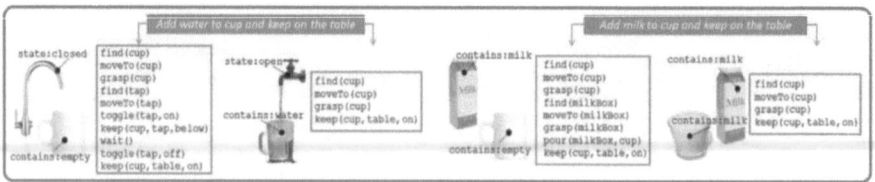

FIGURE 15.6 Representation of the actions to be performed based on natural language instructions (Misra et al. [13]).

currently radio frequency identification (RFID) technology is used for location identification in indoor applications [19]. Other range-based technologies include Wi-Fi access points, visible light, and IoT devices operating on protocols, such as Zigbee, Bluetooth low-energy (BLE), ultra-wideband [19–21], etc. Other authors used a scheme where the robot sends the observation request to the distributed entities as shown in Figure 15.8 [22]. Despite the technological advancements in this domain, self-localization using global positioning system (GPS) technology for indoor applications is still a challenge, especially for the applications where high reliability is required.

3. **Motion Ability of IoRT System**: This is another one of the fundamental abilities of a robotic system that adds value to it as shown in Figure 15.9. The mobility of a robotic system solely depends upon its mechanical design. IoT integration assists the mobile robots in controlling their actuators remotely and reliably, e.g., assistive robots, smart elevators, logistic applications [23] etc. For certain specific applications, e.g., rescue and search, the connectivity may not be present. Then the mobile robots play a crucial role by establishing an ad-hoc network and use the intermediate and forwarding nodes as a communication medium. These mobile robots, when operated using some already-fed lightweight routing protocols, have less overheads, then high efficiency in terms of their movement can be achieved [24].

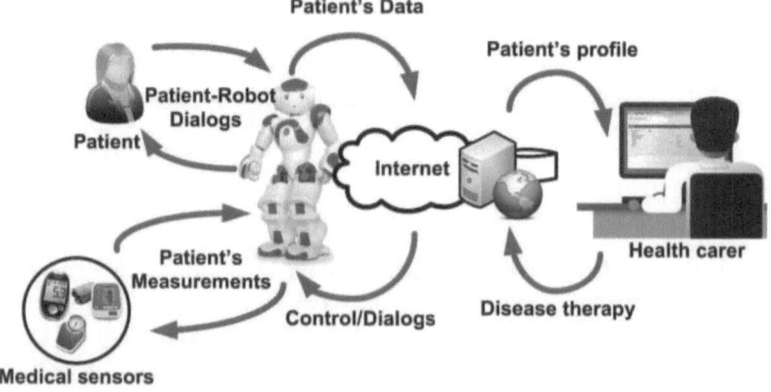

FIGURE 15.7 Support system for diabetes care (Al-Taee et al. [17]).

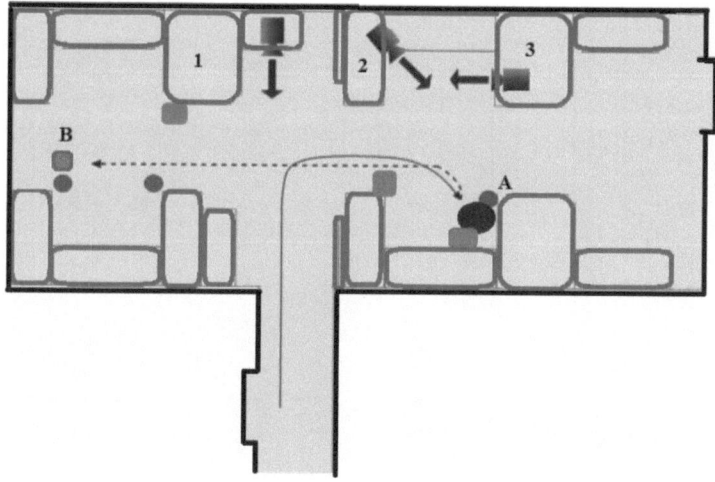

FIGURE 15.8 Distributed cameras located at position 1, 2, 3 help the robot to locate the charging stations marked by alphabet A. The charging station is placed between the brown square and green circle, and visual processing indicates the robot that the correct charging station is located at position A and not at positions B or C [22].

4. **Manipulation Ability of IoRT System**: This is an advanced capability, which sets robots apart from other computerized, automated systems as shown in Figure 15.10. The ability of a robot to interact physically and change its environment, i.e., manipulate it, provides chances for a variety of applications. However, robotics manipulation is a domain that is still in the stages of progress. While results of laboratory research are encouraging and have yielded impressive advances in the deftness and speed of robotic manipulators, many exceed the abilities of a human hand in some aspects.

There are a lot of opportunities for companies willing to produce and commercialize robotic manipulators that will open up new opportunities and applications for the robotic industry. The added advantage of IoT in this scenario is the data acquisition of the object features that are discoverable or non-discoverable with the robot sensors, such as distributed mass, e.g., some

FIGURE 15.9 Representation of a robot's ability to move (roboticsbuisnessreview.com).

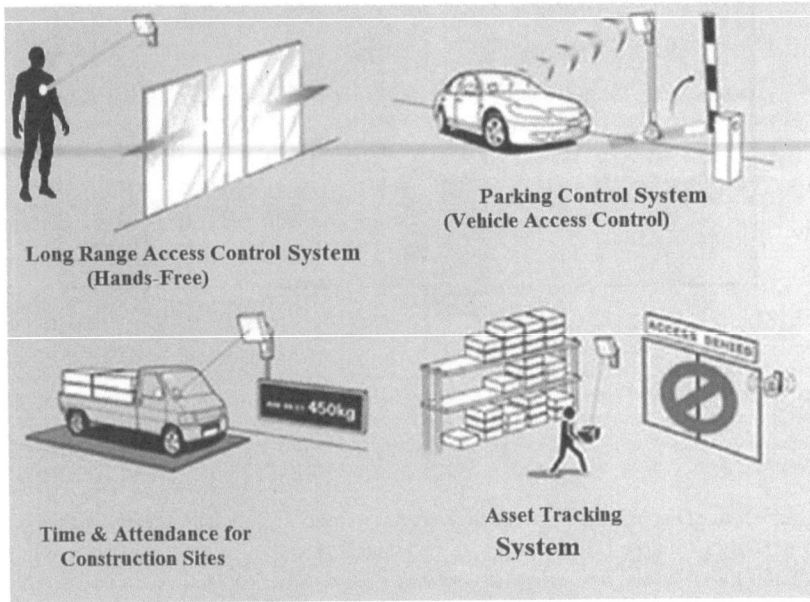

FIGURE 15.10   Representation of long-range RFID application (https://cluster.com).

researchers used RFID technology to detect the size, shape, and grasping points of a filled and empty cup [25]. Long-range RFID technologies were used to locate objects in kitchens [26], factories, toll tax points [27], etc.

5. **Adaptability of IoRT System**: This is the ability of a robot to adapt itself according to the changing environment, conditions, and scenarios, such as change in human behavior, faults, varying tasks, environment, etc. Perception, decisional, and configuration ability as described in the latter part below are the main enablers of adaptability of a robotic system. The major applications and supporting platform are as follows:

a. **Mobile Robots**: These robots are used in the agricultural domain for the deployment of fertilizer, irrigation, etc. These robots adapt themselves according to the variation of the crop sizes and environmental conditions such as light and weather.

b. **Wireless Sensor Networks**: It may be used to acquire the information of the soil moisture, temperature, and humidity of the soil in order to track the path accurately, e.g., Gealy et al. [28] used a robot to regulate the drip rate of individual water sprinklers to control plant water-level control. This also represent the advantage of using IoT in robotic systems.

c. **Building Open Service Gateway Initiative (OSGI)**: This is a home automation platform. AIOLOS (a mobile middleware framework for foraging on the android platform) exposes IoT and robots as shareable services and automatically augments the runtime placement in a distributed environment [29].

6. **Configurability of IoRT System**: This is the ability of a robot to be configured to perform a given task or to reconfigure it to perform a new task. Configuration and reconfiguration deals with both the hardware and software part of a robotic system. In a robotic system almost all the hardware has a software counterpart, such as data acquisition modules, driver software [30], etc. IoT is mainly helpful in supporting the software configurability in order to coordinate with the concerted configurations of numerous devices with each having different capabilities and contributing toward achieving complicated tasks. However, the integration IoT does not clearly address the needs of IoRT systems to uninterruptedly stream the data while interacting physically with the external environment. These requirements are basically needed for applications such as logistics and advance manufacturing as in these applications fast reaction to a disturbance is required along with the ability to adapt itself according to the varying production objectives. In [31], the ubiquitous network robotic platform, as shown in Figure 15.11, is a universal middleware for an IoRT environment; it manages the transfer of functionality to the other layer using real and virtual robots.

Many researchers have developed their own customized architecture-based systems, e.g., Kousi et al. [31, 32] have proposed a service-oriented architecture to support two types of production units, such as mobile production and autonomous production units. Michalos et al. in [33] investigated the flexibility feature of the product that uses high collaborative and autonomous robotic systems proposed a distributed system. In this work, the authors proposed a distributed system for data exchange and coordination of human robot collective processes connected to a centralized task organizer. Configurability can be combined with the decision-making capability, which will enable the systems

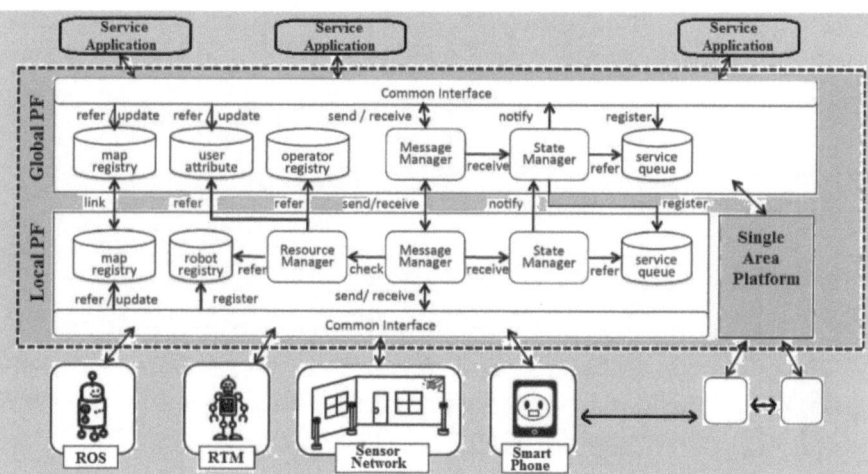

FIGURE 15.11 Representation of a two-layered network robotic platform. The local platform (LPF) configures the robotic system in a sole area. Global platform (GPF) is the middleware layer in between the LPF and service layer (Nishio et al. [31]).

to self-configure according to the desired task. However, self-configuration poses a challenge in IoRT-based systems since the configuration algorithms have to take care of the digital and physical interactions of the system with the real world.

7. **Decisional Autonomy of IoRT System**: This is the ability of a system to find out an optimum way to complete a desired task efficiently. In [34], the authors have presented a decisional system framework where a two-agent model has been proposed, i.e., human and robot, where they share a mutual space for information exchange through several modalities.

The interaction between the two agents occurs as a result of the unambiguous request from the human user to accomplish a certain task or the robot is in such a condition where interaction between the human and robot is useful. The global human–robot interaction framework between the human and robot is shown in Figure 15.12.

Robotics often relies on the artificial intelligence–based predictive models of the external environment, where the superiority of the plan critically depends on the superiority of these models for the estimation of the initial state. In this respect, the IoT environment improves the situational awareness and results in better plans. Task planners [35] based on human awareness use knowledge of human intents inferred through the IoT environment to create plans that admire constraint human interaction as shown in Figures 15.12 and 15.13. IoT also broadens the opportunity of decisional autonomy by making more performers and actions available, such as smart doors, elevators, vehicles, smart home [36], etc. However, IoT-based devices may be accessible or inaccessible,

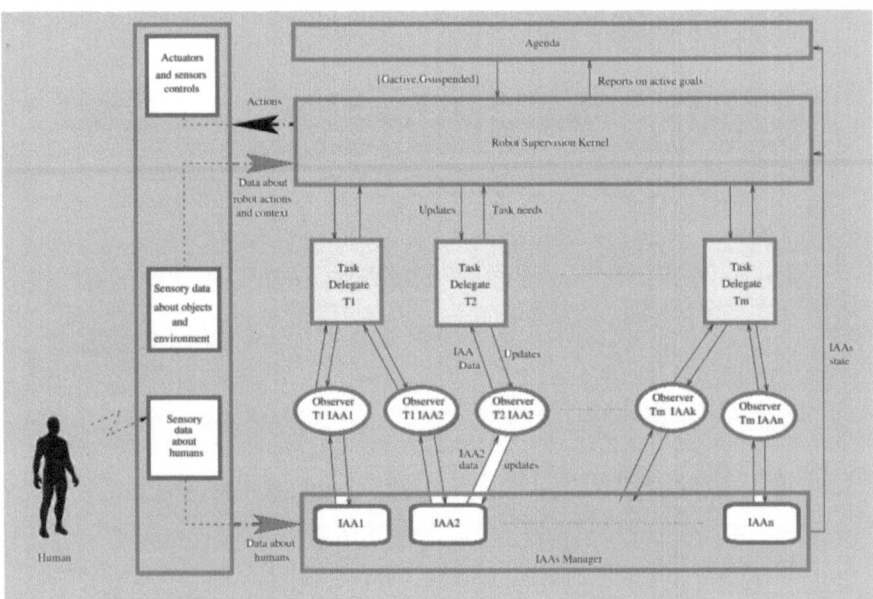

FIGURE 15.12   A global decision framework for human–robot interaction [34].

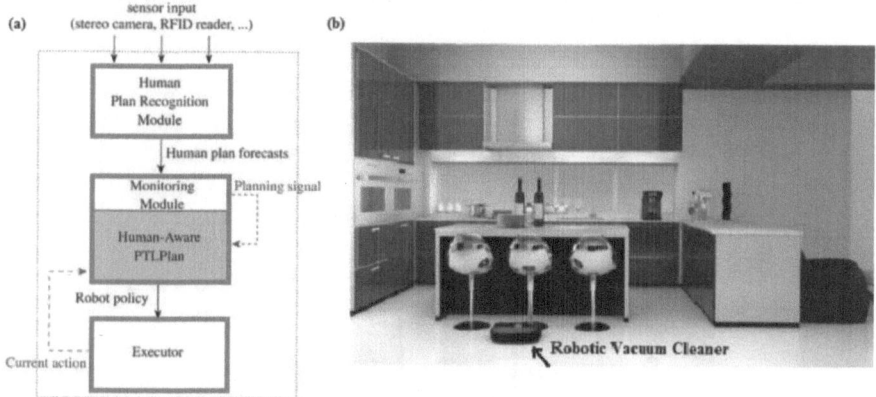

FIGURE 15.13   Vacuum cleaner robot adjusts its plan according to the surrounding environment to avoid the obstacles in the kitchen (Cirillo et al. [35]).

which poses a challenge on the classical multi-agent planning approaches. A possible solution is to plan in terms of intangible services that are mapped to the real device at runtime [37].

8. **Cognitive ability of an IoRT System:** The *cognitive ability of an IoRT system* is basically concerned with providing a robot with intelligent behavior by endowing it with a processing architecture that will enable its learning and reasoning capability about how to react in response to complicated goals in a real world. This part of the discussion is mainly focused on the learning and reasoning of IoRT multi-agent systems. The term "cognitive architectures" is mainly used in artificial intelligence and integrates human intelligence and its cognitive process for developing the system design models. A global cognitive architecture of a robotic system is proposed by Alami et al. in [38] as shown in Figure 15.14 that associates the current processes with the interacting subsystems performing according to the various temporal properties and implementing the cognitive ability of decision-making, communication, interpretation. Recent work shows that the cloud is a powerful tool to extract knowledge from multi-modal data sources like human demonstrations, sensor data observations, and to provide a virtual simulation platform to implement robot control policies [4].

Recent studies show that cognitive techniques are recently proposed in the IoT domain for management of distributive architecture [39]. Here pipelined data analytics have been self-organized by the system on a distributed set of sensor nodes, cloud, etc. To the best of our knowledge the inclusion of the pipelined data analytics has not yet been addressed. If the robots subscribe themselves as an additional agent in the external environment, it will give rise to a new class of problem in distributive context and collaboration for the IoT, as robots have a larger grade of autonomy than the classical IoT based smart devices as they are able to adjust the physical environment leading to complex reliance and interactions.

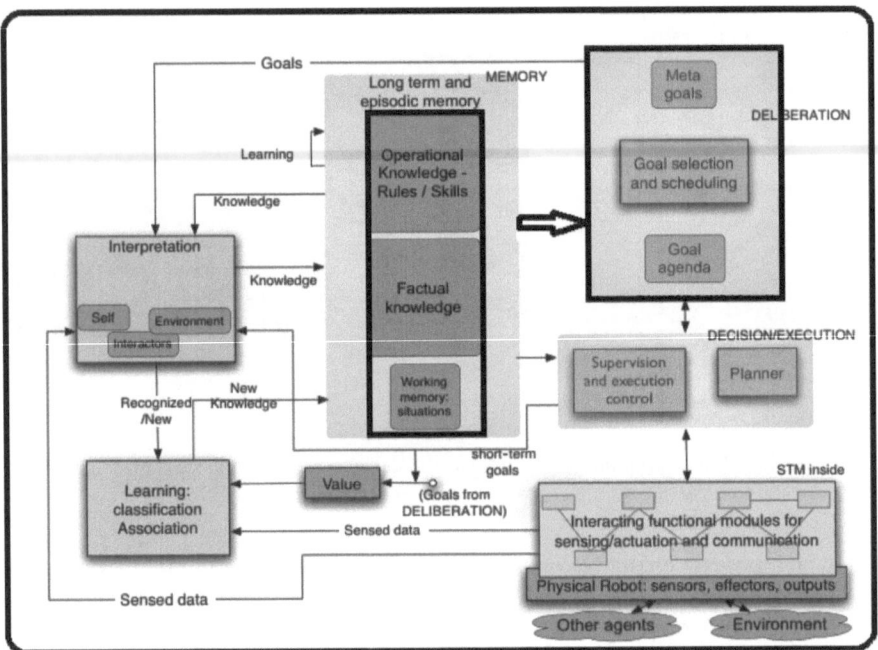

FIGURE 15.14   A global infrastructure cognitive robotics architecture (Alami et al. [38]).

15.2.2   Challenges in the IoRT Domain [2]

As the IoT is a novel area, the research in this field is in its intermediate stage. Many issues that are still unanswered have to yet to be addressed by the scientific community. In this section, we touch on three main challenges in the IoRT domain.

1. **Computational Challenge**: A key feature of IoRT systems is shared-offloading a complicated task to the cloud for its execution. However, this distributive approach requires a more rigorous and unified framework to handle such complex tasks. One possible solution to this problem is that first, a shared pool of robotics and resources shall be influenced together, where the shared-offloading approach should consider factors such as huge data and latency to finish a given task. Second, the IoRT systems should be able to make their decisions about whether it is appropriate to execute a task within the IoRT or not.
2. **Optimization Problem**: Computational challenge fully depends on the optimization. If optimization of a cost function of a robotic algorithm is not done, then the computational challenge gets worse. In a shared-offloading scheme, the computation is done in three steps [40].
   a. Standalone computation done on individual robotic system
   b. Collaborative computation done by a pool of robots in the network
   c. Cloud computation
   Some hybrid cloud models include partial computation considering all the steps together. While developing an IoRT-based optimization framework, it is

advisable to consider all the above computational steps and the communication technologies available and accordingly preset the computational cost [2], which shall lead to optimum computational strategy.

3. **Security Challenge**: This is a main challenge that one may face while working with an IoRT system; especially when a cloud is involved the following two security issues may be faced:

a.   The IoRT-VM system environment should be reliable, otherwise, some other IoRT-based system may interrupt a critical task without the involvement of the actual robot. E.g., in military applications the IoRT-VM system should be able to identify a trustworthy IoRT infrastructure among multiple available networks in order to avoid malicious infrastructure.

b.   Future robotic systems should be able to trust in order to initiate the computational task on the IoRT cloud infrastructure, where the cloud should be self-sufficient to verify the owner of the robotic system.

Here it has to be ensured that no malicious code is running at the background of the surrogated task. In the meantime, confidential data must be permanently saved in the trusted cloud server; on the other hand, cloning of confidential data must be done on private cloud servers. This requires firm methodologies to protect integrity, confidentiality, and trust to protect the IoRT system.

15.3   CONCLUSION

In this chapter, we offered a brief overview of an IoRT-based architectural concept as a progression of the existing cloud concept, a value addition due to the blend of IoT and robotics, and few challenges in IoRT systems so that those interested can get involved in this novel concept. IoRT allows the robotic system to interconnect, share, and distribute the computational resources, such as environmental data, activities related to business, etc., with each other inside a given network. This is basically to acquire accurate knowledge and desired skills that cannot be learned themselves, all below the cover of a sophisticated architectural framework. The advantages of IoT subjugated by the robotic system are mainly of distributive perception and M2M protocols. Present IoRT personifications are almost uniquely found in upright application domains, notably precision agriculture and Industry 4.0. Domain-agnostic solutions, for example, integration of robots in IoT middleware platforms, are only emerging. It is our hope that this chapter will encourage researchers from both disciplines to work to develop IoT, robot, and cloud platform-based ecosystems.

REFERENCES

1. P. Simoens, M. Dragone, and A. Saffiotti, "The Internet of Robotic Things: A review of concept, added value and applications," *Int. J. Adv. Robot. Syst.*, vol. 15, no. 1, pp. 1–11, 2018, doi: 10.1177/1729881418759424

2. P. P. Ray, "Internet of Robotic Things: Concept, technologies, and challenges," *IEEE Access*, vol. 4, pp. 9489–9500, 2016, doi: 10.1109/ACCESS.2017.2647747

3. "IEEE Society of Robotics and Automation's Technical Committee on Networked Robots" 2015. [Online]. Available: https://www-users.cs.%0Aumn.edu/~isler/tc/.

4. Kara D. and Carlaw D., "The Internet of Robotic Things," 2014.
5. P. Simoens *et al.*, "Internet of Robotic Things: Context-aware and personalized interventions of assistive social robots (short paper)," *Proc. 2016 5th IEEE Int. Conf. Cloud Networking, CloudNet 2016*, pp. 204–207, 2016, doi: 10.1109/CloudNet.2016.27
6. S. Bandopadhaya, R. Dey, and A. Suhag, "Integrated healthcare monitoring solutions for soldier using the Internet of Things with distributed computing," *Sustain. Comput. Informatics Syst.*, vol. 26, p. 100378, 2020, doi: 10.1016/j.suscom.2020.100378
7. M. Mazzara, I. Afanasyev, S. R. Sarangi, S. Distefano, V. Kumar, and M. Ahmad, "A reference architecture for smart and software-defined buildings," *Proc. 2019 IEEE Int. Conf. Smart Comput. SMARTCOMP 2019*, no. Section II, pp. 167–172, 2019, doi: 10.1109/SMARTCOMP.2019.00048
8. H. Tran-Dang, N. Krommenacker, P. Charpentier, and D. S. Kim, "Toward the Internet of Things for physical internet: Perspectives and challenges," *IEEE Internet Things J.*, vol. 7, no. 6, pp. 4711–4736, 2020, doi: 10.1109/JIOT.2020.2971736
9. T. Kagaya, "Strategy and efforts for robotics integration aiming at combining information and communications technology with robots," *NTT Tech Rev.*, vol. 10, no. 11, pp. 1–7, 2012.
10. "Oracle Big Data Service." https://www.oracle.com/in/big-data/big-data-service.
11. "Multi-annual roadmap for horizon." 2020.
12. Pieter Simoens, Mauro Dragone, and Alessandro Saffiotti, "A roadmap for US robotics, from internet to robotics." A roadmap for US robotics, from internet to%0Arobotics.
13. D. K. Misra, J. Sung, K. Lee, and A. Saxena, "Tell me Dave: Context-sensitive grounding of natural language to manipulation instructions," *Int. J. Rob. Res.*, vol. 35, no. 1–3, pp. 281–300, 2016, doi: 10.1177/0278364915602060
14. F. Yazdani *et al.*, "Cognition-enabled framework for mixed human-robot rescue teams," *IEEE Int. Conf. Intell. Robot. Syst.*, pp. 1421–1428, 2018, doi: 10.1109/IROS.2018.8594311
15. I. Leite, R. Henriques, C. Martinho, and A. Paiva, "Sensors in the wild: Exploring electrodermal activity in child-robot interaction," *ACM/IEEE Int. Conf. Human-Robot Interact.*, pp. 41–48, 2013, doi: 10.1109/HRI.2013.6483500
16. Y. Liu, R. De Liu, J. R. Star, J. Wang, and H. Tong, "The effect of perceptual fluency on overcoming the interference of the More A-More B intuitive rule among primary school students in a perimeter comparison task: the perspective of cognitive load," *Eur. J. Psychol. Educ.*, 2019, doi: 10.1007/s10212-019-00424-w
17. M. A. Al-Taee, W. Al-Nuaimy, Z. J. Muhsin, and A. Al-Ataby, "Robot assistant in management of diabetes in children based on the Internet of Things," *IEEE Internet Things J.*, vol. 4, no. 2, pp. 437–445, 2017, doi: 10.1109/JIOT.2016.2623767
18. M. Kuga, "Simultaneous," *Sen'i Gakkaishi*, vol. 40, no. 4–5, pp. P393–P395, 1984, doi: 10.2115/fiber.40.4-5_p393
19. A. A. Khaliq, F. Pecora, and A. Saffiotti, "Inexpensive, reliable and localization-free navigation using an RFID floor," *2015 Eur. Conf. Mob. Robot. ECMR 2015 - Proc.*, no. 2, pp. 1–7, 2015, doi: 10.1109/ECMR.2015.7324204
20. L. Luoh, "ZigBee-based intelligent indoor positioning system soft computing," *Soft Comput.*, vol. 18, no. 3, pp. 443–456, 2014, doi: 10.1007/s00500-013-1067-x
21. M. Bonaccorsi, L. Fiorini, F. Cavallo, A. Saffiotti, and P. Dario, "A cloud robotics solution to improve social assistive robots for active and healthy aging," *Int. J. Soc. Robot.*, vol. 8, no. 3, pp. 393–408, 2016, doi: 10.1007/s12369-016-0351-1
22. W. Chamberlain, T. Drummond, and P. Corke, "Distributed robotic vision as a service," *Australas. Conf. Robot. Autom. ACRA*, pp. 2494–2499, 2015.
23. B. Mutlu and J. Forlizzi, "Robots in organizations: The role of workflow, social, and environmental factors in human-robot interaction," *HRI 2008—Proc. 3rd ACM/IEEE Int. Conf. Human-Robot Interact. Living with Robot.*, no. January, pp. 287–294, 2008, doi: 10.1145/1349822.1349860

24. B. Sliwa, C. Ide, and C. Wietfeld, "An OMNeT++ based framework for mobility-aware routing in mobile robotic networks," 2016, [Online]. Available: http://arxiv.org/abs/1609.05351.

25. A. Saffiotti et al., "The PEIS-ecology project: Vision and results," *2008 IEEE/RSJ Int. Conf. Intell. Robot. Syst. IROS*, no. 1, pp. 2329–2335, 2008, doi: 10.1109/IROS.2008.4650962

26. R. B. Rusu, B. Gerkey, and M. Beetz, "Robots in the kitchen: Exploiting ubiquitous sensing and actuation," *Rob. Auton. Syst.*, vol. 56, no. 10, pp. 844–856, 2008, doi: 10.1016/j.robot.2008.06.010

27. R. Y. Zhong, Q. Y. Dai, T. Qu, G. J. Hu, and G. Q. Huang, "RFID-enabled real-time manufacturing execution system for mass-customization production," *Robot. Comput. Integr. Manuf.*, vol. 29, no. 2, pp. 283–292, 2013, doi: 10.1016/j.rcim.2012.08.001

28. D. V. Gealy et al., "DATE: A handheld co-robotic device for automated tuning of emitters to enable precision irrigation," *IEEE Int. Conf. Autom. Sci. Eng.*, vol. 2016-Novem, pp. 922–927, 2016, doi: 10.1109/COASE.2016.7743501

29. T. Verbelen, P. Simoens, F. De Turck, and B. Dhoedt, "AIOLOS: Middleware for improving mobile application performance through cyber foraging," *J. Syst. Soft.*, vol. 85, no. 11, pp. 2629–2639, 2012, doi: 10.1016/j.jss.2012.06.011

30. T. Heikkilä, T. Dobrowiecki, and L. Dalgaard, "Dealing with configurability in robot systems," *MESA 2016—12th IEEE/ASME Int. Conf. Mechatron. Embed. Syst. Appl.—Conf. Proc.*, 2016, doi: 10.1109/MESA.2016.7587120

31. K. Koji, H. Norihiro, and S. Nishio "Ubiquitous network robot platform for realizing integrated robotic applications," *Intell Auton Syst*, vol. 20, pp.278–288, 2013.

32. N. Kousi, S. Koukas, G. Michalos, S. Makris, and G. Chryssolouris, "Service Oriented Architecture for Dynamic Scheduling of Mobile Robots for Material Supply," *Procedia CIRP*, vol. 55, pp. 18–22, 2016, doi: 10.1016/j.procir.2016.09.014

33. G. Michalos, S. Makris, P. Aivaliotis, S. Matthaiakis, A. Sardelis, and G. Chryssolouris, "Autonomous production systems using open architectures and mobile robotic structures," *Procedia CIRP*, vol. 28, pp. 119–124, 2015, doi: 10.1016/j.procir.2015.04.020

34. A. Clodic, V. Montreuil, R. Alami, and R. Chatila, "A decisional framework for autonomous robots interacting with humans," *Proc. - IEEE Int. Work. Robot Hum. Interact. Commun.*, vol. 2005, no. May 2014, pp. 543–548, 2005, doi: 10.1109/ROMAN.2005.1513836

35. M. Cirillo, L. Karlsson, and A. Saffiotti, "Human-aware task planning: An application to mobile robots," *ACM Trans. Intell. Syst. Technol.*, vol. 1, no. 2, 2010, doi: 10.1145/1869397.1869404

36. F. Cavallo et al., "Development of a socially believable multi-robot solution from town to home," *Cognit. Comput.*, vol. 6, no. 4, pp. 954–967, 2014, doi: 10.1007/s12559-014-9290-z

37. C. Goumopoulos, *A middleware architecture for ambient adaptive systems.* 2016.

38. R. Alami, R. Chatila, A. Clodic, S. Fleury, M. Herrb, V. Montreuil, and E. A. Sisbot, Towards Human-Aware Cognitive Robots, 2006.

39. V. Foteinos, D. Kelaidonis, G. Poulios, P. Vlacheas, V. Stavroulaki, and P. Demestichas, "Cognitive management for the internet of things: A framework for enabling autonomous applications," *IEEE Veh. Technol. Mag.*, vol. 8, no. 4, pp. 90–99, 2013, doi: 10.1109/MVT.2013.2281657

40. G. Hu, W. P. Tay, and Y. Wen, "Cloud robotics: Architecture, challenges and applications," *IEEE Netw.*, vol. 26, no. 3, pp. 21–28, 2012, doi: 10.1109/MNET.2012.6201212

16 GaAs Nanostructure-Based Solar Cells with Enhanced Light-Harvesting Efficiency

D. V. Prashant and Dip Prakash Samajdar
Department of Electronics and Communication
Engineering, PDPM Indian Institute of Information
Technology, Design and Manufacturing, Jabalpur, India

CONTENTS

16.1 INTRODUCTION

In the past few decades, photovoltaic solar cells (SCs) have received tremendous attention as a low-carbon energy source and because of their potential to limit the use of fossil fuels. Crystalline Si (c-Si) SCs have dominated the photovoltaic market for years with an efficiency of ~25%, which is quite close to the theoretical limit of 31% (Wright and Uddin 2012). Silicon is abundantly available, non-toxic, and possesses good optoelectronic properties. But fabrication of p-n junctions for Si involves a high temperature (~1000°C) process, which is very expensive and limits the numerous applications of the photovoltaic cells at the domestic level (Chen et al. 2013). Thin-film or second-generation SCs, utilizing much thinner semiconductor absorber layers, are one of the possible solutions for low-cost SCs. But the PCE of thin-film SCs is (~19%), which is quite low compared to conventional cells. Their lower absorption efficiency is mainly attributed to high transmission and reflection losses due to the

DOI: 10.1201/9781003181613-16

thinner absorbing layers (Mavrokefalos et al. 2012). Apart from the thin-film structures, various other solar cell configurations, such as organic polymer-based SCs (Ibrahim et al. 2016), hybrid SCs (Kou et al. 2015), and semiconductor nanostructure-based SCs (Garnett and Yang 2010) have been tested in recent years. Organic solar cells (OSCs) have generated considerable research interest because of their potential qualities like low-cost production, processing at low temperature, great mechanical flexibility, and abundant availability (Khalil et al. 2016; Srivastava, Samajdar, and Sharma 2018). However, the PCE of OSCs is below 15%, and the stability is also very poor, which means they are not used in commercial applications (Yang et al. 2017). On the other hand, hybrid solar cells (HSCs), which are fabricated using a combination of both organic and inorganic semiconducting materials have shown some promising results in terms of PCE as well as other photovoltaic properties (Adikaari, Dissanayake, and Silva 2010). A hybrid heterojunction comprising of a p-type organic polymer and an n-type inorganic semiconductor is fabricated via various techniques, such as printing, coating, and spraying (Ong and Levitsky 2010). In HSCs, the advantageous properties of inorganic semiconductor materials such as superior charge transport, excellent light absorption qualities, quantum confinement effects, and reliability are combined with the simple fabrication methodologies, mechanical flexibility, and cost-effectiveness of organic polymer materials. Some HSC structures have shown PCE as high as ~15%, with excellent optical and charge transport properties, which clearly establishes the fact that these structures have the potential to realize low-cost and high-efficiency SCs (Dayal et al. 2010; Liu 2014; Nicolaidis et al. 2011). However, plenty of theoretical and practical investigations still need to be carried out to find out suitable combinations of organic and semiconductor materials, simpler fabrication techniques, and optimal design of such structures. Similarly, semiconductor nanostructure-based SCs have attracted intense research interest as promising candidates for low-cost, high-efficiency applications in recent years. Various nanophotonic structures, such as nanowires (Fan et al. 2009), nanopyramids (Wang et al. 2018), nancones (Jeong et al. 2012), etc., have been studied by various research groups. It is well-established now that, in comparison to conventional planar SCs and OSCs, substrates with well-arranged and well-designed nanostructures show better light absorption and carrier transport properties (Abujetas, Paniagua-Domínguez, and Sánchez-Gil 2015; Yao et al. 2014). A number of methods are available to synthesize, optimize, and integrate different types of nanostructures on low-cost substrates (Munshi et al. 2014; Yao et al. 2014). In Figure 16.1, we have shown a schematic overview of the structure of a basic nanowire-based SC along with the mechanism of the transport of the photogenerated charge carriers to the respective electrodes.

It is well-known that nearly 40% of the incident sunlight striking the SCs from different angles throughout the day is either reflected back or gets wasted. Consequently, it has been a critical task to minimize the amount of such "wasted photons." Ren et al. 2011 reported that the nanostructured surfaces suppress the omnidirectional broadband reflection and improve the absorption of the incident light (Chen et al. 2013). Additionally, such structures also offer efficient charge transport/separation via increased junction area (Kayes, Atwater, and Lewis 2005). Considering these advantages, tremendous efforts are made to devise the

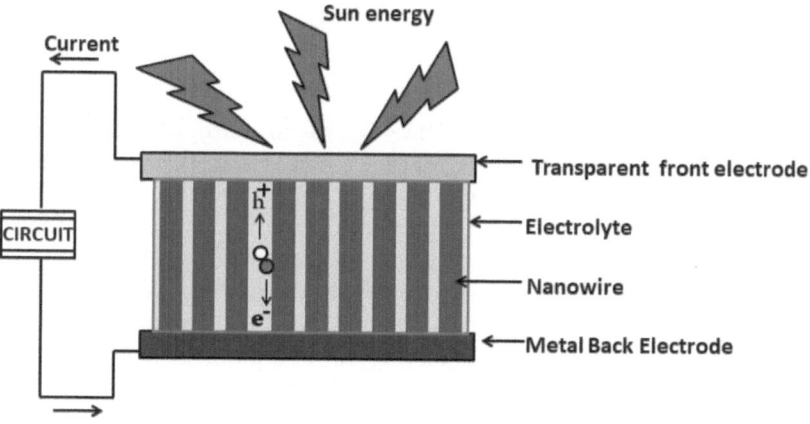

FIGURE 16.1   Pictorial representation of general device structure of nanowire-based SCs.

cost-effective and -efficient SCs by utilizing the light-trapping capabilities of these nanostructures. The light absorption properties of semiconductor nanostructures are quite sensitive to their geometric parameters, such as shape, size, length, and density. The optimization of these parameters results in significant improvement in the photovoltaic performance of the nanostructures (Garnett and Yang 2010; Kelzenberg et al. 2009; Kupec and Witzigmann 2009). Apart from Si, III-V nanostructure arrays have shown exceptional optical and electrical properties, which have paved the pathway for accelerating the research on high-performance photovoltaic applications. To fabricate optimal nanostructure-based SCs, it is necessary to evaluate their geometric designs and optoelectronic properties theoretically (Alamo 2011). In this chapter, we have proposed a fully analytical method to calculate the optimal geometric configuration for GaAs nanostructured-based SCs by using 3D FDTD method. The important geometric parameters like diameter, shape, and density of the nanostructures are optimized in order to achieve highest-possible optical absorption, J_{sc}, and PCE. GaAs nanostructure-based SCs exhibit $J_{sc} > 35$ mA/cm^2 and PCE > 22%, which is ~30% higher in comparison to the conventional planar SC structure. Apart from GaAs, other III-V direct bandgap semiconductors, such as indium phosphide–based (InP) and indium arsenide–based (InAs) nanostructures have shown light-absorption > 90% throughout the incident wavelength regime and $J_{sc} > 40$ mA/cm^2. We also perform FDTD simulations for GaAs NW/poly (3-hexylthiophene) (P3HT) and poly (3, 4-ethylenedioxythiophene) polystyrene sulfonate (PEDOT: PSS) hybrid solar cell structures and observe that optimized thickness of polymer coating over nanostructures can enhance their light-absorption by reducing the fraction of reflected photons. The optoelectronic simulations presented in this article clearly indicate that the geometrically optimized nanostructure-based SCs have much better optical, as well as electrical properties, compared to planar SCs. The obtained results presented here can help generate some novel experimental ideas to realize fabrication of high-efficiency GaAs nanostructure SCs.

16.2   ROLE OF NANOSTRUCTURES IN EFFICIENCY ENHANCEMENT OF SCS

As we have mentioned earlier that for polished Si surface, nearly 40% of the incident photons are reflected back to the air and does not get absorbed, when averaged from all angle of incidence. Thus, to achieve higher PCE in SCs, it is crucial to reduce the number of these wasted photons (Wei et al. 2013).

To address this issue, many efforts have been made, and it was discovered that semiconductor nanostructures with optimized geometrical shapes and sizes exhibit excellent optical as well as electrical properties. In the past few years, many nanostructures like nanowire, nanocone, nanopyramid (NP), and truncated nanopyramid (TNP) have been studied, and by combining intrinsic anti-reflection properties of these novel structures with efficient light absorption properties, we can achieve near perfect light absorption with reduced material cost (Hong et al. 2014; Park et al. 2013; Wu et al. 2018; Zhang et al. 2011). Effective light trapping along the axis of the nanostructures can harvest large portions of the incident photon energy, whereas the radial direction of the nanostructures enhances the carrier transport and collection efficiency by providing short carrier collection length (Wen et al. 2011). To provide a much better picture, we have presented some simulated results, which represents the optical performance of two different SCs structures. In Figure 16.2, the light absorption of the nanostructure array SC is much better compared to conventional planar SC structure throughout the incident wavelength region (300–1100 nm). The superior optical properties of nanostructure-based SCs are attributed to their superior anti-reflection and light-harvesting qualities. We also calculate the short-circuit current density (J_{sc}) for both the structures only to find out that nanostructure array–based SC achieved J_{sc} of 38 mA/cm^2, which is ~15 mA/cm^2 higher compared to the conventional structure.

Apart from superior electrical and optical properties, nanostructure array–based SCs have other benefits over conventional SCs, such as low-cost fabrication, low-temperature processing, and scalability (Takagahara and Takeda 1992). Over the

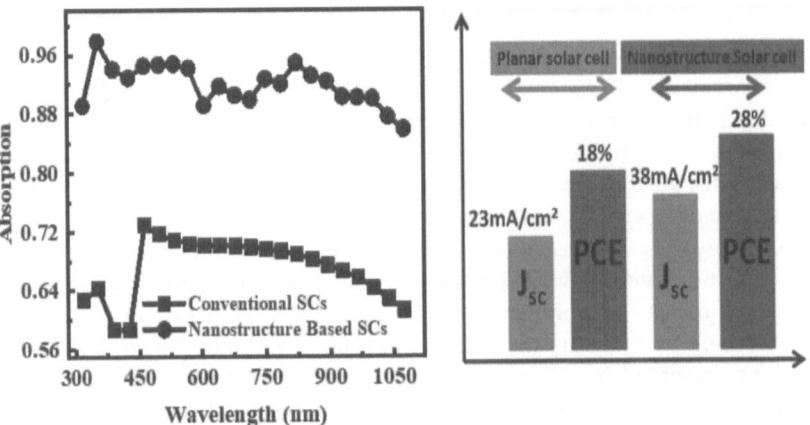

FIGURE 16.2   Comparative plot of optical absorption of conventional planar SCs and nanostructure-based SCs.

past few years, many theoretical, as well as experimental, works have been published in the field of high-efficiency nanostructure-array SCs, and much of the research is mainly focused on improving the design of these SCs. In the majority of reports, the performance of the device is enhanced by optimizing the geometrical parameters and design of the nanostructures, and it is found that the photovoltaic properties of such structures largely depend on their geometric configuration (Wu et al. 2017). To provide a clear-cut idea of the role of nanostructures and optimization of its geometrical dimensions in the efficiency enhancement of SCs, we have listed some of the recent theoretical results published in this domain (Table 16.1).

TABLE 16.1

Recent Experimental and Theoretical Works in the Field of Nanostructure Array-Based SCs

Structure	Experimental Details	J_{sc} (mA/cm²)	V_{OC} (Volts)	FF (%)	PCE (%)
A unit cell of GaAs nanowire array (NWA) with fixed length (L) and variable diameter (D) and filling ratio (FR) (Wen et al. 2011).	The influence of geometrical parameters on the optical absorption of the GaAs NWA is thoroughly analyzed using the **FDTD** method. To achieve maximal light absorption and PCE, the diameter D and FR of NWs are optimized. The NWA shows broadband absorption throughout the wavelength region (300–900 nm).	28.7	0.96	-	22.3
A hybrid solar cell (HSC) structure containing c-Si nanopyramid (NP) structures coated with PEDOT: PSS polymer, embedded with metal nanoparticles (MNPs) (Sachchidanand and Samajdar 2019).	The length, base-size, and FR of the NPs are optimized. The NPs are coated with optimized thickness of organic PEDOT:PSS to improve the absorption characteristics of the structure. MNPs with optimized diameters are also embedded in between the NPs.	25.91	0.56	78.2	11.32
A unit cell of GaAs nanowire-coated with PEDOT: PSS polymer (hybrid structure) (Prashant, Samajdar, and Sharma 2019).	Superior anti-reflection and light-absorption properties are achieved by optimizing the geometric parameters (diameter and FR) of the GaAs nanowire. By optimizing the polymer PEDOT, PSS thickness broad absorption spectra (**<85%**) is achieved.	36.6	0.96	86.2	27.8

(Continued)

TABLE 16.1 *(Continued)*

Recent Experimental and Theoretical Works in the Field of Nanostructure Array-Based SCs

Structure	Experimental Details	J_{sc} (mA/cm^2)	V_{OC} (Volts)	FF (%)	PCE (%)
HSC composed of Si nanocones and conductive PEDOT: PSS polymer (Jeong et al. 2012).	The aspect ratio (height/diameter) of the nanocones is optimized and conformal coating of PEDOT: PSS polymer is provided via spin coating. The fabricated structure has shown excellent antireflection and light-trapping properties.	31	0.55	67.7	11.1
HSC composed of silicon NP arrays and conductive PEDOT: PSS polymer (Chen et al. 2013).	The nanopyramid structures fabricated on the Si substrate improves the light absorption and carrier transport properties by providing anti-reflection properties and radial-junction architecture.	32.5	0.48	69.6	10.8
HSC composed of GaAs NW arrays and conductive PEDOT: PSS polymer (J. Chao, Shiu, and Lin 2010).	By controlling the etching time, the length of the NWs is varied. The light-trapping properties of the structure is strongly affected by the length of the NWs, and superior optical properties can be achieved by optimizing their value.	10.24	0.61	55	3.46
HSC composed of Si nanopyramids and conductive PEDOT: PSS polymer (X. Wang et al. 2018).	To achieve excellent light-trapping the periodicity of the nanopyramids and conformal polymer-coating thickness is optimized. Similarly, by improving the morphology of nanopyramids reduced carrier recombination rate is achieved.	32	0.57	74.3	13.8
HSC composed of GaAs NW array/ PEDOT: PSS/ P3HT conductive polymers (J. J. Chao, Shiu, and Lin 2012).	A new type of hybrid structure GaAs NWs and PEDOT:PSS incorporating P3HT as an electron-blocking layer is demonstrated. A large conduction band offset b/w the GaAs, and P3HT favors the electron transport towards the electrode and improves the carrier collection efficiency. Additionally, by optimizing the thickness of the polymer coating, superior carrier transport and light absorption is achieved.	20.2	0.65	70	9.2

Various novel designs have been proposed in recent years to enhance the performance of the SCs. We have used a variety of optimization techniques and implemented them on different shapes of nanostructures to propose the best possible design. Meanwhile, plenty of the research work is done with Si material. However, in this chapter, we have focused GaAs as our core material because of its promising optical as well as electrical properties and its room temperature band gap of 1.43 eV, which is very close to the ideal value of band gap for inorganic material of 1.5 eV for maximum photoabsorption (Bi and LaPierre 2009; Chao, Shiu, and Lin 2010).

16.3   RESULTS AND DISCUSSIONS

16.3.1   Optical Simulation of GaAs Nanostructure-based SCs

The light absorption in nanostructures like nanowires (NWs) is controlled by two dominant mechanisms: (a) diameter-dependent mode resonance within the NWs, and (b) periodic nature of reflection and transmission of the incident light (Wen et al. 2011).

To achieve the maximum possible light absorption, we have to find out the relationship between the geometrical parameters and each light-absorbing process of such nanostructures. Similarly, for other nanostructures like nanopyramids (NPs), nanocones (NCs), and truncated Pyramid (TPs), their base size and FR govern the optical properties. To understand the effect of base size on the absorption mechanism, first we have varied the diameter (D) of the NWs, and during this process, we have kept period (P) constant. The diameter of the GaAs NWs is varied from 120 nm–300 nm, keeping the FR (defined as base-size to period ratio) fixed at 0.5. From Figure 16.3(a), it can be seen that as the base-size of the NWs is increased above 120 nm, the light absorption in the longer wavelength region (>750 nm) is significantly enhanced. This improvement in light absorption of NWs can be understood from a report presented by (Colombo et al. 2012; Mariani et al. 2013), where the concept of

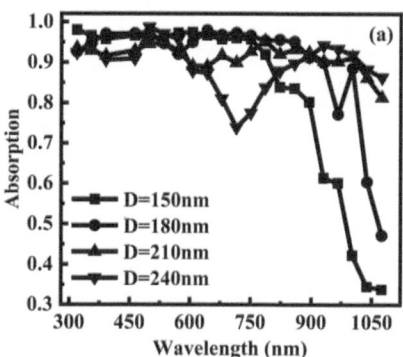 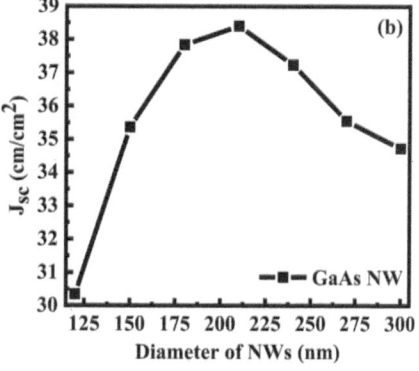

FIGURE 16.3   (a) Optical absorption spectra, (b) J_{sc} of the GaAs NWA-based SCs as a function of their base diameter.

guided resonance modes to explain the process of light absorption in nanostructures is used. For nanowires (NWs), when their diameter is too small (80 nm–140 nm), the incident light cannot be guided into the NWs properly as they are able to support only a few modes of resonance. Conversely, if the D of the NWs is too large, there will be an increase in surface reflection, which will result in loss of incident photons (Wu et al. 2017).

Thus, to guarantee maximal photon absorption, there should be an optimal intermediate value of base size. In our case, we have observed a similar phenomenon for GaAs NW arrays. When the D of the NWs is small (120 nm or 180 nm), the light absorption in the higher-wavelength region is poor, which is due to the feeble support of guided resonance modes. However, as the D of NWs is increased to 210 nm, significant increase in optical absorption especially in higher-wavelength regions is observed, which is attributed to the contribution of the guided resonance modes and reduced reflection and transmission of incident light. Further, as the D of the NWs increased above 240 nm, the light absorption reduces drastically in the 850 nm–1100 nm wavelength region, which is due to increased reflection of incident light. To verify our obtained results, we have calculated the J_{sc} as a function of the D of the GaAs NW array shown in Figure 16.3(b), where it can be seen that, as the base-size is increased above 120 nm, the value of J_{sc} increases and reaches its maximum value at an optimal point. This increment in J_{sc} is attributed to improved light absorption and the supporting role of different resonance modes throughout the incident wavelength regime. However, as the base size is increased after a certain point, J_{sc} starts decreasing, which is due to increased reflection and transmission losses of incident light (Sturmberg et al. 2014; Wu et al. 2016).

To achieve broad absorption spectra and to reduce reflection and transmission losses, we have evaluated the optimal FR or base-size to period ratio for nanostructure arrays. First, we have fixed the D of the NWs at their optimal value 210 nm (from simulated results) and then varied their period (P), which results in changes in FR.

In Figure 16.4(a), one can notice that, for lower value of FR like 0.4, the light absorption of NWs is unstable or oscillating in nature for the whole wavelength

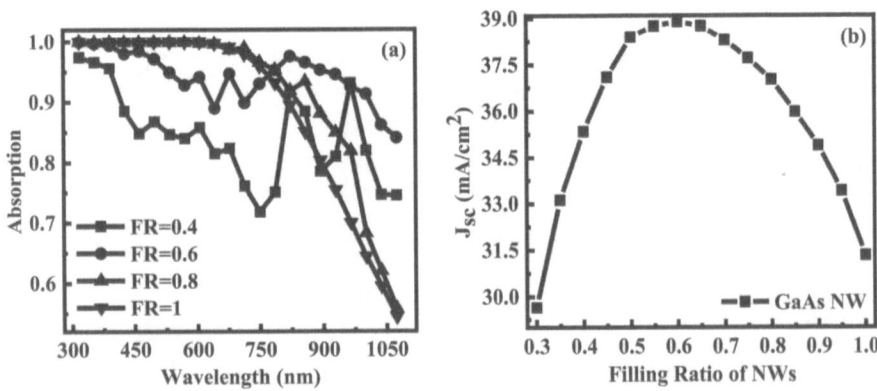

FIGURE 16.4   (a) Optical absorption, (b) J_{sc} of GaAs NWA-based SCs as a function of their FR.

region, which is due to very large reflection from the flat surface of the substrate between the NPs. However, as the FR of the NWs is increased above 0.4, the oscillations starts decreasing and NWs achieve a stable and broad absorption spectra at FR = 0.6, which is attributed to reduced surface reflection and transmission losses throughout the incident wavelength region. Although, further increase in the value of FR results in degradation of absorption in the wavelength region ranging from 700 nm–1100 nm because of the larger coverage of NWs in a unit cell. It is also responsible for reflection from the top of the NW arrays, which can be observed from Figure 16.4(a) for FR = 1. In Figure 16.4(b), we have shown the J_{sc} of the NWs as a function of their FR, and it is evident that, as the FR is increased, J_{sc} of the NWs increases and reaches a maxima, which is attributed to the reduced surface reflection of the incident light. Further, an increment in the FR of NWs results in a decrement of J_{sc}, which is due to an increase in light reflection from the top of the nanostructure arrays.

After optimizing the geometrical properties of NWs, we have carried out the same analysis to find the optimal geometrical configuration for other semiconductor nanostructures like truncated nanopyramid (TNP) and nanopyramid (NP) structures.

In Figure 16.5, we have presented the optical properties of all three nanostructures for their optimal geometrical parameters and one can notice that all nanostructures absorb nearly 90% of the incident light throughout the entire incident wavelength region, which is attributed to superior light absorption and minimal reflection/transmission losses exhibited by the nanostructures after the optimization of their geometric parameters. However, compared to NW and NP structures, TNP structures have shown they are a stable and broad light absorption spectrum, which is due to the uniqueness in their design. The tilted side surfaces of the TNP structures allow them to capture the photons incident from different angles owing to multiple reflections in between the adjacent TNPs, whereas in the case of NW structures, it is not possible because of their cylindrical shape and only their top flat surface is effective for light absorption. Similarly, the flattened top surface of the TNP structures also helps to absorb the incident photons unlike NP structures where

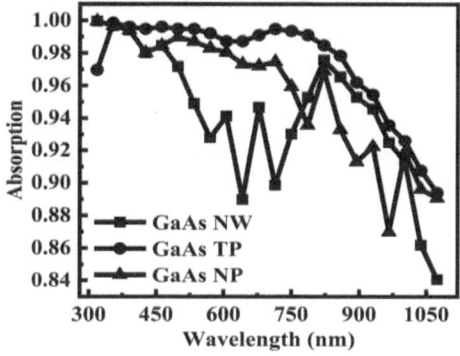

FIGURE 16.5 Optical absorption of nanostructure array-based SCs for their optimized geometrical parameters.

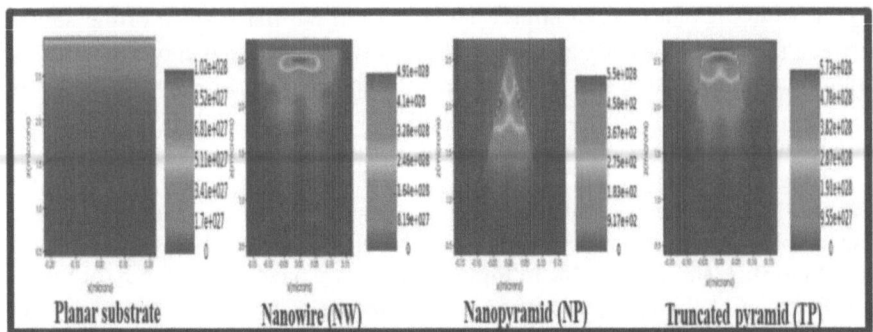

FIGURE 16.6 Optical generation rates (cm⁻³s⁻¹) of planar substrate and different nanostructures for their optimized geometric configuration.

sharp top edges could hardly absorb the incident photons(Prashant, Samajdar, and Arefinia 2021).

To verify this, we have generated the optical photon absorption profiles of all three structures (refer to Figure 16.6). For GaAs NW, high optical generation rates are concentrated at the top surface of the NWs, and for NP structure, almost zero photon absorption is observed at its top edges, while for the TNPs, high optical generation rates are obtained at both top and side walls. In Table 16.2, we have listed some of the important photovoltaic parameters that we have obtained from the FDTD simulations. A base size of 240 nm and FR of 0.7 are found to be the optimal value for both TNP and NP structures. In comparison to NP and TNP structures, slight differences in the optimal value of geometric parameters of NW structures is observed, which is due to the difference in the shape of NW in comparison to the other two structures.

TABLE 16.2

Important Photovoltaic Parameters Obtained from Optical Simulations of Different Nanostructure-Based SCs

Structure	Photon Generation rate (cm⁻³s⁻¹)	Optical J_{sc} (mA/cm²)	Average Light Absorption (%)	Optimal Configuration
Nanowire (NW) (Prashant et al. 2020)	3.1×10^{28}	38.41	94.2	D = 210 nm, FF = 0.6
Nanopyramid (NP) (Prashant et al. 2020)	3.8×10^{28}	39.81	95.5	D = 240 nm, FF = 0.7
Truncated Pyramid (TNP) (Prashant et al. 2020)	4.3×10^{28}	41.69	97.3	D = 210 nm, FF = 0.7

16.3.2 OPTICAL SIMULATION OF III-V SEMICONDUCTOR NANOWIRE NWS BASED SCS

Apart from GaAs, various other III-V semiconductor materials have been used in a number of new cutting-edge classes of electronic and optoelectronic devices, such as photodetectors, high-electron mobility transistors, and other optoelectronic devices (Li et al. 2016; Raychaudhuri et al. 2009). III-V materials like InP and InAs are also direct bandgap materials with bandgap of 1.34 eV and 0.33 eV, respectively, and have great technological significance (Vurgaftman and Meyer 2001). Like GaAs, these materials also possess attractive properties like high-electron mobility, infrared emission, non-toxicity etc., and can be used in photovoltaic cells application. To get a better understanding, we have performed optical simulations of InP and InAs NW array-based SCs and found that, like GaAs nanostructure-based SCs, these structures also show properties like scalability and quantum confinement effects, and by optimizing their geometrical configurations, one can achieve superior optical and electrical properties.

In Figure 16.7, the light absorption properties of the three different III-V semiconductor NW array-based SCs (for their optimized geometric configurations) are presented, and it can be observed that all three structures have achieved light absorption above 80% throughout the incident wavelength region. Further, we have also calculated J_{sc} and other important photovoltaic parameters of these structures, which are listed in Table 16.3. Out of the three materials, InAs NWs have shown excellent optoelectronic properties and the highest J_{sc} of 41.8 mA/cm² , which is attributed to broad absorption spectra (> 90%) throughout the incident wavelength region.

16.3.3 OPTICAL SIMULATION OF SEMICONDUCTOR NW/POLYMER HSC

In recent years, a lot of research has been done on semiconductor nanostructures/polymer-based HSC structures, and it is reported that polymer coating onto the semiconductor nanostructures can further improve their light absorption by reducing the

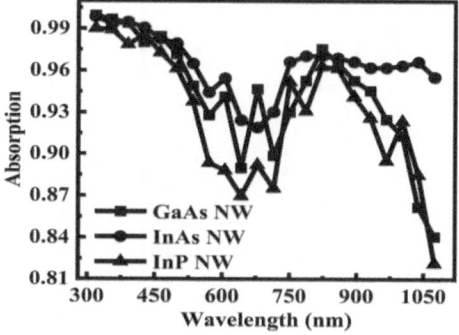

FIGURE 16.7 Optical absorption of III-V semiconductor NW-based SCs at their optimized geometrical configuration.

TABLE 16.3

Important Photovoltaic Parameters Obtained from Optical Simulations of Different III-V Semiconductor NW-Based SCs

Structure	Photon Generation rate (cm^{-3}s^{-1})	Optical J$_{sc}$ (mA/cm²)	Average Light Absorption (%)	Optimal Configuration
GaAs NW (Prashant et al. 2020)	3.1×10^{28}	38.41	94.2	D = 210 nm, FF = 0.6
InP NW (Prashant et al. 2020)	2.9×10^{28}	36.53	92.1	D = 210 nm, FF= 0.6
InAs NW (Prashant et al. 2020)	4.2×10^{28}	41.82	96.5	D = 210 nm, FF = 0.6

amount of reflected and transmitted light (Wang et al. 2014). In addition, by optimizing the thickness of polymer coating onto the nanostructures, properties such as optical absorption, carrier transport, and J$_{sc}$ can be significantly enhanced (Chao, Shiu, and Lin 2010; Chao, Shiu, and Lin 2012; Yong et al. 2012). We perform optical simulation of GaAs NW/PEDOT: PSS HSC structure.

In Figure 16.8, the optical absorption of both the structures at their optimized geometric configuration is presented, and we can see that the light absorption of the GaAs NWs is significantly enhanced after the polymer coating, which is attributed to efficient light trapping provided by the coated polymer due to reduced reflection losses. The light absorption of the GaAs NW structure is improved from >84% to >94% throughout the incident wavelength region with the introduction of the polymer coating. In our case, we have provided 50 nm thick polymer coating onto the NWs, which is an optimal value that we have already evaluated in our recent work (Prashant, Samajdar, and Sharma 2019). We have calculated the optical J$_{sc}$ of the hybrid structure as a function of polymer coating thickness in Figure 16.8(b), where it can be noticed that, as the thickness

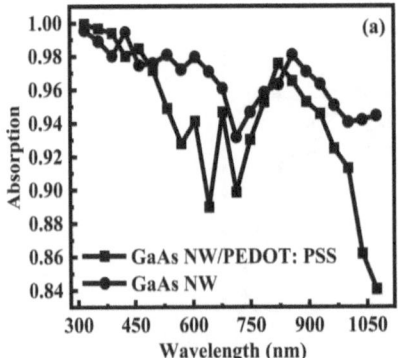
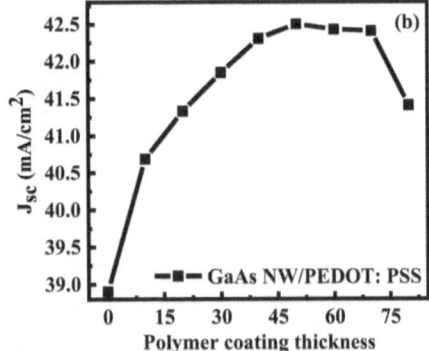

FIGURE 16.8 (a) Normalized absorption plot of GaAs NW SCs and GaAs NW/PEDOT: PSS HSCs at their optimal geometric configuration, (b) J$_{sc}$ as a function of polymer coating thickness.

TABLE 16.4

Important Photovoltaic Parameters Obtained from Optical Simulations of GaAs NW SCs and GaAs NW/PEDOT: PSS Hybrid SCs

Structure	Photon Generation rate (cm⁻³s⁻¹)	Optical J$_{sc}$ (mA/cm²)	Average Light Absorption (%)	Optimal Configuration
GaAs NW (Prashant et al. 2021)	3.1×10^{28}	38.41	94.2	D = 210 nm, FF = 0.6
GaAs NW/PEDOT: PSS (Prashant et al. 2021)	4.8×10^{28}	42.53	97.3	D = 210 nm, FF = 0.6, Coating thickness of 50 nm

of the polymer coating onto the NWs is increased, the J$_{sc}$ of the structure is enhanced and reaches a maximum value at 50 nm. However, further increase in coating thickness results in decrease of J$_{sc}$, due to poor light trapping by the structure. Thicker polymer coating obstructs the incident light to reach effectively to the NWs. In Table 16.4, we have listed some of the important geometric parameters that we have obtained from the FDTD simulation. The optical J$_{sc}$ of the GaAs NWs SC is increased by 4 mA/cm² after the optimal PEDOT: PSS polymer coating.

16.3.4 Device Simulations of Planar and Nanostructure-based SCs

Finally, we have obtained the J-V and P-V characteristics of the planar and nanostructure-based solar cell (shown in Figure 16.9).

GaAs nanostructure-based SC have achieved much higher open-circuit voltage (V$_{oc}$) and J$_{sc}$ compared to its planar counterpart, which is attributed to its superior light-capturing and -harvesting properties. The PCE of nanostructure-based SC is 28%, which is nearly 1.65 times much better than the planar structure. The nanostructure-based SC shows J$_{sc}$ of 33 mA/cm², which is ~10 mA/cm² higher than the

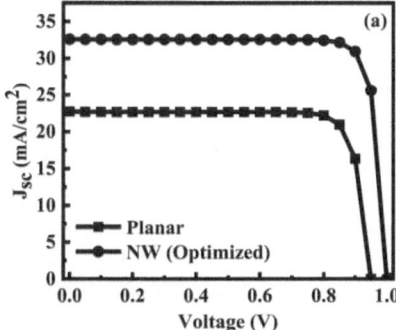
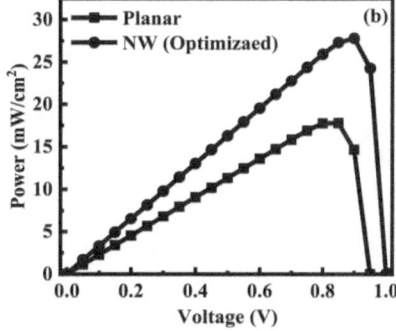

FIGURE 16.9 (a) J-V and (b) P-V curves of conventional and nanostructure-based SCs.

TABLE 16.5

Important Photovoltaic Parameters of Both the Structures Obtained from the Optical and Electrical Simulations

Structures	Photon Generation (cm^{-3}s^{-1})	Average Light Absorption (%)	Optical J_{sc} (mA/cm^2)	Electrical J_{sc} (mA/cm^2)	V_{oc} (V)	Fill Factor (%)	PCE (%)
Planar structure (Prashant et al. 2021)	1.3×10^{28}	67.57	27.48	22.78	0.94	84.10	17.97
Nanostructure-based SC (Prashant et al. 2021)	3.1×10^{28}	91.96	38.91	32.60	0.99	86.29	27.85

planar SC (Table 16.5). However, electrical J_{sc} obtained from the DEVICE simulation is much less than the optical J_{sc} computed using FDTD. In the case of FDTD simulation, 100% IQE is assumed, where all photogenerated electron and hole pairs are collected at the electrodes and contribute to device photocurrent.

16.4 SUMMARY

To prove that the semiconductor nanostructure-based SCs can be a new pathway for low-cost and high-efficiency SCs, we have carried out the optical and electrical simulations of different nanostructure-based SCs using Lumerical Inc. commercial software and compared the results with the corresponding planar structures. The geometrical parameters of such nanostructures are optimized using the FDTD simulation results to improve the overall performance of the SCs. In our optical simulations: first, we have evaluated the optimal base diameter and FR for the GaAs NW structures as 210 nm and 0.6, respectively. NPs with base-size = 240 nm and FR = 0.7 can reach J_{sc} as high as 38.4 mA/cm^2, which is nearly 37% better than the planar structure. Similarly, TNP structures at their optimal base size of 210 nm and FR of 0.6 exhibit J_{sc} of 41.7 mA/cm^2, which is almost 50% better than the planar structure and comparatively higher than the other structures. Amongst all three nanostructures, GaAs TNP structures have shown much better light absorption and J_{sc}, which is attributed to their better geometric shape allowing them to absorb the incident photons effectively. We have also carried out optical simulation for two other III-V direct bandgap materials (i.e., InP and InAs NWs) and found that like GaAs NWs, these structures also possess confinement effects, and superior light trapping can be achieved by optimizing their geometric configuration. Out of these three different semiconductor NWs, InAs NWs have shown superior optical properties and J_{sc} attributed to their low bandgap properties, which results in broad and stable absorption spectra. Further, to improve the J_{sc} and optical absorption of GaAs NW SCs, we have provided PEDOT: PSS polymer coating (at optimized thickness) onto the

NWs, which reduces the amount of reflection and transmission of the incident light and improves the J_{sc} by 4 mA/cm². The FDTD simulations reveal that, despite low volume-filling ratio and material usage, nanostructure array-based SCs can achieve light absorption (~94%) throughout the incident wavelength region. Finally, we have carried out the electrical simulations in Lumerical DEVICE software to show that the NW-based SC has delivered PCE of 27.85%, which is 1.65 times better than the PCE of planar structure. The work we have carried out and the obtained results for the GaAs nanostructure array-based SCs could shed some light to carry out some experiments on GaAs nanostructure SCs.

16.5 SIMULATION METHODOLOGY

In our analysis, we have fixed the length of the nanostructures to 2 μm, considering the fabrication difficulty. However, it has been reported in many works that the optical as well as the electrical properties of the nanostructures have a logarithmic dependence on their length (Anthony 2014; Chao, Shiu, and Lin 2012). The FDTD method is used to perform the optical simulations. Periodic boundary conditions are applied the along the x- and y-directions, whereas to avoid parasitic reflection perfectly matched layer (PML), boundary condition is assumed along the z-direction (refer to Figure 16.10).

A plane wave source with a wavelength ranging from 300 nm–1100 nm is incident along the nanostructures perpendicular to the GaAs substrate. The reflection, transmission, and absorption are calculated using frequency domain power monitors. J_{sc}, optical generation rates, and other important parameters are determined with the entire structure excited by the AM 1.5 solar spectrum. A photogeneration

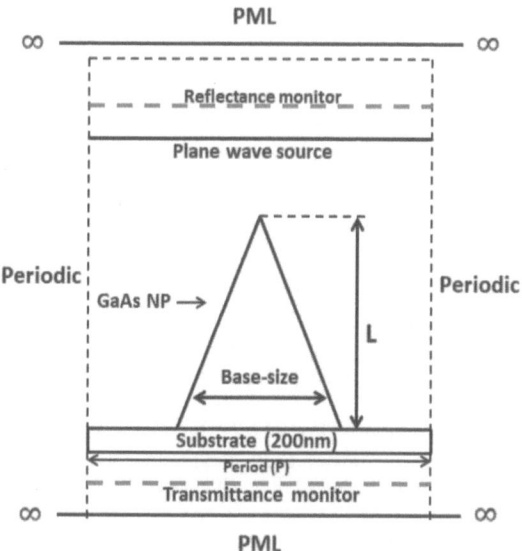

FIGURE 16.10   Simulated unit of GaAs nanostructure-based SC.

rate computed using FDTD solutions is imported to the Lumerical DEVICE software, to calculate other important electrical parameters, such as V_{oc}, J_{sc}, and PCE. The required refractive index data for III-V semiconductor materials is obtained from the Palik material data base of Lumerical FDTD solutions (Tang et al. 2018).

AKNOWLEDGMENTS

We would like to acknowledge **Science and Engineering Research Board, Department of Science and Technology, Government of India (ECR/2017/002369)** for providing financial support to accomplish this research work. This work is a part of our research project titled "Analytical Modeling and Simulation of III-V Nanostructure-based Hybrid SCs."

REFERENCES

1. Abujetas, Diego R., Ramón Paniagua-Domínguez, and José A. Sánchez-Gil. 2015. "Unraveling the Janus Role of Mie Resonances and Leaky/Guided Modes in Semiconductor Nanowire Absorption for Enhanced Light Harvesting." *ACS Photonics* 2 (7): 921–29. doi:10.1021/acsphotonics.5b00112
2. Adikaari, A. A. D. T., D. M. N. M. Dissanayake, and S. R. P. Silva. 2010. "Hybrid Organic-Inorganic Solar Cells : Recent Developments and Outlook." *IEEE Journal of Selected Topics in Quantum Electronics* 16 (6): 1595–1606. doi:10.1109/JSTQE.2010.2040464
3. Alamo, Jesús A. Del. 2011. "Nanometre-Scale Electronics with III-V Compound Semiconductors." *Nature* 479 (7373): 317–23. doi:10.1038/nature10677
4. Anthony, Michael. 2014. "Electronic Thesis and Dissertations Los Angeles Hybrid Solar Cells Based on Gallium Arsenide Nanopillars A Thesis Submitted in Partial Satisfaction of the Requirements for the Degree Master of Science in Electrical Engineering by Michael Anthony Haddad."
5. Bi, H., and R. R. LaPierre. 2009. "A GaAs Nanowire/P3HT Hybrid Photovoltaic Device." *Nanotechnology* 20 (46). doi:10.1088/0957-4484/20/46/465205
6. Chao, Jiun-jie, Shu-chia Shiu, and Ching-fuh Lin. 2010. "GaAs Nanowire/PEDOT: PSS Hybrid Solar Cells." *Advanced Functional Materials* 1 (c): 2–3.
7. Chao, Jiun Jie, Shu Chia Shiu, and Ching Fuh Lin. 2012. "GaAs Nanowire/Poly(3,4-Ethylenedioxythiophene): Poly(Styrenesulfonate) Hybrid Solar Cells with Incorporating Electron Blocking Poly(3-Hexylthiophene) Layer." *Solar Energy Materials and Solar Cells* 105. Elsevier: 40–45. doi:10.1016/j.solmat.2012.05.021
8. Chen, Jheng-yuan, Ming-hung Yu, Shun-fa Chang, Kien Wen Sun, Jheng-yuan Chen, Ming-hung Yu, Shun-fa Chang, and Kien Wen. 2013. "Hybrid Solar Cells with Imprinted Nanopyramid Structures" 133901. doi:10.1063/1.4822116
9. Colombo, C., M. Heiß, M. Grätzel, A. Fontcuberta Morral, C. Colombo, M. Hei, M. Grätzel, and A. Fontcuberta Morral. 2012. "Gallium Arsenide p - i - n Radial Structures for Photovoltaic Applications" 173108 (2009): 13–16. doi:10.1063/1.3125435
10. Dayal, Smita, Nikos Kopidakis, Dana C. Olson, David S. Ginley, and Garry Rumbles. 2010. "Photovoltaic Devices with a Low Band Gap Polymer and CdSe Nanostructures Exceeding 3% Efficiency." *Nano Letters* 10 (1). American Chemical Society: 239–42. doi:10.1021/nl903406s
11. Fan, Zhiyong, Daniel J. Ruebusch, Asghar A. Rathore, Rehan Kapadia, Onur Ergen, and W. Paul. 2009. "Challenges and Prospects of Nanopillar-Based Solar Cells" 1. doi:10.1007/s12274-009-9091-y

12. Garnett, Erik, and Peidong Yang. 2010. "Light Trapping in Silicon Nanowire Solar Cells." *Nano Letters* 10 (3): 1082–87. doi:10.1021/nl100161z

13. Hong, Lei, Xincai Wang, Hongyu Zheng, Hao Wang, Xu Xiaoyan, and Hongyu Yu. 2014. "Light Trapping in Hybrid Nanopyramid and Nanohole Structure Silicon Solar Cell beyond the Lambertian Limit Light Trapping in Hybrid Nanopyramid and Nanohole Structure Silicon Solar Cell beyond the Lambertian Limit" 074310: 112–16. doi:10.1063/1.4893707

14. Ibrahim, M. L. Inche, Zubair Ahmad, Khaulah Sulaiman, M. L. Inche Ibrahim, Zubair Ahmad, and Khaulah Sulaiman. 2016. "Analytical Expression for the Current-Voltage Characteristics of Organic Bulk Heterojunction Solar Cells Analytical Expression for the Current-Voltage Characteristics of Organic Bulk Heterojunction Solar Cells" 027115 (2015). doi:10.1063/1.4908036

15. Jeong, Sangmoo, Erik C. Garnett, Shuang Wang, Zongfu Yu, Shanhui Fan, Mark L. Brongersma, Michael D. Mcgehee, and Yi Cui. 2012. "Hybrid Silicon Nanocone – Polymer Solar Cells." *Nano Letters* 12 (6): 2971–76. doi:https://dx.doi.org/10.1021/nl300713x

16. Kayes, Brendan M., Harry A. Atwater, and Nathan S. Lewis. 2005. "Comparison of the Device Physics Principles of Planar and Radial P-n Junction Nanorod Solar Cells." *Journal of Applied Physics* 97 (11): 114302-1-114302–11. doi:https://dx.doi.org/10.1063/1.1901835

17. Kelzenberg, M. D., M. C. Putnam, D. B. Turner-Evans, N. S. Lewis, and H. A. Atwater. 2009. "Predicted Efficiency of Si Wire Array Solar Cells." In *Conference Record of the IEEE Photovoltaic Specialists Conference*, 001948–53. doi:10.1109/PVSC.2009.5411542

18. Khalil, Asma, Zubair Ahmed, Farid Touati, and Mohamed Masmoudi. 2016. "Review on Organic Solar Cells." In *13th International Multi-Conference on Systems, Signals and Devices, SSD 2016*, 342–53. doi:10.1109/SSD.2016.7473760

19. Kou, Yanlei, Kong Liu, Zhijie Wang, Dan Chi, Shudi Lu, Shizhong Yue, Yanpei Li, Shengchun Qu, and Zhanguo Wang. 2015. "Hybrid Silicon Nanocone – Polymer Solar Cells Based on a Transparent Top Electrode." *RSC Advances* 5 (53). Royal Society of Chemistry: 42341–45. doi:https://dx.doi.org/10.1039/C5RA04222D

20. Kupec, J., and B. Witzigmann. 2009. "Dispersion, Wave Propagation and Efficiency Analysis of Nanowire Solar Cells." *Optics Express* 17 (12): 10399–410. doi:10.1364/oe.17.010399

21. Li, Qiang, Maojun Zheng, Miao Zhong, Liguo Ma, Faze Wang, Li Ma, and Wenzhong Shen. 2016. "Engineering MoSx/Ti/InP Hybrid Photocathode for Improved Solar Hydrogen Production." *Scientific Reports* 6 (June): 1–9. doi:10.1038/srep29738

22. Liu, Ruchuan. 2014. "Hybrid Organic/Inorganic Nanocomposites for Photovoltaic Cells." *Materials* 7 (4): 2747–71. doi:10.3390/ma7042747

23. Mariani, Giacomo, Ramesh B. Laghumavarapu, Bertrand Tremolet De Villers, Joshua Shapiro, Andrew Lin, Benjamin J. Schwartz, and Diana L. Huffaker, et al. 2013. "Hybrid Conjugated Polymer Solar Cells Using Patterned GaAs Nanopillars Hybrid Conjugated Polymer Solar Cells Using Patterned GaAs Nanopillars," 013107 (2010): 3–6. doi:10.1063/1.3459961

24. Mavrokefalos, Anastassios, Sang Eon Han, Selcuk Yerci, Matthew S. Branham, and Gang Chen. 2012. "Efficient Light Trapping in Inverted Nanopyramid Thin Crystalline Silicon Membranes for Solar Cell Applications." *Nano Letters* 12 (6): 2792–96. doi:10.1021/nl2045777

25. Munshi, A. M., D. L. Dheeraj, V. T. Fauske, D. C. Kim, J. Huh, J. F. Reinertsen, L. Ahtapodov, et al. 2014. "Position-Controlled Uniform GaAs Nanowires on Silicon Using Nanoimprint Lithography." *Nano Letters* 14 (2): 960–66. doi:10.1021/nl404376m

26. Nicolaidis, Nicolas C., Ben S. Routley, John L. Holdsworth, Warwick J. Belcher, Xiaojing Zhou, and Paul C. Dastoor. 2011. "Fullerene Contribution to Photocurrent Generation in Organic Photovoltaic Cells." *The Journal of Physical Chemistry C* 115 (15). American Chemical Society: 7801–5. doi:10.1021/jp2007683

27. Ong, Pang Leen, and Igor A. Levitsky. 2010. "Organic/IV, III-V Semiconductor Hybrid Solar Cells." *Energies* 3 (3): 313–34. doi:10.3390/en3030313

28. Park, Hyesung, Sehoon Chang, Joel Jean, Jayce J. Cheng, Paulo T. Araujo, Mingsheng Wang, and Moungi G. Bawendi, et al. 2013. "Graphene Cathode-Based ZnO Nanowire Hybrid Solar Cells." *Nano Letters* 13 (1): 233–39. doi:10.1021/nl303920b

29. Prashant, D. V., D. P. Samajdar, and Dheeraj Sharma. 2019. "Optical Simulation and Geometrical Optimization of P3HT/GaAs Nanowire Hybrid Solar Cells for Maximal Photocurrent Generation via Enhanced Light Absorption." *Solar Energy* 194 (July). Elsevier: 848–55. doi:10.1016/j.solener.2019.11.027

30. Prashant, D. V., Suneet Kumar Agnihotri, and Dip Prakash Samajdar. 2021. "Geometric Optimization and Performance Enhancement of PEDOT: PSS/GaAs NP Array Based Heterojunction Solar Cells." *Optical Materials*. doi:10.1016/j.optmat.2021.111080.

31. Prashant, D. V., Dip Prakash Samajdar, and Zahra Arefinia. 2021. "FDTD-Based Optimization of Geometrical Parameters and Material Properties for GaAs-Truncated Nanopyramid Solar Cells." *IEEE Transactions on Electron Devices*, 1–7. doi:10.1109/ted.2021.3055190.

32. Prashant, D. V., Dip Prakash Samajdar, and Sachchidanand. 2020. "Optical Simulation of III-V Semiconductor Nanowires/PEDOT: PSS-Based Hybrid Solar Cells: Influence of Polymer Coating Thickness and Geometrical Parameters on Light Harvesting and Overall Photocurrent." In *Energy Systems, Drives and Automations*, edited by Afzal Sikander, Dulal Acharjee, Chandan Kumar Chanda, Pranab Kumar Mondal, and Piyush Verma, 361–68. Singapore: Springer.

33. Prashant, D. V., Dip Prakash Samajdar, and Dheeraj Sharma. 2020. "Optical Simulations of High Efficiency GaAs Nanostructure Array Based Solar Cells: Significance of Geometrical Parameter Optimization on Light Harvesting Properties and Overall Photocurrent." *AIP Conference Proceedings* 2265 (November): 5–9. doi: 10.1063/5.0016773.Raychaudhuri, Sourobh, Shadi A. Dayeh, Deli Wang, and Edward T. Yu. 2009. "Precise Semiconductor Nanowire Placement Through Dielectrophoresis." *Nano Letters* 9 (6): 2260–66. doi:10.1021/nl900423g

34. Ren, Shenqiang, Ni Zhao, Samuel C. Crawford, Michael Tambe, Vladimir Bulovi, and Silvija Gradečak. 2011. "Heterojunction Photovoltaics Using GaAs Nanowires and Conjugated Polymers." *Nano Letters* 11 (2): 408–13. doi:10.1021/nl1030166

35. Sachchidanand, and D.P. Samajdar. 2019. "Light-Trapping Strategy for PEDOT:PSS/ c-Si Nanopyramid Based Hybrid Solar Cells Embedded with Metallic Nanoparticles." *Solar Energy* 190 (August). Elsevier: 278–85. doi:10.1016/j.solener.2019.08.023

36. Srivastava, A., D. P. Samajdar, and D. Sharma. 2018. "Plasmonic Effect of Different Nanoarchitectures in the Efficiency Enhancement of Polymer Based Solar Cells: A Review." *Solar Energy* 173 (July): 905–19. doi:10.1016/j.solener.2018.08.028

37. Sturmberg, Bjo C. P., Kokou B. Dossou, Lindsay C. Botten, Ara A. Asatryan, Christopher G. Poulton, Ross C. Mcphedran, and C. Martijn De Sterke. 2014. "Optimizing Photovoltaic Charge Generation of Nanowire Arrays: A Simple Semi-Analytic Approach." *ACS Photonics* 1 (8): 683–89. doi:10.1021/ph500212y

38. Takagahara, Toshihide, and Kyozaburo Takeda. 1992. "Theory of the Quantum Confinement Effect on Excitons in Quantum Dots of Indirect-Gap Materials the Conduction-Band Minimum Is Located at Ko =2m/a(0, 0,0.85) and Another Five Equivalent Points in the Case of Si and At." *The American Physics Society* 46 (23): 15578–81. https://journals-aps-org.ezproxy.lib.monash.edu.au/prb/pdf/10.1103/PhysRevB.46.15578

39. Tang, Quntao, Honglie Shen, Hanyu Yao, Kai Gao, Ye Jiang, and Youwen Liu. 2018. Dopant-Free Random Inverted Nanopyramid Ultrathin c-Si Solar Cell via Low Work Function Metal Modified ITO and TiO2 Electron Transporting Layer. *Journal of Alloys and Compounds*. 769. Elsevier B.V. doi:10.1016/j.jallcom.2018.08.072

40. Vurgaftman, I., and J. R. Meyer. 2001. "Band Parameters for III – V Compound Semiconductors and Their Alloys." *Journal of Applied Physics* 89 (11): 5815–75. doi:10.1063/1.1368156

41. Wang, Wenbo, Xinhua Li, Long Wen, Yufeng Zhao, Huahua Duan, Bukang Zhou, Tongfei Shi, Xuesong Zeng, Ning Li, and Yuqi Wang. 2014. "Optical Simulations of P3HT/Si Nanowire Array Hybrid Solar Cells." *Nanoscale Research Letters* 9 (1): 1–6. doi:10.1186/1556-276X-9-238

42. Wang, Xixi, Zhaolang Liu, Zhenhai Yang, Jian He, Xi Yang, Tianbao Yu, Pingqi Gao, and Jichun Ye. 2018. "Heterojunction Hybrid Solar Cells by Formation of Conformal Contacts between PEDOT : PSS and Periodic Silicon Nanopyramid Arrays." *Small* 14 (15): 1704493-1–1704493–97. doi:10.1002/smll.201704493

43. Wei, Wan-rou, Meng-lin Tsai, Shu-te Ho, Shih-hsiang Tai, Cherng-rong Ho, Shin-hung Tsai, Chee-wee Liu, Ren-jei Chung, and Jr-hau He. 2013. "Above-11%-E Ffi Ciency Organic – Inorganic Hybrid Solar Cells with Omnidirectional Harvesting Characteristics by Employing Hierarchical Photon-Trapping Structures."

44. Wen, Long, Zhifei Zhao, Xinhua Li, Yanfen Shen, Haoming Guo, and Yuqi Wang. 2011. "Theoretical Analysis and Modeling of Light Trapping in High Efficiency GaAs Nanowire Array Solar Cells." *Applied Physics Letters* 99 (14): 2009–12. doi:10.1063/1.3647847

45. Wright, Matthew, and Ashraf Uddin. 2012. "Organic-Inorganic Hybrid Solar Cells: A Comparative Review." *Solar Energy Materials and Solar Cells* 107. Elsevier: 87–111. doi:10.1016/j.solmat.2012.07.006

46. Wu, Dan, Xiaohong Tang, Kai Wang, and Xianqiang Li. 2016. "Effective Coupled Optoelectrical Design Method for Fully Infiltrated Semiconductor Nanowires Based Hybrid Solar Cells." *Optics Express* 24 (22): A1336–48. doi:10.1364/oe.24.0a1336

47. Wu, Dan, Xiaohong Tang, Kai Wang and Xianqiang Li. 2017. "An Analytic Approach for Optimal Geometrical Design of GaAs Nanowires for Maximal Light Harvesting in Photovoltaic Cells." *Scientific Reports* 7 (March). Nature Publishing Group: 1–8. doi:10.1038/srep46504

48. Wu, Yao, Xin Yan, Xia Zhang, and Xiaomin Ren. 2018. "Photovoltaic Performance of a Nanowire/Quantum Dot Hybrid Nanostructure Array Solar Cell." *Nanoscale Research Letters* 13 (62). Nanoscale Research Letters: 1–7. doi:10.1186/s11671-018-2478-5

49. Yang, Ziyan, Ting Zhang, Jingyu Li, Wei Xue, Changfeng Han, Yuanyuan Cheng, and Lei Qian. 2017. "Multiple Electron Transporting Layers and Their Excellent Properties Based on Organic Solar Cell." *Scientific Reports* 7 (9571). Springer US: 1–9. doi:10.1038/s41598-017-08613-7

50. Yao, Maoqing, Ningfeng Huang, Sen Cong, Chun-Yung Chi, M. Ashkan Seyedi, Yen-Ting Lin, Yu Cao, Michelle L. Povinelli, P. Daniel Dapkus, and Chongwu Zhou. 2014. "GaAs Nanowire Array Solar Cells with Axial p–i–n Junctions." *Nano Letters* 14 (6): 3293–3303. doi:10.1021/nl500704r

51. Yong, Chaw Keong, Keian Noori, Michael Gao, Hannah J. Joyce, H. Hoe Tan, Chennupati Jagadish, Feliciano Giustino, Michael B. Johnston, and Laura M. Herz. 2012. "Strong Carrier Lifetime Enhancement in GaAs Nanowires Coated with Semiconducting Polymer." *Nano Letters* 12 (12): 6293–6301. doi:10.1021/nl3034027

52. Zhang, Genqiang, Scott Finefrock, Daxin Liang, Gautam G. Yadav, and Haoran Yang. 2011. "Nanoscale MINIREVIEW Semiconductor Nanostructure-Based Photovoltaic Solar Cells," 2430–43. doi:10.1039/c1nr10152h

17 Advanced Technologies in IoT for Development of Smart Cities

Radhika G. Deshmukh
Shri Shivaji College of Arts, Commerce &
Science, Akola, India

CONTENTS

17.1 INTRODUCTION

In this modern age, our life is impacted by various IoT devices creating innovative and advantageous new services. Globalization has pushed this forward, and everyone is connected by means of technology. The IoT connected the Internet to the physical world via sensors. By the year 2025, the devices connected to the Internet are roughly expected to be as much as 50 billion. One hundred smart cities are being planned by the Indian central government in the coming five years all across the country. We can define smart city as a city equipped with the latest and most modern technologies such as IoT; sensor-based networks; big data; machine-based learning; 2G, 3G, and 4G modes of interaction; and computing machines of higher and superior performance so that citizens' overall lives

DOI: 10.1201/9781003181613-17

are improved. The Internet of Things gadgets are interlined generally by the networking foundation all across the globe [1–5]. The creation of a smart city is the main focus with ideal monitoring of power systems and smart homes. The new industries making use of information and communication technologies are referred to in the term smart cities, along with an eco-friendly environment, energy-related technologies, and other supporting benefits inside the environments of cities and residences. People are moving from rural areas to cities because of the drastic increase in population and unemployment concerns. One of the main reasons for migration from rural to urban areas is to enrich educational opportunities for children. This is the root cause of various important concerns taking place in cities [6–11]. To eliminate some problems in more crowded cities, the concept of the smart city has been introduced by various governments. To successfully implement smart cities, the IoT is considered the most important aspect so that many devices are linked to each other and can be significantly used in these smart cities.

17.2 WHAT IS IoT?

The electronic devices connected to IoT and networks are used to connect various industrial equipment allowing the gathering of data and device management with the help of software so that the efficiency increases, allowing modern services to get benefits related to the health and safety of environment. The physical Internet is the term used for the IoT [12–16]. It is the wireless network between objects and is a worldwide infrastructure for data or an information-related society allowing modern and highly advanced features with the help of connected items.

Main features of IoT:

- The linkage and inter-connection:
  The global information is used for the connection of everything, especially infrastructure and communication.
- Services related to things:
  Inside the things' consent it is offered modern technologies in the data board and physical environment.
- Heterogeneity:
  In the Internet of Things the gadgets are linked with each other and other devices with various kinds of networks.
- Dynamic changes:
  There are dynamic changes to the state and working of modern devices. The gadgets are sometimes sleeping; they can be disconnected or cancelled, which includes speed changes and location changes along with devices.
- Enormous scale:
  There are billions of devices, which are supposed to be administered ideally so that they can communicate with each other.

17.2.1 THE WHY OF IoT

The IoT is essential for various reasons, such as:

- For industrial and everyday life; it offers dynamic control
- Improving utilization of resources

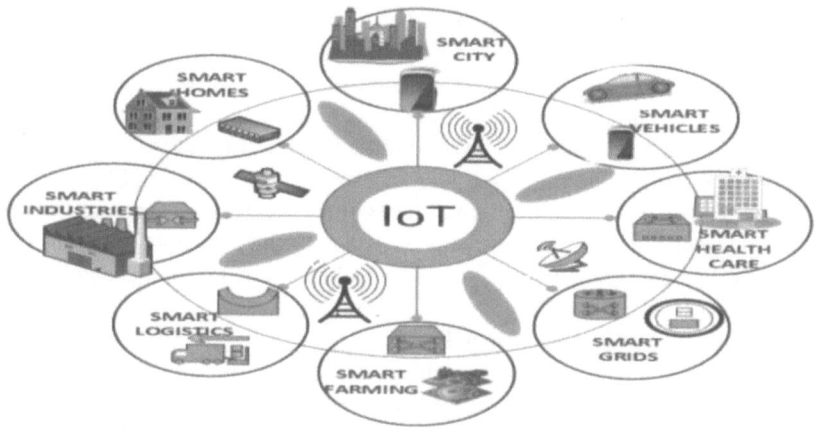

FIGURE 17.1 IoT and its applications.

- Creating better relations between humans and nature
- The integration of human society and physical systems by forming an intellectual entity

17.2.2 Applications of IoT

The following are some of the applications of IoT (Figure 17.1):

- **IoT makes the physical world more attractive**
  It is a matter of common observation that by spending time with friends and family, people experience greater satisfaction; they love to interact with their loved ones even when they are outdoors and surrounded by Mother Nature [17–20]. With the help of IoT, the real world is given more emphasis as compared to the virtual or computer-based world.
- **IoT for the sustainability of environment**
  Traffic management is also helped by IoT. The technology can be used to reduce the congestion of traffic, idling, and environmental contamination. In addition, for ecological health or tracking of special and endangered species, IoT can be used in wild places and this technology can prevent poaching by checking long-range sensors, which are linked to each other by IoT. This also ensures that the environment remains cleaner and greener.
- **It improves our health**
  Health is also improved, as many medical devices are connected by IoT. Fitbit types of personal fitness trackers help us to considerably reduce the intake of calories and sleep in a much better way. There are various kinds of mood trackers, which are used to show as emotional life patterns that can keep us cool and calm. In the same way, the IoT can considerably reduce the time and inefficiencies of doctors and other paramedical teams. Health care professionals can save their time while at work and can work more efficiently to save lives.

- **Companies and the use of IoT**
  Since there is a lower cost for IoT, this newer technology allows various activities to be monitored and managed in innovative ways. Human labor and its inherent costs can be replaced with the help of innovative technology and hence supplies can be maintained and monitored so that this technology is affordable in terms of finances. It is more beneficial to make use of symbiotic infrastructure rather than going for massive sensor networking deployment. In this way, one can rely on the widely adopted current standards. For entrepreneurships and innovation, the IoT offers greater opportunities [21, 22].
- **Government objectives for IoT policy**
  India will account for 5–6 percent of the global IoT industry. To undergo research and development for all the surrounding technology ecosystems, IoT-related products are to be designed to meet the needs of Indian consumers in multiple business domains, such as water quality, health, natural disasters, agriculture, security, automobile, waste management, smart cities, transportation, automated metering and monitoring of utilities, supply chain management, etc..

17.3   THE SMART CITY CONCEPT

A framework of technology known as a smart city generally comprises information and communication technology to address growing urbanization challenges. The technology can be developed, deployed, and promoted for sustainable developmental means. The data-transmitting network of machines and other connected objects making use of cloud technology and wireless technology is used. To receive data, analyze it, and manage it in the real time is the key role of cloud-based IoT applications, and this can considerably improve the life standards of urban residents.

17.3.1   THE NEED FOR SMART CITIES

In modern times, as much as 54 percent of the people across the world live in cities. It is expected that by the end of 2050, this percentage will hike to 66 percent. Environmental, social, and economical sustainability is there along with the quick expansion of resources in our cities in order to keep and maintain peace. One never ending phenomena is urbanization. IoT and secure wireless connectivity are transforming city elements.

For the development of smart cities, there are four essential elements.

- Enveloping wireless connectivity
- Open information
- Trustworthy security
- A monetization scheme and its flexibility

For any smart city application, pervasive wireless connectivity is the first building block. This wireless connectivity must be reliable and pervasive as well. For the applications of smart cities, low-power wide area networking (WAN) technologies are considered best due to the cost efficiency. The LTE Cat M, NB-IoT, LoRa, and Bluetooth are some of the technologies that are supposed to play a role in the fabric of connected cities. The upside of 5G innovation is a turning point that impels brilliant city innovation into the standard and quickens new arrangements. This IoT-based present-day innovation includes connected cameras, current street frameworks, and different public-checking frameworks that can give an additional layer of insurance and crisis backing to help residents when required.

17.3.2   FOUR CORE SECURITY OBJECTIVES FOR SMART CITY SOLUTIONS

For smart city solutions, all ecosystem partners integrate ecosystem partners, which are likely to have an impact on:

- Availability and impact: Without reliable, actionable, and real-time data, the concept of a smart city cannot thrive. To collect and share data is critical, and it is also important to find the security solutions so that adverse effects can be avoided regarding Internet availability.
- Integrity: From the manipulation, the smart cities are generally free and accurate data is needed in the most reliable form for the smart cities.
- Confidentiality: So as to prevent the sensitive information disclosure in an unauthorized manner, there is a need to take various steps. All the details about consumers are included in the collected, sorted, and analyzed data.
- Accountability: With the specific user, ideally the interaction with sensitive networking should be logged and linked. For these logs the forging is difficult, and the integration protection is kept reliable. For the proper thriving of smart cities, establishing sustainable commerce models is essential for the facilitation of success of all ecosystem players. The software for Internet of Things should act like the contributor of ecosystems. The integrators and government are included in the ecosystem along with developers and OEMs. Intellectual property should be given value and rewards. New business models can be developed with the help of subscription software so that everyone can extract value from their own part in the sustainability of the ecosystem.
- Fleets of vehicles shared between owners soon will offer affordable subscriptions. Owners will be able to choose from the custom option arrays. Traffic problems can be reduced and patterns of traffic can be optimized as well. Urban areas continue to grow and increase in size. To enhance sustainability, smart city technology is developed so that humanity can be served better. Smart cities can improve by leveraging enveloping connection, open information, security end to end, and solutions of software monetization.

17.4   APPLICATIONS OF IoT AND SMART CITIES

There are number of applications of IoT in developing smart cities.

17.4.1 TRAFFIC CONGESTION

For larger smarter cities, there are many challenges to optimizing the congestion of traffic.

An urban city's traffic congestion can be monitored with the help of IoT.

By installing GPS, monitoring can be done on newer motor vehicles. The traffic needs to be controlled by city officials and citizens. They can also plan a shopping trip in advance, or in case of urgency, they can find a short route to arrive at work. Whenever there is an accident and traffic is going slow, IoT systems re-plan routes, thus helping to keep safe distances between vehicles. IoT systems monitor the function of vehicles, i.e., the speed of vehicles in dangerous situations.

17.4.2 AIR QUALITY MANAGEMENT

The tool for the air management of cities is software-based on the cloud, and contamination information is captured inside the real time and emissions are forecasted. Air pollution with all needed measurements is predicted by it for the efficacy and technological betterment so that air management can be managed more easily in the modern cities. On a dashboard, all the needed information about the quality of air is displayed and sensors are used to keep it protected in the larger and smart cities. In urban cities, the IoT can track the air quality. All across such cities, the pollution sensors can check the quality of air, and all the information is made ready for the citizen.

17.4.3 SMART HEALTH

For patients there are certain critical parameters, such as the condition of heart, body temperature, pulse rate, breathing rate, etc. The IoT can be used to monitor all these parameters. In remote locations, the needed information about hospitals and ambulances is sent to the patient. IoT is often used to support patients who are unhealthy mentally, neonates and infants, and toddlers making use of relevant data from various areas.

17.4.4 SMART ENERGY

The energy of a city can be monitored as well with the help of IoT. The IoT can also offer the details of consumed energy in buildings by the public, lighting, traffic lights, camera control, transport, etc. The main energy consumption sources can be isolated with the help of this.

17.4.5 SMART INFRASTRUCTURE

There should be a proper planning of city infrastructures and important buildings. Efficient planning is needed for the reduction of carbon dioxide emission. The emission of carbon dioxide should be kept as low as possible. Investments in electrical cars are an example. Along with planning of self-propelled vehicles, C smart lighting should be used for pedestrians, and these lights should be manufactured so that their brightness can be adjusted and daily use of these lights can be tracked so as to lessen the regulation and monitoring of electrical power for the building's real conditions. Various areas are needed to be identified by external agents with the aim being to use the distributed database for structural management in buildings with the help of sensors. The main examples include atmospheric sensors for the tracking of pollution, sensors for the detection of humidity and warmth, and deformation sensors to study vibration and building stress. Seismic readings can be possibly added so as to acknowledge the city buildings' infrastructure and the way these buildings are affected by lights and earthquakes. In buildings as well as nearby areas, the sensors can be installed. The IoT contains an ecosystem, communication systems, information, city residents, and other smart gadgets, which all play a significant role.

17.4.6 SMART PARKING

IoT has an advantage in smart parking. Available parking spaces can be identified by intelligent parking services including sensors. The driver can be informed about open parking via a smart phone. Today smart parking is reality and a complicated infrastructure is not needed; there is also no need for higher investments. For the smart cities, smart parking is considered ideal.

17.4.7 MANAGING WASTES IN A SMART WAY

The efficiency of waste collection can be optimized by managing wastes in smart ways. This also helps reduce the cost of operation, and it is considered an ideal solution for various environmental concerns linked with waste collection in inefficient ways. On filling up of a container, a message is sent to the truck driver at their cell phone (smart phone) as a level sensor is obtained at the waste container. In the same way, to empty the full container a message is sent to the smart phone thus avoiding half-empty pickups.

17.4.8 SMART HOMES

IoT helps to maintain temperature in houses. Mobile devices collaborate with house sensors and control elements like AC. It helps to control and monitor household devices, such as smart cookers, smart washing machines, smart egg trays, smart garbage cans, smart gardening, etc.

17.5 INTERNET OF THINGS IN TERMS OF THE FUTURE

IoT has huge potential in all sectors. Some future developments include:

1. **Vision Van of Mercedes-Benz**
   For urban areas with various innovative technologies, the vision of Mercedes-Benz is a van concept containing onboard technology for delivery systems.

This includes autonomous drones which are supposed to autonomously deliver inside a 10 km radius of the van. Time is saved during the loading and processing of delivery, and this is considered an additional benefit. On the last mile, the time of delivery is shortened by as much as 50 percent as the drones deliver parallel to the deliverer.

2. **Smart Eye**

Google's best project is the Glass. Sensors, Wi-Fi, and Bluetooth are the equipped devices of smart eyes and provide essential components in front of eyes in such a way that no distraction is caused. The messages can be read with the help of innovative technology and in the same way Internet can be surfed as well.

Challenges in IoT space:

Though IoT has various applications, it has challenges also

- Technological standardization in most areas is still fragmented
- Connectivity
- Power management
- Security

We need multiple areas of expertise and should cross interdisciplinary boundaries to overcome the challenges.

17.6   CONCLUSIONS

The roadmap of IoT is very exciting. IoT has unlimited sectors and provides solutions in industrialization, fashion, health care, education, etc. The cloud-based IoT is the best solution for smart cities. All these smart cities are controlled with the help of a cloud platform in such a way that simpler city ecosystems can be formed. From intelligent and smart technologies, cities can benefit. There is a huge impact on IoT-based smart city solutions on some independent variables, such as management of health care and safety, accessibility of information, management of energy, traffic management, parking management, and management of air quality, etc. IoT-based smart city solutions are available, and citizens need to have a positive mindset and need to acknowledge the emerging technologies and their impact on everyday life. A strong IoT and its revenue model is needed these days along with superior data security and government support. IoT offers smart city solutions in terms of management of better health care safety, superior accessibility of information to the populace, and better and advanced management of energy. Several millions have already been invested by industries in India to set up an e-infrastructure based on IoT. This can also be used to develop other similar and related applications. The government of India and other related organizations are supposed to study the implementations of IoT according to the requirements of its citizens. India is well-positioned to grow economically by keeping a strong focus on technological development in the modern IoT era.

REFERENCES

1. Rajguru, S., Kinhekar, S. and Pati, S., April 2015. Analysis of Internet of Things in a smart environment. *International Journal of Enhanced Research in Management and Computer Applications*, 4(4), pp. 40–43.
2. Alsamhi, S., Ma, O and Ansari, M. 2018. Artificial Intelligence-Based Techniques for Emerging Robotics Communication: A Survey and Future Perspectives. arXiv 2018, arXiv:1804.09671.
3. Vermesan, O. and Friess, P., 2013. *Internet of Things – Converging Technologies for Smart Environments and Integrated Ecosystems*. River Publishers. Denmark.
4. Gomez, K., Hourani, A., Goratti, L., Riggio, R., Kandeepan, S. and Bucaille, I., June 2015. Capacity evaluation of aerial LTE base-stations for public safety communications. In *Proceedings of the 2015 European Conference on Networks and Communications (EuCNC)*, Paris, France, 29 June–2 July 2015, pp. 133–138.
5. Rana, A.K. and Sharma, S., 2021. Contiki Cooja Security Solution (CCSS) with IPv6 Routing Protocol for Low-Power and Lossy Networks (RPL) in Internet of Things applications. In *Mobile Radio Communications and 5G Networks* (pp. 251–259). Springer, Singapore.
6. Rana, A.K. and Sharma, S., 2019. Enhanced Energy-Efficient Heterogeneous Routing Protocols in WSNs for IoT Application.
7. Mohamed, N., Al-Jaroodi, J., Jawhar, I., Idries, A. and Mohammed, F., 2018. Unmanned Aerial Vehicles Applications in Future Smart Cities. Technological Forecasting and Social Change/ 2018, in press.
8. Kolios, P., Pitsillides, A., Mokryn, O. and Papdaki, K., 2016. 7—Data dissemination in public safety networks. In *Wireless Public Safety Networks 2*; Câmara, D., Nikaein, N., Eds.; Elsevier, Amsterdam, The Netherlands, pp. 199–225.
9. Kumar, A., Salau, A.O., Gupta, S. and Paliwal, K., 2019. Recent trends in IoT and its requisition with IoT built engineering: A review. In *Advances in Signal Processing and Communication* (pp. 15–25). Springer, Singapore.
10. Charmonman, S., Mongkhonvanit, P., Dieu, V. and Linden, N., 2015. Applications of Internet of Things in e-learning. *International Journal of the Computer, the Internet and Management*, 23(3), pp. 1–4.
11. Veeramanickam, M.R.M. and Mohanapriya, M., 2016. Iot enabled futurus smart campus with effective e-learning: i-campus. *GSTF Journal of Engineering Technology (JET)*, 3(4), pp. 8–87.
12. Vharkute, M. and Wagh, S., 2015, April. An architectural approach of Internet of Things in e-learning. In *2015 International Conference on Communications and Signal Processing (ICCSP)* (pp. 1773–1776). IEEE.
13. Li, H., Ota, K. and Dong, M., 2018. Learning IoT in edge: Deep learning for the Internet of Things with edge computing. *IEEE Network*, 32(1), pp. 96–101.
14. Dalal, P., Aggarwal, G. and Tejasvee, S., 2020. Internet of Things (IoT) in Healthcare System: IA3 (Idea, Architecture, Advantages and Applications). Available at SSRN 3566282.
15. Rana, A.K. and Sharma, S., 2021.Industry 4.0 manufacturing based on IoT, cloud computing, and big data: Manufacturing purpose scenario. In *Advances in Communication and Computational Technology* (pp. 1109–1119). Springer, Singapore.
16. Wang, Q., Zhu, X., Ni, Y., Gu, L. and Zhu, H., 2020. Blockchain for the IoT and industrial IoT: A review. *Internet of Things*, 10, p. 100081.
17. Saloni, M., et al., 2017. Lasso: A Device-to-device Group Monitoring Service for Smart Cities. Smart Cities Conference (ISC2), 2017 International. IEEE.
18. Dua A., Kumar N., Das A.K., Susilo W., 2017. Secure Message Communication Protocol among Vehicles in Smart City. IEEE Transactions on Vehicular Technology. Dec 12.

19. Kumar, K., Gupta, E.S. and Rana, E.A.K., 2018. Wireless Sensor Networks: A review on "Challenges and Opportunities for the Future world-LTE".

20. Sachdev, R., 2020, April. Towards security and privacy for edge AI in IoT/IoE based digital marketing environments. In *2020 Fifth International Conference on Fog and Mobile Edge Computing (FMEC)* (pp. 341–346). IEEE.

21. Loriot, M., Aljer, A., and Shahrour, I., 2017. Analysis of the use of LoRaWan technology in a large-scale smart city demonstrator. *Sensors Networks Smart and Emerging Technologies (SENSET)*, IEEE.

22. Kim, T., Cho, J. Y., & Lee, B. G., 2012, July. Evolution to smart learning in public education: a case study of Korean public education. In *IFIP WG 3.4 International Conference on Open and Social Technologies for Networked Learning* (pp. 170–178). Springer, Berlin, Heidelberg.

18 Implementation of Soft Skills for Humanoid Robots Using Artificial Intelligence

Saira Banu Atham*, Rakesh Ahuja#, and Mohammed Farooq Abdullah†
*Presidency University, Bangalore, Karnataka, India
#Chitkara University Institute of Engineering and Technology, Chitkara University, Punjab, India
†Kumaraguru College of Technology, Coimbatore, India

CONTENTS

DOI: 10.1201/9781003181613-18

18.1 INTRODUCTION

The basic idea of robotics is to perform difficult and repetitive tasks and to reduce the burden of human beings. During the 19th century our ancestors used devices called automata, which means human-like devices. The word robot is encrypted from the Czech word "robota", which means "forced labour "or "heavy work".

18.1.1 THE FIRST GENERATION OF INDUSTRIAL ROBOTS (1950–1967)

The robots in this span were program-based machines that did not have the capability to change the mode of the task and had no communication with the outer world. These robots are equipped with low-tech assistive technology and without servo controllers. A unique feature of this robots is the huge noise they produce during the movement of the arm axes. Most of the first-generation robots are equipped with actuators and controlled by logic gates. These robots are capable of loading and unloading simple items.

Many developments in industrial robots [1] took place, for example, programmables, manipulators, etc. The first true industrial robot is called Unimate, shown in Figure 18.1. It was developed using the design of George Devol. These Unimate robots were used in automotive factories for spot welding cars and managing the workpieces. Companies like General Motors and Ford understood the huge importance of robots in automotive factories, Thus, a sudden increase in the demand for robots was initiated, which caused robot manufacturers to thrive. One among them

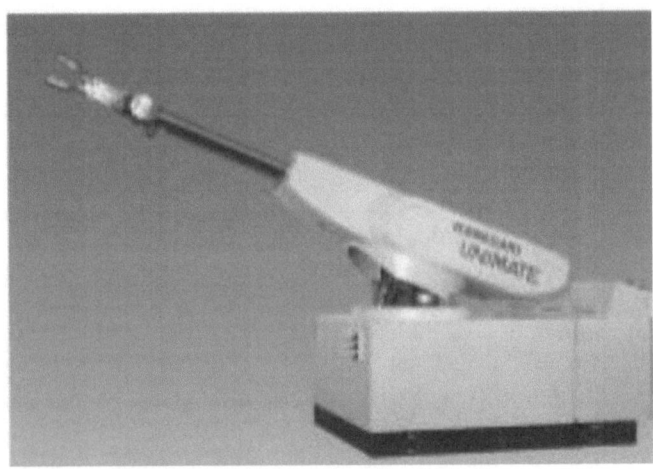

FIGURE 18.1   The Unimate robot.

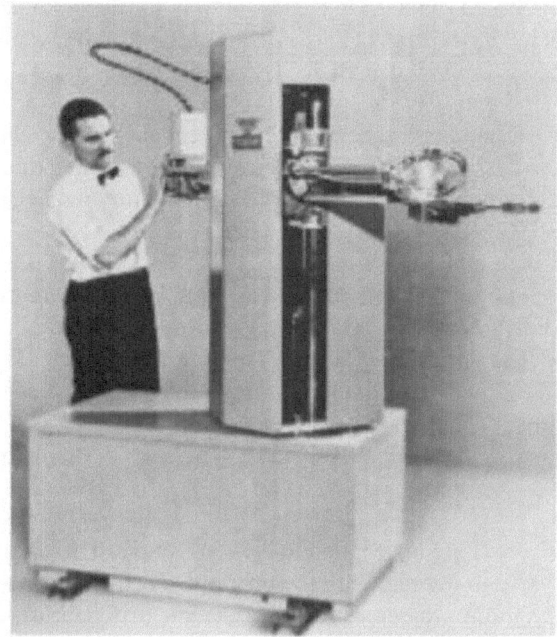

FIGURE 18.2 Versatran robot.

is the reputed American Machine & Foundry company. They manufactured a new type of robot called Versatran, shown in Figure 18.2. This type of robot is an automatic handling equipment with an artificial arm and hand. These are cylindrical-shaped robots used in car companies to carry large bobbins and packing them in a box. The Japanese understood the importance of robots in the 1960s. Robots played an important role in Japanese factories, and creators started to concentrate on the development of robots. Japan took the lead in the world in manufacturing industrial robots. The first national-level robot association, named JIRA, was initiated in 1971.

Europe turned towards the use of robots in the 1960s.The installation of robots was initiated in Europe in 1967 through the Swedish company Svenska Metallverken. It used robots for pick and place jobs and for welding. The company Unimation manufactured welding robots, and the company General Motor used robots for their automotive plants in Italy.

18.1.2 THE SECOND GENERATION OF ROBOTS (1968–1977)

Second-generation robots are called industrial robots. These robots are featured with basic programmable self-adaptive behavior. Microprocessors or PLCs are used to do point-to-point or continuous motions. These robots are capable of doing a specific task, i.e., application-specific or dedicated tasks. The functioning of the robots is a shift from hydraulic actuators to electronic components as more cost-effective devices. After the Yom Kippur war in 1973, the oil crisis resulted in lowering the usage of hydraulic-based robots. This leads the way for the popularity of electricity-driven robots. Victor Scheinman built the first prototype of the Stanford Arm in 1969. This type of robot is equipped with six DC PDP motors and six microprocessors. Tachometers and potentiometers are used to measure the position and velocity of robot joints. After four years, Scheinman, a student at Stanford University came up the idea of the next design of robot called Vicam for assembling the parts. Unimation developed the new design of robot called PUMA (Programmable Universal Machine for Assembly) in 1978. The KUKA, which is an industrial robot, was introduced in 1973. In 1974, Cincinnati Milacron released the T3 Robot (The Tomorrow Tool). These robots are used in Volvo plants and control the Swedish minicomputer. In 1974, the Swedish company ABB developed the IRB-6 robots for performing complex tasks such as machining and arc-welding. The Japanese company Hitachi built the HI-T-HAND, which is an expert robot used to insert mechanical parts with clearance of 10 μm.

18.1.3   THIRD-GENERATION ROBOTS (1978–1999)

Third-generation robots are also called ubiquitous robots. These robots are managed with vision and voice to interact with the operator. This generation of robots are in-build with self- programming capabilities to execute different tasks. And these programs can be either offline or online. High-level programming like Python is used for motion programming. This generation of robots interface with CAD or databases. The robot's movements are developed based on the data read from the sensor. This type of robot is built with diagnostic capabilities, which can report location and type of failure. Japanese scientist Hiroshi Makino from Yamanashi University proposed the SCARA (Selective Compliance Assembly Robot Arm) which is used for assembling small objects. Carnegie Mellon University developed the CMU Direct Drive Arm robot named AdeptOne. It has higher accuracy and faster operations because the motors connect directly to the arms. It has cameras, force sensors and laser scanners, and artificial intelligence. In 1998, ABB developed Flex-Picker, the world's fastest-picking robot. A robot control system named MRC was developed by Motoman in 1994, which can control two robots synchronously.

18.2   APPLICATION OF ROBOTS IN DIFFERENT AREAS

The importance of robotics is increasing in all aspects of human life in the work environment and home. Robotics is expected to spread positivity, increase efficiency, improve the level of service and the safety level of life. Robots play a key role in many fields. The following subsections discuss the applications of robots in various fields.

18.2.1   HEALTH CARE

The usage of robots in the medical field, shown in Figure 18.3, is increasing day by day. Robot medical assistants are used to monitor the vital statistics of the patients and give alerts to the nurse when a human presence is needed. They are also used

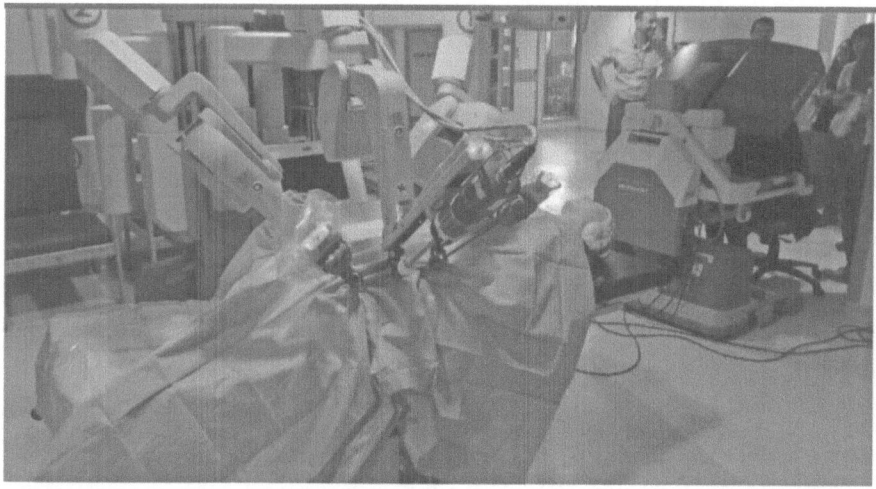

FIGURE 18.3   Robots in the medical world.

to enter medical-related information of the patient into the electronic health record. We can also see robots moving down the corridors of hospitals carrying supplies for the patients. Instead of an inches-long opening, robots are used to assist the doctors in performing tiny incision surgery. Thus, robots are part of a large evolution in the areas of medicines.

18.2.2   AGRICULTURE

The introduction of robots in the agriculture field, shown in Figure 18.4, has increased production with low overall cost. The government has come up with different projects to use the different forms of robots in the agricultural industry, such as various operations like thinning, pruning, spraying, weed removal and mowing. Sensor devices are used to track disease and pests that affect the crops.

18.2.3   FOOD INDUSTRY

The advancement of robotics technology will soon decorate the kitchen in every house. Intelligent robots invented by Moley are used for preparing meals for hundreds of people. This robotic chef is controlled by smartphones. The controller selects the recipe, and if the ingredients are ready the robots will automatically start the cooking efficiently and quickly shown in Figure 18.5.

18.2.4   MANUFACTURING INDUSTRY

Robotics are used in manufacturing, shown in Figure 18.6, to improve the productivity and lower the production cost. The robots are used in the industry to perform tasks like monotonous, repetitive, or intricate jobs. These machines have increased the production and lowered overall cost.

FIGURE 18.4   Robots in agriculture.

FIGURE 18.5   Robots in the food industry.

FIGURE 18.6   Robots in industry.

18.2.5   MILITARY

Unmanned drones in the war field show the importance of robots in the military sector (Figure 18.7). These devices are used for observation and support operations on the war field. Military drones are used during disaster situation to provide the soldiers with real- time information.

FIGURE 18.7   Robots in the military.

18.3 DIFFERENT TYPES OF ROBOTS

18.3.1 PREPROGRAMMED ROBOTS

Preprogrammed robots operate in a controlled environment where they do simple, monotonous tasks. An example of a preprogrammed robot, shown in Figure 18.8, would be a mechanical arm on an automotive assembly line. The arm serves one function — to weld a door on, to insert a certain part into the engine, etc. — and its job is to perform that task longer, faster and more efficiently than a human.

18.3.2 HUMANOID ROBOTS

Humanoid robots, shown in Figure 18.9, are robots that look like and/or mimic human behavior. These robots usually perform human-like activities (like running, jumping and carrying objects), and are sometimes designed to look like us, even having human faces and expressions. Two of the most prominent examples of humanoid robots are Hanson Robotics' Sophia and Boston Dynamics' Atlas.

18.3.3 AUTONOMOUS ROBOTS

Autonomous robots, shown in Figure 18.10, operate independently of human operators. These robots are usually designed to carry out tasks in open environments that do not require human supervision. An example of an autonomous robot would be the Roomba® vacuum cleaner, which uses sensors to roam freely throughout a home freely.

18.3.4 AUGMENTING ROBOTS

Augmenting robots, shown in Figure 18.11, either enhance current human capabilities or replace the capabilities a human may have lost. Some examples of augmenting robots are robotic prosthetic limbs or exoskeletons used to lift hefty weights.

FIGURE 18.8   Preprogrammed robots.

FIGURE 18.9 Humanoid robots.

18.3.5 TELEOPERATED ROBOTS

Teleoperated robots, shown in Figure 18.12, are mechanical bots controlled by humans. These robots usually work in extreme geographical conditions, weather, circumstances, etc. Examples of teleoperated robots are human-controlled submarines used to fix underwater pipe leaks during the BP oil spill or drones used to detect landmines on a battlefield.

FIGURE 18.10 Amazon wheeled autonomous robot.

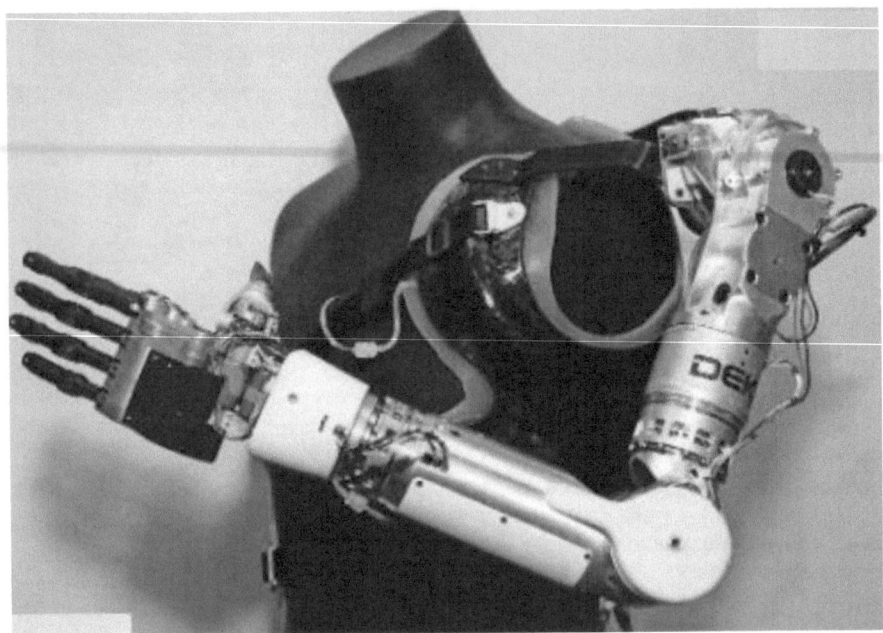

FIGURE 18.11   Augmenting robots.

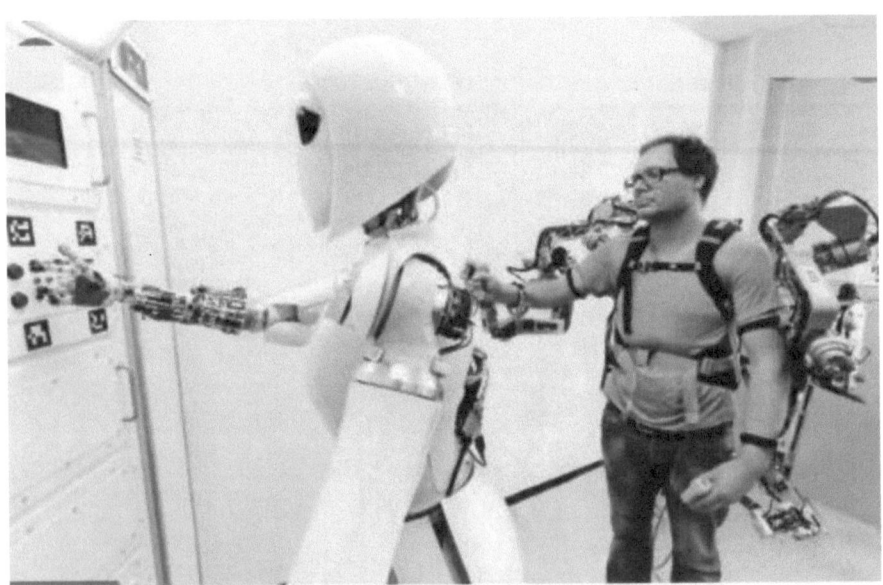

FIGURE 18.12   Teleoperated robots.

18.4   DIFFERENT SKILLS OF CURRENT ROBOTS

18.4.1   ADAPTABILITY AND FLEXIBILITY

According to the 4th Industrial Revolution, robots are expected to have cognitive flexibility, which is considered an in-demand skill. This is the ability to change the cell of the robot to adapt to changes that occur in the product and the business. The following factors will affect the demand in the market:

a. Change in fashion
b. Change in technology
c. Competition
d. Development of new products

Future robots are expected to have adaptability skills to absorb the changes in business and make decisions or update the modifications according to the demands.

18.4.2   CRITICAL THINKING SKILLS

The way of analyzing and evaluating an issue to arrive at a judgment is called critical thinking. Robots used for answering questions asked of them and giving proper solutions for the questions will improve the critical thinking skill robot.

18.4.3   SPEECH RECOGNITION SKILLS

Speech is an important parameter for creating the network and for improving the business. Training robots to understand speech and respond back with the proper answer is a challenging task for the researcher and developer. Robots that can recognize speech are used in hospital to attend patients, schools to interact and educate children, and in the customer-care industry for responding to customer queries.

18.4.4   PEOPLE MANAGEMENT SKILLS

Robotization of the workplace is gaining importance. Robots are trained to play five key roles during people management. Social relationship, self-development, self-esteem, appreciation, and recognition are the interpersonal challenges faced by the robots when they deal with people management in organizations. Today, robots are used for applications like picking orders in warehouse, delivering products to a university campus, and disposing of bombs.

18.4.5   PROBLEM-SOLVING SKILLS

Problem solving is an important pillar in the field of education. Robots are appointed in schools for teaching problem-solving skills to students. Students learn more from the robots than in schools. This is because the robots can give answers in more ways, thus providing more options to the students to think. But still more learning tools are needed to utilize robots in an effective way in the education system.

18.5   TEACHING HUMAN SKILLS TO ROBOTS

18.5.1   Speaking Skills for the Home Robot

18.5.1.1   Skill

Speaking is one of the important skills for progressing in life. Without verbal communication, networking becomes very difficult, which in turn decreases the productivity or performance of the system. Usage of proper words at the appropriate situation is a challenging task even for the humans. Incorporating speaking skills for human-like robots is a real challenge for the robotic industry [4–6].

Robots with voice recognition modules are used for controlling automotive vehicles. Speech recognition systems are designed based on the hidden Markov model (HMM).

18.5.1.2   Algorithm

Researchers started using the HMM algorithm for training robots with speech recognition skills. This statistical model is designed with hidden state information. Though the state information is hidden, the output derived from the state is visible to the user.

The basic components of the speech recognition system are shown in Figure 18.13. The voice from the mic will reach the analog-to-digital conversion (ADC) in the form of analog signals. The ADC will convert these analog signals to digital signals. The input waveform signal, shown in Figure 18.14, from the microphone is transferred into fixed-sized acoustic vectors represented by Y 1:T = y1,...,yT, which is a procedure of feature extraction. This extracted feature attempts to find the words represented as w1: L = w1,...,wL. Here Y is the generator, and the decoder represents wˆ = arg max w {P(w|Y)}. The Bayes Rule transforms this wˆ into equivalent findings wˆ = arg max w {p(Y|w)P(w)}. P(y)/w) is designed by the acoustic model, and P(w) is designed by the language model.

Feature extraction is the process of converting the speech waveform to the parametric representation at the lower data rate for analyzing and processing. Linear

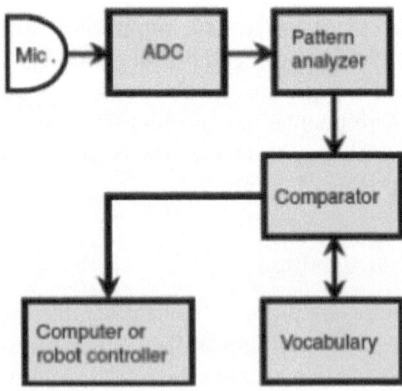

FIGURE 18.13   Speech recognition system.

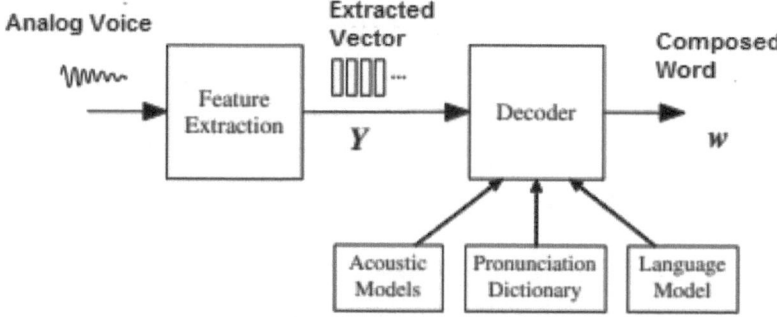

FIGURE 18.14   HMM-based speech recognition system—architectural diagram.

prediction cepstral coefficients (LPCC), mel-frequency cepstral coefficients (MFCC), linear prediction coefficients, discrete wavelet transforms (DWT), line spectral frequencies and perceptual linear predictions (PLP) are the commonly used techniques for speech extraction in voice recognition systems.

18.5.2   OBSTACLE AVOIDANCE SKILLS

Robots are equipped with the capability to avoid obstacles using infrared sensors implemented on microcontroller-embedded devices. The bubble band algorithm [3] is used to detect the obstacle in a particular area. The motion of the robots is adjusted according to the obstacle present with the help of the bubble present in the device.

Quinlan and Khatib [18] proposed the bubble band algorithm. This method contains a bubble with maximum free space around the robot, which allows the robot to travel in any direction without collision as shown in Figure 18.15.

18.5.3   SINGER MODULE FOR HUMANOID ROBOTS

The research in the robotic field is a trending topic in different areas of the world. Making a singing robot using the speech synthesizer software called Vocaloid Editor will generate the songs for the robot to sing. A synthesizer is a tool designed to generate digital music sounds artificially. Physical modeling synthesis, subtractive

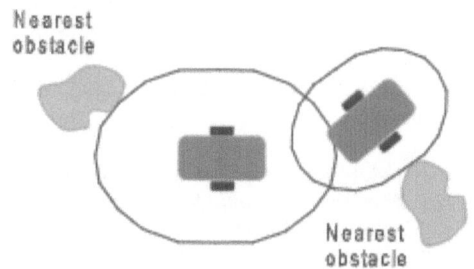

FIGURE 18.15   Obstacle avoidance.

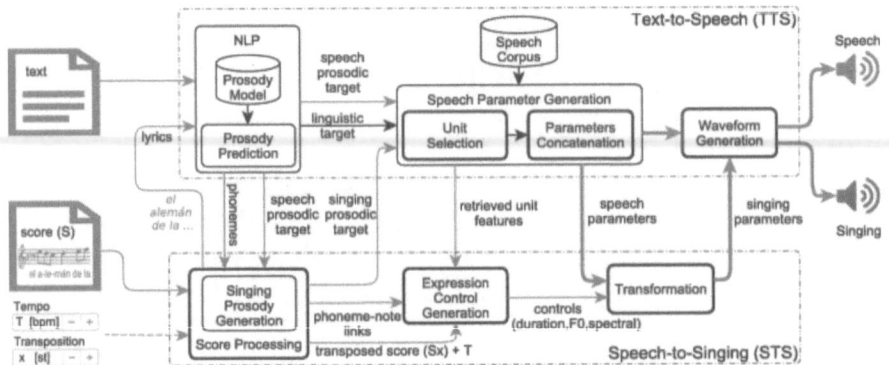

FIGURE 18.16   Text-to-speech synthesis framework.

synthesis, and addictive synthesis are the different techniques used by the synthe-sizer shown in Figure 18.16. The algorithms used for synthesizing the text to speech is called text-to-speech.

> **Prosody model**: The intonation, duration, and patterns are generated associ-ated with the prosody words, syllables, and phrases of the speech.
> **Prosody prediction:** In this block, each word will be given a value based on the importance the user gives for the word while reading.
> **Speech corpus:** This is the database of the speech audio files and transcriptions.
> **Singing prosody generation:** This block transfers the node duration and the FOs into prosodic representation.

REFERENCES

1. Gasparetto, A., & Scalera, L. (2019). A Brief History of Industrial Robotics in the 20th Century. *Advances in Historical Studies*, 8(1), 24–35. https://doi.org/10.4236/ahs.2019.81002
2. Munawar, A., & Fischer, G. (2016). A Surgical Robot Teleoperation Framework for Providing Haptic Feedback Incorporating Virtual Environment-Based Guidance, *Frontiers in Robotics and AI*, 3, DOI: 10.3389/frobt.2016.00047
3. Susnea, I., Minzu, V., & Vasiliu, G. (2014). Simple, Real-Time Obstacle Avoidance Algorithm for Mobile Robots, *Applied Mechanics and Materials*, 571–572, 1068–1075.
4. Joshuva, A., & Sugumaran, V. (2015). Speech Recognition for Humanoid Robot, *International Journal of Applied Engineering Research*, ISSN 0973-4562 10(68) , pp. 280–296.
5. Srivastava, N. et al. (2012). Speech Recognition Using Mel-Frequency, Cepstrum Coefficients (MFCC) and Neural Networks.
6. Razak, Z., Ibrahim, N. J., Tamil, E. M., & Idris, M. Y. I. (2012) Quarnic Verse Recitation Feature Extraction Using Mel-Frequency Cepstral Coefficient (MFCC). *Department of Al-Quran & Al-Hadith, Academy of Islamic Studies*, University of Malaya.
7. Kabir, A., & Ahsan, S. M. M. (2007). Vector Quantization in Text Dependent Automatic Speaker Recognition using Mel-Frequency Cepstrum Coefficient, 6th WSEAS International Conference on Circuits, Systems, Electronics, Control & Signal Processing, Cairo, Egypt, December 29–31, pp. 352–355.

8. Dharmale, N. S., & Mahamune, R. S. (2013). Robotic Automation Using Speech Recognition.

9. Gaikwad, S. K., Gawali, B. W., & Yannawar P. (November 2010). A Review on Speech Recognition Technique, *International Journal of Computer Applications (0975 – 8887)*, 10(3), pp. 1120–1135.

10. Muda, L., Begam, M., & Elamvazuthi, (March 2010). Voice Recognition Algorithms Using Mel Frequency Cepstral Coefficient (MFCC) and DTW Techniques, *Journal of Computing*, 2(3), pp. 380–395.

11. Lumelsky, V., & Skewis, T. (1990). Incorporating Range Sensing in the Robot Navigation Function. *IEEE Transactions on Systems, Man, and Cybernetics*, 20, 1058–1068.

12. Lumelsky, V., & Stepanov, A. (1990). Path-Planning Strategies for a Point Mobile Automaton Moving Amidst Unknown Obstacles of Arbitrary Shape, in *Autonomous Robot Vehicles*. New York, Spinger-Verla.

13. Kamon, I., Rivlin, E., & Rimon, E. (April 1996). A New Range-Sensor Based Globally Convergent Navigation Algorithm for Mobile Robots, in *Proceedings of the IEEE International Conference on Robotics and Automation*, Minneapolis, pp. 1480–1490 .

14. Khatib, O. (1985). Real-Time Obstacle Avoidance for Manipulators and Mobile Robots. *IEEE International Conference on Robotics and Automation*, March 25–28, St. Louis, pp. 500–505.

15. Koren, Y., & Borenstein, J. (August 1988). High-Speed Obstacle Avoidance for Mobile Robotics, in *Proceedings of the IEEE Symposium on Intelligent Control*, Arlington, VA, pp. 382–384.

16. Borenstein, J., & Koren, Y. (1991). The Vector Field Histogram – Fast Obstacle Avoidance for Mobile Robots. *IEEE Journal of Robotics and Automation*, 7, pp. 278–288.

17. Ulrich, I., & Borenstein, J. (May 1998). VFH+: Reliable Obstacle Avoidance for Fast Mobile Robots, in *Proceedings of the International Conference on Robotics and Automation (ICRA'98)*, Leuven, Belgium, pp. 270–285.

18. Quinlan, S., & Khatib, O. (1993). Elastic bands: connecting path planning and control, in *Proceedings of the IEEE International Conference on Robotics and Automation*, Atlanta.

Index